CLINICAL NEURORADIOLOGY
100 MAXIMS

Leena M. Ketonen, MD PhD

Professor of Radiology, and Head,
Division of Neuroradiology,
University of Texas Medical Branch,
Galveston, Texas, USA

Michel J. Berg, MD

Assistant Professor of Neurology,
University of Rochester,
Comprehensive Epilepsy Program,
Rochester, New York, USA

A member of the Hodder Headline Group
LONDON • SYDNEY • AUCKLAND
Co-published in the USA by Oxford University Press, Inc., New York

First published in Great Britain in 1997 by
Arnold, a member of the Hodder Headline Group,
338 Euston Road, London NW1 3BH

Co-published in the United States of America by
Oxford University Press, Inc.,
198 Madison Avenue, New York, NY 10016
Oxford is a registered trademark of Oxford University Press

Whilst the advice and information in this book is believed to be true and
accurate at the date of going to press, neither the authors nor the publisher
can accept any legal responsibility or liability for any errors or omissions
that may be made. In particular (but without limiting the generality of the
preceding disclaimer) every effort has been made to check drug dosages;
however it is still possible that errors have been missed. Furthermore,
dosage schedules are constantly being revised and new side-effects
recognized. For these reasons the reader is strongly urged to consult the
drug companies' printed instructions before administering any of the drugs
recommended in this book.

British Library Cataloguing in Publication Data
A catalogue record for this book is available from the British Library

Library of Congress Cataloging-in-Publication Data
A catalog record for this book is available from the Library of Congress

ISBN 0 340 60878 1

Typeset in 9/11pt Palatino by Gray Publishing, Tunbridge Wells
Printed and bound in Great Britain by The Bath Press

To our children for their inimitable support and patience

Emily, Irene, and Rose
Ben and Matt

And to all our students, house officers, colleagues, and teachers

Contents

Chapter 11   Infections and Inflammatory Conditions

Chapter 12   Pediatric Neuroradiology

Chapter 13   Trauma

Chapter 14   Spine and Spinal Cord: Congenital Malformations

Chapter 15   Spine and Spinal Cord: Tumors

Chapter 16   Spine and Spinal Cord: Infections, Back Pain, and Myelopathy

Chapter 17   Epilogue

Foreword to the 100 Maxims in Neurology Series

The *100 Maxims in Neurology* series was originated by Roger Porter, MD. Based on his success with *Epilepsy: 100 Elementary Principles,* he recognized that this new approach in conceptualizing and presenting clinical information could extend well beyond epilepsy.

The basic principle of the *100 Maxims* is to bridge the gap between the didactic presentation of information in a review article or text with the clinical wisdom and pearls presented on clinical rounds or a phone consultation with a colleague. Experienced physicians make clinical decisions based on data from the literature, discussions with colleagues, and their own observations and lessons. The nuances of clinical neurology are thus often absent from the texts that provide careful and critical academic presentations.

Each maxim is intended to provide the reader with a clinical rule, a warning, an observation, or a therapeutic principle. From the declarative statement, each maxim evolves as a brief clinical discussion, focusing on practical issues of pathophysiology, diagnosis or therapy. The goal is not a reference book, but a book that can be read from cover to cover and will confer a solid foundation of clinical knowledge. Selected references allow the interested reader to delve deeper into topics of interest.

As the baton of editorship passes from Roger Porter to me, the maxim series remains a new and poorly defined terrain in the ever expanding universe of neurology books. The maxim format holds great promise to teach neurology. Although the books are not "needed" by most neurologists – who will already have books in their libraries that cover the specific topics – the maxim books will provide an important tool for neurologists to practice their art. Standard textbooks summarize and present the "party line", often at the cost of avoiding controversy and making specific recommendations about difficult clinical decisions. Idealized patients with black and white diagnoses exist mainly in textbooks. Doctors most often face gray zones and patients that fit into neither the square nor the circular hole. The maxim texts are intended to address the gray zone where doctors practice medicine.

Creating and executing a new series of neurology books has been time consuming and challenging. My deepest thanks go to my family – Deborah, Janna, and Julie – for providing the support and allowing the time.

Orrin Devinsky

Foreword to this Volume

Clinical neurology has evolved. Although the clinical history and neurologic examination remain the foundation of neurologic diagnosis, neuroimaging has become the standard for clinical localization. A generation ago, lesions were identified using careful neurologic examination. In some of the best residency programs, house officers admitted several patients per week and performed detailed, laborious sensory, parietal lobe, visual field, and other examinations. Although neuroradiology has been rumored to threaten the very existence of both the neurologic examination and the neurologist, neuroimaging is the neurologist's partner and right hand.

As neuroradiology becomes an indispensable component in the practice of neurology, neurologists must become very comfortable with neuroimaging. From the neuroanatomy on CT and MRI to the differential diagnosis of abnormal signals, neurologists must be as skillful with an image as they are with a hammer.

Drs Ketonen and Berg have brought the fundamentals and nuances of neuroradiology into a concise, clear presentation. Their comprehensive, well-balanced text will provide the reader with invaluable and practical information. The authors have done a masterful job. There is no fat in this text. The book's contents are part of the modern foundation of neurology. Any practicing neurologist should be familiar with the clinical pearls of wisdom contained in *Clinical Neuroradiology: 100 Maxims*.

Orrin Devinsky

Preface

The goal of this volume, *Clinical Neuroradiology: 100 Maxims*, is to provide the reader with a solid knowledge base of neuroradiology. This text is designed to teach and is not intended to be a comprehensive reference. The radiologic appearances of common, important esoteric, and classic neurologic entities are described. In addition, both time-honored and new "pearls of wisdom" are emphasized to make the reader skilled in clinical neuroradiology.

In the mid-1970s, the introduction of computed tomography (CT) caused a revolution in the fields of neurology, neurosurgery, and neuroradiology. Since then, magnetic resonance imaging (MRI) and interventional angiography have resulted in a further explosion in neuroradiology. Functional imaging techniques, including positron emission tomography (PET) and single photon emission computed tomography (SPECT), have moved from the laboratory to the clinic. These tools have resulted in a tremendous, yet evolving, advancement in our understanding of pathophysiology, and have improved our ability to treat many diseases effectively. However, with this technologic explosion comes greater complexity. These tests have many overlapping indications and can be performed with numerous variations making it critical, especially in this day of financial constraints, for physicians to understand how optimally to order and correlate the findings of these studies to best serve the patient. In addition, each of these tests has its own unique contraindications and limitations with which physicians must be familiar.

This volume is designed to enhance the basic level of knowledge of neuroradiology that many clinicians possess. Major emphasis is placed on MRI because it is generally the most versatile and accurate single neuroradiologic test. It has largely supplanted CT and contrast myelography and has modified the use of catheter angiography in many neurologic diseases. The popularity of MRI is rational: MRI does not use ionizing radiation, is essentially noninvasive, and gives superb anatomic and pathologic information about brain and spinal cord parenchyma, individual cranial nerves, vessels, CSF spaces, and surrounding tissues. However, important findings on all neuroimaging modalities are discussed in this text.

As the level of comfort with the current technology improves, several new imaging techniques on the horizon will present a fresh challenge. Functional neuroimaging, especially functional MRI, offers a new dimension, which may dramatically alter imaging strategies in the future. Computer hardware and software advances will modify how scans are displayed and interpreted. Technology already exists to present these tomographic studies in three-dimensional displays. In the near future it is probable that computers will automatically be able to extract almost any structure, enabling highly detailed quantitative examination. It is likely that all difficult neurosurgical operations will be planned at computer workstations before and during surgery. Directed-beam therapy is one application of this technology that is already in clinical use. Because neuroradiology is a highly dynamic field, it is impossible to predict what the next few years will hold. This book will be limited to established facts, with a few comments on new innovations.

A detailed understanding of physics and engineering principles is neither required nor expected of physicians ordering neuroimaging studies such as CT and MRI. However, a basic knowledge of the principles that govern these imaging techniques will lead to more effective use of these tests. Thus, this book begins by describing the key principles of neuroimaging. Throughout the text, emphasis is placed on the values of the various acquisition techniques of MRI, as this is confusing.

Following the introduction, maxims are given for major neurologic disorders. In the discussion, the salient features of each neurologic disease are reviewed. The major thrust of each discussion focuses on neuroimaging. When relevant, details of the study technique are given, especially for MRI.

The reader is encouraged to read this book from beginning to end. However, each maxim is designed to stand alone, as is each figure. Once the reader has completed this text he or she should have a solid working knowledge of modern neuroradiology and an improved ability to judiciously order neuroimaging studies and correctly correlate imaging findings to symptoms and pathology.

Leena M. Ketonen
Michel J. Berg

Acknowledgements

The authors are deeply indebted to Roger Porter, MD, a great teacher, for initiating the 100 Elementary Principles series and for first asking us to participate as writers. We owe a great deal of gratitude to Georgina Bentliff, Director of Health Science Publishing, and to Marianne Kirby, Senior Production Editor at Arnold, for their encouragement and patience. We are also very thankful for the guidance we received from the series editor, Orrin Devinsky, MD.

Special thanks to Allen Pettee, MD and Linda Williams, MD for reviewing the entire manuscript and to James Power, MD, Bruce Beesley, MD, Rebecca Myerson, MD and Daniel Rifkin, MD for reviewing portions of the manuscript. All of these reviewers made valuable suggestions that have made this book more powerful.

We greatly appreciate the contribution of Leena Valanne, MD who collected several cases that were included in this text.

No neuroimaging would be possible without highly qualified technicians. We sincerely appreciate all of our technologists and especially the superb leadership provided by our chief technologists, Ms Connie White, Ms Deborah Caskey, Mr Dale Hultz, and Mr Dennis Dornfest. Credit is also due to our secretaries, Valerie Bates, Rebecca Müller and Deborah Selke.

We also wish to thank David Goldblatt, MD and John Bodensteiner, MD for reviewing our initial proposal and giving strong support for the concept of this book.

Finally, we deeply appreciate the ongoing encouragement provided by our respective institutions and departments. Special thanks to James L Burchfiel, PhD, Giuseppe Erba, MD, Robert C Griggs, MD, and Arvin Robinson, MD at the University of Rochester and to Eric VanSonnenberg, MD at the University of Texas Medical Branch.

Michel J Berg
Rochester, NY

Leena M Ketonen
Galveston, TX

1
Introduction

1. *MRI produces images of the radio-frequency responses from tissue; thus, we speak of **signal**. CT and conventional plain radiographs are images of the attenuation of x-rays; thus, we speak of **density**. PET and SPECT are images of radioactive decay; thus, we speak of **activity***

MRI is based on the physical property of nuclear magnetic resonance (NMR). Routine MRI results in an image of certain properties of hydrogen nuclei (often referred to as protons). The protons that contribute to MRI in tissue reside predominantly in water and fat. Thus, simplistically, a conventional MR image is a measure of the amount of fat and water within different regions. In reality the situation is more complex.[1-3]

An MR image (Fig. 1.1) is acquired by placing the tissue to be imaged (in this text, the head or spine) into a strong, homogeneous magnetic field (usually 0.3–1.5 T). This constant magnetic field aligns the nuclear spins of a portion of the hydrogen nuclei within the tissue. A coil that either surrounds the region (e.g. head or body coil) or overlies the region (surface coil) produces a radiofrequency (energy) pulse (approximately in the FM radiofrequency band range) to excite the tissue. The coil also serves as an antenna to detect the tissue response (the MR signal). The initial radiofrequency pulse perturbs the nuclear spins (exciting the protons into a higher energy state) in the tissue. Over a short period of time (milliseconds to seconds), the excited protons relax from the higher energy state to the lower energy state. As they relax, they each give off energy in the form of a radiofrequency response (the echo).

In a spin-echo sequence the echo is controlled by specialized radiofrequency pulses. In a gradient-echo sequence the echo is controlled by rapidly changing magnetic field gradients. The echo is detected (by the coil) and then processed by computers into an image. The strength of the radiofrequency response from each point is dependent on various machine parameters, the imaging technique, and specific tissue properties (maxim 2).

The strength of the signal that a substance produces is specifically related to the surrounding environment. The signals of an identical tissue are often different in different parts of an image due to magnetic field inhomogeneities (Fig. 2.1). Thus, MRI signals cannot be directly compared between studies or even within the same study. (This is not the case with CT.)

The key to obtaining spatial information (that is, which radiofrequency response came from where) lies in having known gradients within the magnetic field. A set gradient is present in the (large) constant magnetic field along the long axis, and other gradients are created by the coil. The frequency of a tissue's radiofrequency response (the Larmor frequency) is directly proportional to the precise magnetic field at the location of the individual proton at the time it relaxes. Thus, by having a known gradient within the magnetic field, the computer can calculate which frequency came from which location. All of the frequency-direction pixel information is acquired during a single repetition at each phase of the signal. To obtain an entire image (not just one line), a repetition is required for each different phase view. Thus, the number of repetitions required to collect all the information depends on the number of views (subdivisions) in the phase direction. The frequency-encoded information is displayed in one direction (usually vertical) and the phase-encoded information in an orthogonal direction (usually horizontal). Knowing these directions is important for understanding certain artifacts (maxim 4). The frequency direction is identified with the Greek letter v on most MR images (Fig. 1.1).

CT and conventional radiographs are easier to understand. They are produced by x-rays that are directed through tissue and then detected. The image depends on how much of the original x-ray beam reaches the detector. Dense tissue, like bone, attenuates (primarily by scatter) more x-rays than does less dense tissue. The amount of attenuation of a given substance is absolute and measured in Hounsfield units on CT (Table 1.1). Iodinated contrast medium attenuates x-rays primarily by absorption. On plain radiographs, the x-rays activate an intensifying screen that emits ultraviolet and visible light rays, which expose the film. Bone, which is dense and scatters most of the x-rays, appears white. Less dense tissues are progressively darker. CT employs thin x-ray beams and scintillation or ionization chamber detectors.

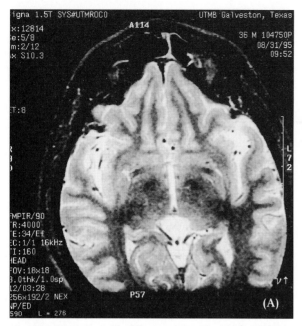 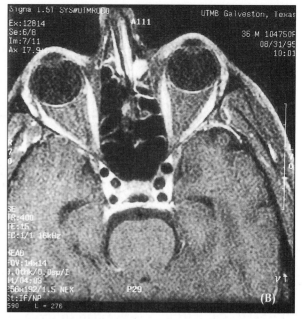

Fig. 1.1 MR image annotation. Each manufacturer has a different scheme for displaying image and patient information. It is essential that you are familiar with the notation on all the images that you review. The first detail to check on every image (or sheet of film) is the patient's name!

Typical information on MR images includes the acquisition technique, TR, TE, slice thickness, and field of view (FOV). The field strength is not always given, although it is important in certain circumstances (maxim 62). The axial images are generally viewed "from the feet" looking to the vertex of the head. That is, the patient's right side is on the viewer's left. Unfortunately this rule has exceptions. The second detail to check on every image is the orientation! Look for R (right) and L (left) on axial and coronal images. Inspection of the scout image (Fig. 3.1) should clarify which sagittal images are from the right and which are from the left. When information about the orientation is not given, complain loudly (it is hard to believe that this occurs, but it does). The medical community anxiously awaits the day when all manufacturers have standardized image annotation and image formats.

Most images have a relative position scale that is very useful for determining the size of lesions or anatomic structures. An arbitrary zero plane is typically set near the center of the field and each image's position relative to this plane is given. For example, the bottom line in the left upper corner in Fig. 1.1A is "Ax S10.3". This indicates that this image is 10.3 mm superior to the zero plane. For coronal images, A (anterior) and P (posterior) positions; for sagittal images, R (right) and L (left) positions; and for axial images, S (superior) and I (inferior) positions are given.

Most neuroimages have a cm scale. In Figs. 1.1A and B, a 5 cm line with 1 cm gradations is printed along the right margin (near the L). Use this scale for measuring objects on the film. (More accurate measurements can be made on display workstations.) Sometimes the distance to the center of the magnet bore is given; e.g. Fig. 1.1A, the L72 adjacent to the cm scale is printed 72 mm to the left of the bore's center.

The image annotation shown here is specific for one software version of one manufacturer (GE Signa 5.x). It is not possible to give examples of all the different MR sequences and image orientations, but these two examples illustrate the important points. This particular annotation is organized with patient and general information in the upper right, scanning sequence information in the lower left, and image, log, and machine information in the upper left.

In the lower right corner note the υ indicating that the frequency-encoded information is displayed in the vertical

direction and, by default, the phase-encoded information is displayed in the horizontal direction.

A: Upper right (patient information):
- Name of institution (UTMB Galveston, Texas)
- Patient's name (blacked out here)
- Patient's age (36), gender (M) and hospital identification number
- Date of study (8/31/95)
- Time of the study (09:52).

Lower left (technical information):
- Acquisition sequence (FMPIR/90 = fast multiplanar inversion recovery)
- Repetition time in milliseconds (TR:4000)
- Time to echo (effective) in milliseconds (TE:34/Ef)
- Which echo in a series of echoes (EC 1/1), bandwidth (16 kHz)
- Inversion time in milliseconds (TI:160)
- Structure imaged (HEAD)
- Field of view in centimeters (FOV:18 × 18)
- Slice thickness in millimeters (3.0 thk), interslice gap (1.0sp)
- Slice count (12), time to acquire sequence (03:28)
- Matrix size (256 × 192), number of excitations (2 NEX)
- No phase wrap (NP), extended dynamic scale (ED)
- Window width (590), level (L=276).

Upper left (image and machine information):
- Machine (Signa 1.5T SYS#UTMROCO): 1.5T refers to the field strength of the magnet
- Examination number in the log book (it is a running number) (Ex:12814)
- Series number (here 5 out of 8) (Se:5/8)
- Image number (here 2 out of 12) (Im:2/12)
- Orientation and location of scan (see above) (Ax S10.3). Notation about contrast will be on this line (+C) as in Fig. 1.1B.

B: Everything as in Fig. 1.1A except:

Upper left (last line):
- +C = intravenous contrast administered

Lower left corner information:
- SE = spin echo
- TR:400 = repetition time
- TE:15 = time to echo
- EC:1/1 16 kHz = which echo in a series of echoes; bandwidth
- 3.0thk/0.0 sp/I = slice thickness/space between the slices/ interleaved
- St: IF/NP = inferior saturation pulse (St: I); fat suppression (F)/no phase wrap (NP)

Table 1.1 CT density values

Tissue	Density (Hounsfield units)
Metal	1000
Calcium	100–1000*
Blood	80–85
Gray matter	35–40
White matter	25–30
Water (CSF)	0
Fat	–100
Air (vacuum)	–1000

*Depending on density; cortical bone approaches 1000.

The spatial information for CT is obtained by using a series of x-ray beams, each passing through a slightly different block of tissue. Images are reconstructed by a computer and presented in the same fashion as plain radiographs.[4]

Positron emission tomography (PET) and single photon emission computed tomography (SPECT) images are created by detecting the effects of radioactive decay of certain short-lived radioactive substances. SPECT involves the direct detection of γ-particles as the decay product. In PET, emitted positrons collide with electrons resulting in annihilation of the particles and the production of two γ-rays directed at 180° to each other. Nearly simultaneous recordings by detectors located on opposite sides of the patient are required for the event to be counted. Spatial information for image reconstruction is obtained by making measurements with detectors at many different known positions, in a fashion similar to that of CT. The activity detected is a measure of specific tissue properties. The property measured depends on the specific radioactive substance used. The most common substances used with SPECT – [^{123}I]iodoamphetamine ([^{123}I]IMP) and [^{99m}Tc]hexametazime ([^{99m}Tc]HMPAO) – reflect regional cerebral blood flow (rCBF).[5] With PET, regional cerebral glucose metabolism is measured with [^{18}F]fluorodeoxyglucose (FDG), regional cerebral blood volume with carbon [^{15}O]monoxide, regional cerebral blood flow with [^{15}O]water, and regional cerebral oxygen consumption with [^{15}O]oxygen.[6] Because the amount of blood flow, glucose utilization, and oxygen consumption within a given region of the brain reflects the metabolic activity of that region, these studies indirectly measure brain function (hence the term "functional imaging"). A variety of other radioactive substances are available or being developed to measure metabolic, receptor binding, and other tissue properties.

References

1. Cox IH, Roberts TP, Moseley ME. Principles and techniques in neuroimaging. In: Kucharczyk J, Moseley M, Barkovich AJ (eds), *Magnetic resonance neuroimaging*. Boca Raton: CRC Press, 1994: 1–103.
2. Edelman RR, Warach S. Magnetic resonance imaging. *N Engl J Med* 1993; **328**: 708–16.
3. Council on Scientific Affairs; Jacobson HG section editor. Fundamentals of magnetic resonance imaging. *JAMA* 1987; **258**: 3417–23.
4. Kinkel W, Bates V. Computerized tomography in clinical neurology. In: Joynt RJ (ed.), *Clinical Neurology*. Philadelphia: JB Lippincott, 1994: 1–134.
5. Masdeu JC, Brass LM, Holman BL, Kushner MJ. Brain single-photon emission computed tomography. *Neurology* 1994; **44**: 1970–7.
6. Herscovitch P. Radiotracer techniques for functional neuroimaging with positron emission tomography. In Thatcher RW, Hallet M, Zeffiro T, John ER, Huerta M (eds), *Functional neuroimaging technical foundations*. New York: Academic Press, 1994: 29–42.

2. *MRI displays the T1 and T2 properties of tissue*

- On T1-weighted MRI, a tissue with a short T1 will be bright and with a long T1 will be dark.
- On T2-weighted MRI, a tissue with a short T2 will be dark and with a long T2 will be bright.
- Proton density (PD) images are set to minimize both the T1 and the T2 effects, giving a "map" of the density of the protons.
- MRI contrast is a paramagnetic agent that shortens the T1 of tissues near the contrast. Thus, contrast appears bright on T1-weighted images.

The T1 and T2 relaxation time constants are measures of certain physical properties of the tissue imaged (or, for that matter, any substance that demonstrates NMR). They depend on the nature of the specific molecule and the local magnetic field environment around it.[1–4]

By varying specific MRI machine image acquisition parameters (for instance, the time to repetition (TR) and time to echo (TE)* in conventional spin-echo acquisitions) images may be weighted to display predominantly either the T1 or the T2 tissue property. A long TR minimizes T1 effects and a short TE minimizes T2 effects. Thus, a T1-weighted image has a short TR (≈500 ms) and a short TE (≈20 ms), and a T2-weighted image has a long TR (≈2000 ms) and a long TE (≈80 ms). A proton density (also termed intermediate, spin density, balanced, or rho) image uses a long TR (≈2000 ms) and short TE (≈20 ms) to minimize both the T1 and T2 effects, resulting in a signal that is proportional to the density of the protons. Images with each weighting may be acquired in any plane (sagittal, axial, coronal, or oblique) and with a variety of techniques including conventional spin-echo and gradient-echo. The "double spin-echo" technique gives a PD image on the first echo and a T2-weighted image on the second echo.

On spin-echo images, contrast is primarily governed by the TR and TE as discussed above. Contrast is influenced by other machine parameters in other acquisition techniques. For instance, in gradient-echo (GRE) the

---

*The TR is the (operator-selected) time between repetitions of the pulse cycle. It is usually fixed but with cardiac gating varies with the heart rate. The TE is the (operator-selected) time from the center of the excitation pulse to the peak of the measured tissue response (echo).

contrast is controlled primarily by the flip angle, with less contribution from the TR and TE. The fast spin-echo (FSE) technique introduces other factors that modify image contrast such as echo trains and interecho spacing. Furthermore, TR and TE are derived differently in FSE than in the conventional spin-echo technique.[5] FSE images are susceptible to different artifacts and are not as sensitive as conventional spin-echo images to magnetic susceptibility from paramagnetic material.[6] Inversion recovery images have an inversion time (TI) parameter.

Paramagnetic substances, such as gadolinium (used in intravenous contrast), disrupt the local magnetic field, shortening the T1 relaxation time and result in high signal on T1-weighted images. Paramagnetic substances (such as iron present in old blood) also shorten the T2 relaxation time, resulting in low signal on T2-weighted and gradient-echo images (maxim 62). This effect is sometimes referred to as magnetic susceptibility (maxim 4). Ferromagnetic substances have the same effect, but are infrequently encountered in biological tissue.

Fat and water have characteristic patterns and are primarily responsible for the signal on MRI neuro-imaging. Fat has a short T1 and short T2, and thus is high signal on T1-weighted images and low signal on T2-weighted images. Water has a long T1 and long T2 and thus is low signal on T1-weighted images and high signal on T2-weighted images. Densely calcified tissues such as bone generally give off very little signal because no fat or water is present. Blood within vessels is low signal on spin-echo sequences because it flows out of the region between the series of excitation pulses (flow void). Flowing blood is high signal on gradient-echo sequences because only one radiofrequency pulse is given and no outflow signal loss occurs.

High signal on T1-weighted images is either fat, gadolinium contrast, another paramagnetic compound resulting in magnetic susceptibility (e.g. particulate calcium (maxim 13) or manganese (maxim 67)), or blood (e.g. methemoglobin). Fat suppression* is used to differentiate subacute blood and other T1 high signal lesions from fat (Fig. 2.1). Low signal on T2-weighted images may be due to a region devoid of fat and water (e.g. air), a flow void, a substance with different proton binding (calcified regions including bone), or magnetic susceptibility from a paramagnetic substance (e.g. gadolinium contrast, iron (hemosiderin), manganese, deoxyhemoglobin or methemoglobin within red blood cells; Table 60.1).

Magnetic field inhomogeneities are illustrated in both the white and the gray matter in Fig. 2.1. The 1s are all within white matter and the 2s are all within gray matter. Note how different the signal of each of these tissues is

*Fat suppression is an imaging enhancement technique that suppresses the signal from fat or water. This suppression can be accomplished with a variety of methods including application of a frequency-selective presaturation pulse, the use of a short T1 inversion–recovery sequence, the Dixon/Chopper method or a hybrid method. Fat suppression is most commonly used to optimize visualization of contrast enhancement in fat-rich areas such as the orbit, skull base, pituitary region, and spine.

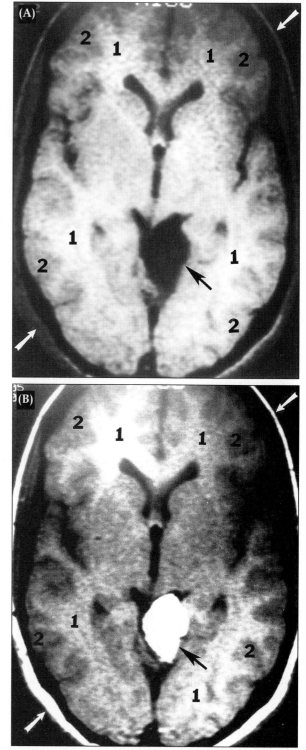

Fig. 2.1 Use of fat suppression and magnetic field inhomogeneity. An incidental quadrigeminal plate region lesion is present on these T1-weighted images. **A:** T1-weighted axial image reveals a high signal lesion (black arrow). Note that the subcutaneous fat has a similar high signal (white arrows). **B:** T1-weighted axial image with fat saturation sequence. The signal from the lesion is suppressed, demonstrating that it is a lipoma (black arrow). Also note that the signal of the subcutaneous fat is suppressed (white arrows).

in different parts of the image. For example, in Fig. 2.1B the gray matter in the lower left of the image surprisingly has a higher signal than does the white matter in the upper right. Also note the quantity of noise within each tissue. For example, within the white matter there are numerous pixels with a gray-scale value as dark as the typical gray-matter pixel. This phenomenon makes automated tissue segmentation difficult (maxim 100).

References

1. Pomeranz J. *Craniospinal magnetic resonance imaging*. London: WB Saunders, 1989.
2. Atlas S. *Magnetic resonance imaging of the brain and spine*. New York: Raven Press, 1991.
3. Osborn A. *Diagnostic Neuroradiology*. St Louis: CV Mosby, 1994.
4. Elster AD. *Questions and answers in magnetic resonance imaging*. St Louis: CV Mosby, 1994.
5. Brown MA, Semelka RC. *MRI: Basic principles and applications*. Chichester: John Wiley, 1995.
6. Tice HM, Jones KM, Mulkern RV, *et al*. Fast spin-echo imaging of intracranial neoplasms. *J Comput Assist Tomogr* 1993; **17**: 425–31.

3. *Order CT to evaluate calcium (bone) and acute blood; use MRI for soft tissue and vascular disease*

During the past several decades there has been an explosion in neuroimaging technology. CT, MRI, conventional contrast arteriography, and ultrasonography (US) are the most frequently used. In a growing number of institutions, MR spectroscopy* and functional neuroimaging using MRI, PET, and SPECT are available for both research and clinical use. Occasionally plain radiographs, contrast myelography and nuclear medicine or contrast cisternography are indicated.

Although no standard guidelines exist about which test to perform in every condition, there is general agreement in most circumstances. The usual clinical decision is between MRI and CT. The general principle is that CT is best for acute blood and calcium (including bone) and MRI for soft tissue abnormalities (gray and white matter) and vascular disease. Often CT and MRI are complementary and both should be performed.

Comparing CT and MRI must be done with caution. For instance, although the axial images appear similar, they are acquired in substantially different planes (Fig. 3.1). In addition, with CT, it is important to examine the soft tissue (brain) windows, blood (subdural) windows,

*Magnetic resonance spectroscopy (MRS) is possible because hydrogen, carbon, fluorine, sodium, and phosphorus display measurable spectra with magnetic resonance. That is, these elements have slightly different resonance frequencies depending on the type of molecule and their location on that molecule. The signals produced are small, and thus large voxel sizes (compared with conventional MRI) are currently required. High field strengths (4 T or more) produce higher resolution MRS images in shorter times but, overall, MRS requires long acquisition times.

and bone windows, or information may be lost (maxims 61, 85 and 87).

CT is the appropriate initial study in the following circumstances:

1. *Acute nontraumatic brain deficit*
   • Subarachnoid or intracerebral hemorrhage
   • Bleed into neoplasm
   • Stroke
2. *Trauma*
   • Subdural or epidural hematoma

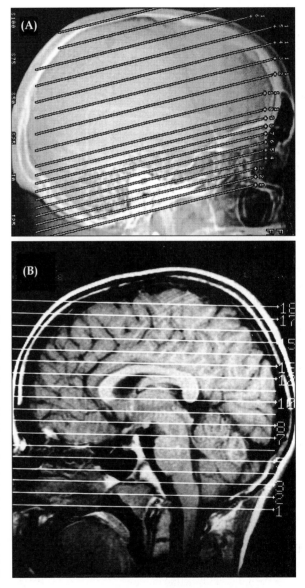

Fig. 3.1 Typical scout images for head CT and MRI. The difference in slice orientation must be taken into account when comparing axial MRI and CT studies. **A:** CT scout image. Lines denoting 5 mm thick sections through the posterior fossa and 10 mm sections through the rest of the brain are superimposed on a lateral skull image. The anatomical slices are oriented at 25° to the orbitomeatal line. **B:** MRI scout image. Lines denoting the axial sections are superimposed on the T1-weighted midline sagittal image. The MRI axial images are oriented in a significantly different plane than are axial CT images.

- Contusion
- Fracture

3. *Ventricle size assessment*
   - Hydrocephalus
   - Shunt malfunction
4. *Myelopathy/radiculopathy* (with contrast myelography)
   - Spinal stenosis with cord compression
   - Other compressive spinal cord lesions
   - Vertebral pathology (contrast not necessary)
   - Delayed contrast imaging to detect syrinx
5. *MRI contraindicated*
6. *Evaluation for calcium within lesion*
7. *Bone pathology*
   - Skull base
   - Temporal bone
   - Sutures: synostosis
   - Bony erosion from tumor.

MRI is the appropriate initial study for:

1. *Subacute or chronic brain deficit*
   - Stroke
   - Tumor
   - Abscess
   - Malformation
2. *Degenerative disease*
   - Alzheimer dementia
   - Cerebrovascular dementia
   - Huntington disease
   - Multiple system atrophy
3. *Developmental delay*
   - Delayed myelination or dysmyelination
   - Congenital malformation
4. *New-onset seizure*
   - Tumor
   - Infection
   - Vascular malformation
   - Subdural hematoma
5. *Epilepsy (chronic)*
   - Indolent tumor
   - Cortical dysplasia
   - Vascular malformation
   - Hippocampal sclerosis
6. *Multiple sclerosis* (with contrast for acute lesions)
7. *Myelopathy*
   - Spinal cord lesion
   - Spinal cord atrophy
   - Syrinx
   - Transverse myelitis
8. *Vascular lesions*
   - Arteriovenous malformation
   - Aneurysm
   - Cavernous or venous malformation
   - Venous thrombosis
   - Vasculitis
9. *Meninges*
   - Chronic basilar meningitis
   - Carcinomatous meningitis
10. *MR angiography.*

Cranial ultrasonography is essentially restricted to infants before fontanelle closure. The choice between conventional angiography, CT angiography, MR angiography, and carotid ultrasound/Doppler depends more on local expertise than on the specific modality. Transcranial Doppler may be used to determine whether vasospasm or ongoing emboli are present. Plain radiographs are indicated as the initial study to evaluate for bony pathology in the spine.

4. *Artifacts may mimic or obscure lesions*

Both CT and MRI use complicated electronic hardware and computer processing to produce images. Imperfections in the images created by hardware, software, and the environment are termed artifacts. Artifacts must be distinguished from valid imaging features. In practice, CT and MRI hardware is tuned periodically to achieve optimal performance and minimize hardware-related artifacts.

CT has been in widespread use and remains essentially unchanged, except for improved resolution and faster acquisition times, since its introduction in the mid-1970s. The latest development, helical (spiral) CT, enables very rapid image acquisition (on the order of one second per image), and has resulted in the major advance of CT angiography. CT angiography of the neck and intracranial vessels (mainly the circle of Willis) has a resolution comparable to or better than that of MR angiography. Nevertheless, CT images are density maps and most physicians are familiar with CT artifacts.

In contrast to CT, MRI is a rapidly developing technology undergoing frequent modifications. Its complex nature results in a wide variety of artifacts that are often sequence and machine specific. This is a very important point, as MRI may easily be misinterpreted by a reader unfamiliar with the system's idiosyncrasies. In addition, every software upgrade (sometimes performed several times per year) has the potential to produce new and different artifacts. Technologists and physicians who program, perform, supervise, and read MRI studies need to be highly skilled at recognizing artifacts and applying techniques to minimize them. All physicians should be familiar with the appearance of the most typical MRI and CT artifacts.

Metal is a common environmental cause of CT and MRI artifacts, sometimes making all or a portion of the study uninterpretable (Fig. 4.1). Sources that can be easily eliminated with proper patient preparation include metal in clothing (buttons, zippers, studs), certain cosmetics (e.g. metallic glitter in eye make-up, hair colors, and hairspray), and jewelry. Metallic objects within the patient's body such as dental braces, surgical clips and wires, bullets, and metallic debris may cause artifacts. In this situation, before performing the study, the potential benefit should be weighed against the risk of an inadequate result. In MRI, the risk of injury must be

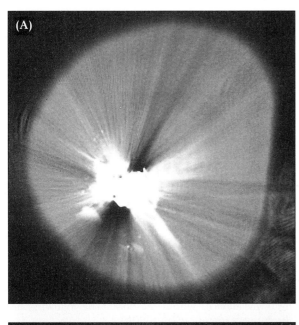

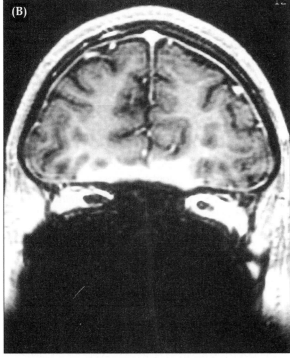

Fig. 4.1 Artifacts due to metal. **A:** Axial CT with metal artifacts from bullet fragments completely obscuring the brain tissue. **B:** Coronal gradient-echo image of a 12-year-old child with braces. The braces distort the orbital area and cause a "black hole" in the facial area.

considered: ferromagnetic materials can dislodge under the influence of the strong magnetic field, or heat within the changing magnetic field (maxim 5).

CT-specific artifacts

Beam-hardening artifacts are always present to some extent on head and spine CT (Figs. 7.5A and 43.1A). They consist of alternating bright and dark lines at high contrast interfaces (e.g. where soft tissue is adjacent to bone). Beam hardening is the result of the polychromatic nature of the x-ray beam. Lower energy photons are disproportionately attenuated by dense tissue, resulting in a more highly penetrating beam than can be compensated for by the software reconstruction. This artifact is most marked at the level of the thick skull base within the posterior fossa, especially at the pons ("Hounsfield lines") and middle cranial fossa. This artifact may also be present along the skull–cerebral convexity interface, obscuring the epidural and subdural spaces. Thinner sections or an altered imaging angle may minimize this artifact. A repeat study or MRI is sometimes required if pathology is suspected in these regions.

Partial volume artifact is the result of averaging voxels* that have heterogeneous composition. For example, a voxel that is filled partially with bone and partially with soft tissue will have an intermediate density value that may simulate pathology. Partial volume artifact increases with pixel size and section thickness.

Patient motion artifact appears as streaks in the direction of motion. It may be minimized with high-speed image acquisition. This is a major application of helical (spiral) CT in which individual images can be acquired within 1 second.

MRI-specific artifacts

Pulsatile blood and CSF flow, respiratory motion, and patient movement produce *ghost artifacts*.[1] They appear primarily in the phase-encoded direction (usually horizontal) because the sampling rate in this direction is in the range of periodic patient and physiologic motion (Fig. 41.2B). The phase-encoded data is obtained significantly more slowly than the (usually vertical) frequency-encoded data. To minimize this artifact, strict patient immobilization and motion-suppression protocols may be utilized. Additionally, the phase-encoded direction can be changed if the ghosting obscures or raises the possibility of pathology.

Chemical shift artifact occurs at tissue interfaces, usually between fat and water.[2] The signal emitted by fat protons has a slightly lower frequency than water. This results in spatial mismapping in the frequency-encoded direction (remember, the spatial location is encoded by the specific response frequency – maxim 2). The artifact is a black or white band at the interface between the two tissues and is often present in the orbit at the junction of the globe (water) and retrobulbar fat, and in the vertebral body between the disc space (water) and vertebral body (fat in the bone marrow) (Fig. 4.2). Chemical shift also occurs at water–iophendylate (a lipid-containing contrast material used for myelograms) and water–silicon interfaces.

*A voxel is the three-dimensional unit–volume of a digitally imaged object. A pixel is the unit picture element, the two-dimensional, smallest component of a digital image.

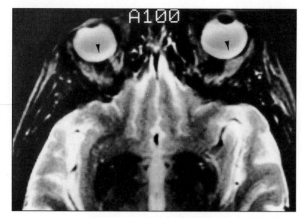

Fig. 4.2 Chemical shift on MRI. A T2-weighted axial image through the eye globes and basal ganglia demonstrates crescent-shaped high-signal areas in the posterior aspects of the globe (black arrowheads) due to chemical shift artifact. This artifact is common at water (globe)–fat (orbital fat) interfaces.

Focal distortion or a bright band at interfaces between substances with different magnetic properties is called *susceptibility artifact*. This occurs at tissue interfaces and is caused by local distortion of the magnetic fields. The magnetic field distortion creates dephasing and frequency shifts of nearby protons resulting in T1 and T2 shortening (high signal on T1-weighted images and low signal on T2-weighted images). The susceptibility artifacts are typically seen in the paranasal sinus region, skull base, and pituitary gland. Short TE values, spin-echo sequences, and broader band widths minimize this artifact. The magnetic susceptibility effect is greatest on gradient-echo sequences and is used to detect para-magnetic and ferromagnetic substances.

Alternating parallel light and dark rings immediately adjacent to high contrast interfaces are called *truncation artifact*, *Gibbs artifact*, or *spectral leakage*. Truncation artifact arises as a fundamental consequence of the Fourier transforms used to process MR signals to the images.[3] They appear in both phase and frequency encoding directions. Truncation errors can be minimized by increasing the number of phase-encoding steps (up to 256) or reducing the field of view. However, this artifact can never be completely eliminated. Truncation artifact is especially disruptive in images of the spinal cord, where it can mimic a syrinx. It often appears in the brain periphery as alternating light and dark bands, giving an onion-skin appearance. Use of different pulse sequences and the axial orientation can help to differentiate this artifact from pathology.

References

1. Henkelman RM, Bronskill MJ. Artifacts in magnetic resonance imaging. *Rev Magn Reson Med* 1987; **2**: 121–6.
2. Szumowski J, Simon JH. Proton chemical shift imaging. In: Stark DD, Bradley WG Jr (eds), *Magnetic resonance imaging*, 2nd edn. St Louis: CV Mosby, 1991: 488–90.
3. Levy LM, DiChiro G, Brooks RA, *et al*. Spinal cord artifacts from truncation errors during MR imaging. *Radiology* 1988; **166**: 479–85.

5. MRI can be dangerous for patients with metallic objects within their bodies

There are no known adverse effects on human tissue from recurrent exposure to MRI. The following devices can safely be scanned with 1.5 T MRI:

- Shunts, Omaya reservoir
- Central venous catheters
- Porta catheters
- Ventriculoperitoneal and atrial shunts
- Coronary artery washers and markers
- Surgical skin staples
- Wires and fixation screws
- Harrington rods
- Silverstone carotid clamps
- Orthodontic braces
- Intrauterine devices (all).

All imaging contraindications are related to metallic or electronic devices within the body. The following are absolute contraindications to MRI:

- Cerebral aneurysm clips (unless the clip was verified as MRI compatible before implantation)
- Cardiac pacemakers
- Neurostimulators, implanted hearing aids.

Several other objects are relative contraindications:

- Insulin pumps
- Some metallic vena cava occlusive devices
- Some cardiac valve prostheses
- Ball-valve-type penile prostheses
- Recently placed surgical clips adjacent to great vessels.

There are three major concerns:

- Impairment of function of a life-sustaining device (such as a pacemaker)
- Deflection of a metal object in a critical location (such as a metal filing within the eye or an aneurysm clip on a cerebral vessel)
- Heating resulting in tissue damage (electric currents, which dissipate energy in the form of heat, are generated in wire loops within fluctuating magnetic fields).

Before any MRI study, the patient must be screened carefully. When the safety of scanning an object or device is in question, the device can be tested for magnetic properties with a hand-held magnet. The review by Shellock and Curtis[1] provides information concerning the behavior of more than 260 metallic implants and foreign bodies exposed to magnetic fields and is an excellent reference for neuroradiologists and MRI technologists.

Intracranial aneurysm clips can move in a strong magnetic field because, even years after surgery, little scar tissue is formed around the clips. A recent patient death caused by a cerebral aneurysm clip torquing with ensuing intracranial hemorrhage resulted in the FDA issuing a warning that extreme caution must be exercised when scanning a patient with known aneurysm clips unless the same clip has tested as compatible with MRI.[2,3] The FDA recommends that, in patients with intracranial aneurysm clips, MRIs should be performed only when the medical benefit is substantial and cannot be achieved otherwise and when the risks can be minimized by positive identification of the implanted clip and verification of its probable magnetic properties. Positive identification of the magnetic properties of an aneurysm clip is not always reliable because the compounds used in the manufacture of a specific clip model can change. In contrast, vascular clips in other parts of the body are usually encased in scar tissue, which prevents deflection after a sufficient time of healing (usually 6 weeks is considered safe).

Intraorbital metallic foreign bodies present a special problem because patients are sometimes unaware of their presence. An intraorbital metal fragment caused hemorrhage leading to blindness in one patient during MRI.[4] If the patient's history suggests an ocular foreign body, plain radiographs or screening CT is needed to identify and locate these objects to determine whether MRI is safe.

References

1. Shellock FG, Curtis JS. MR imaging and biochemical implants, materials and devices: an updated review. *Radiology* 1991; **180**: 541–50.
2. Klucznik RP, Carrier DA, Pyka R, *et al.* Placement of a ferromagnetic intracerebral aneurysm clip in a magnetic field with fatal outcome. *Radiology* 1993; **187**: 855–6.
3. *Caution needed when performing MRI scans on patients with aneurysm clips.* FDA Medical Bulletin 23, 1993.
4. Kelly WM, Paglen PG, Pearson JA, *et al.* Ferromagnetism of intraocular foreign body causes unilateral blindness after MR study. *AJNR Am J Neuroradiol* 1986; **7**: 243–5.

6. *Pregnancy is a relative, but never an absolute, contraindication to radiologic imaging*

Performing either CT (with proper shielding) or MRI during pregnancy is probably safe. However, because of concerns about the unknown risk of these procedures on a developing fetus, pregnancy is a relative, but never an absolute, contraindication to radiologic imaging. If possible, the imaging procedure should be postponed until the second or third trimester, and preferably delayed until after delivery. In women of child-bearing age, an attempt should be made to perform radiologic examinations during the first 10 days of the menstrual cycle to avoid early pregnancy.[1]

If imaging the mother during pregnancy is necessary, MRI is preferable to CT in most circumstances because CT invariably delivers some ionizing radiation to the developing fetus. Even though *in vivo* and *in vitro* studies have not shown any evidence of developmental defects from MRI, pregnant patients should not be examined routinely with MRI.[2] CT is appropriate only when an intracranial bleed (in any compartment) or trauma is suspected or when it is necessary to determine whether a lesion detected on MRI is calcified.

Cervical spine CT and cervical or cerebral angiography pose little risk to the fetus. However, radiographic procedures involving the thoracic or lumbosacral spine (such as conventional myelography) may deliver a substantial amount of radiation to the fetus and should be avoided. MRI should be used if thoracic or lumbosacral spine evaluation is necessary.

Both iodinated and gadolinium contrast media cross the placenta. Although no adverse effects on the human fetus have been identified with either of these agents, their use should be minimized. Thus, for example, contrast should be reserved for angiography and not be administered with CT in the pregnant patient.

If a CT, radiographic, or conventional angiographic procedure is performed, the uterus should be shielded with a leaded apron, although this blocks only a small amount of the scattered radiation. No shielding is possible during MRI.

References

1. Schwartz RB. Neurodiagnostic imaging of the pregnant patient. In: Devinsky O, Feldmann E, Hainline B (eds), *Neurological complications of pregnancy*. New York: Raven Press, 1994: 243–8.
2. Shellock FG. Biological effects of MRI: a clean safety record so far. *Diagn Imaging Clin Med* 1987; February: 96.

7. *Normal radiographic anatomy must be mastered*

Every point in the central nervous system has a name (or multiple names). At a macroscopic level, many connections and functions are ascribed to all the different sites. Following is a series of standard, normal radiographic images with the major anatomical features identified. This abbreviated atlas is intended as a review. All practicing physicians who deal with neurologic diseases should be completely familiar with all of the labeled structures (Figs. 7.1–7.10).

Fig. 7.1 Normal coronal T2-weighted head MRI.

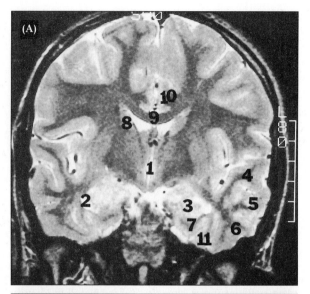

A: Normal Coronal T2-weighted MRI through the third ventricle:
1. third ventricle;
2. temporal horn;
3. hippocampus;
4. superior temporal gyrus;
5. middle temporal gyrus;
6. inferior temporal gyrus;
7. parahippocampal gyrus;
8. head of the caudate nucleus;
9. genu of corpus callosum;
10. cingulate gyrus;
11. fusiform gyrus.

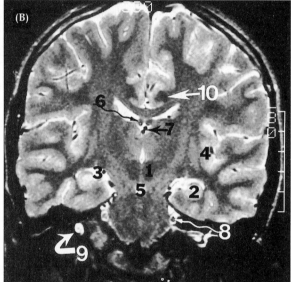

B: Normal coronal T2-weighted MRI at the level of the red nuclei:
1. red nucleus;
2. hippocampus;
3. choroidal fissure;
4. insula;
5. interpeduncular cistern;
6. fornix (wavy black arrow);
7. internal cerebral veins (arrow);
8. fifth cranial nerve (wavy white arrow);
9. cochlea (curved white arrow);
10. cingulum – white matter fasciculus within the cingulate gyrus (straight white arrow).

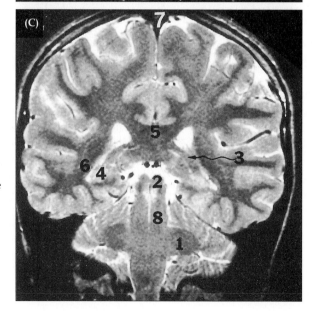

C:. Normal coronal T2-weighted MRI at the level of the quadrigeminal plate:
1. cerebellar white matter – middle cerebellar peduncle;
2. colliculi (quadrigeminal plate);
3. fornix (wavy black arrow);
4. tail of the hippocampus;
5. splenium of corpus callosum;
6. temporal stem white matter;
7. superior sagittal sinus (flow void);
8. pons.

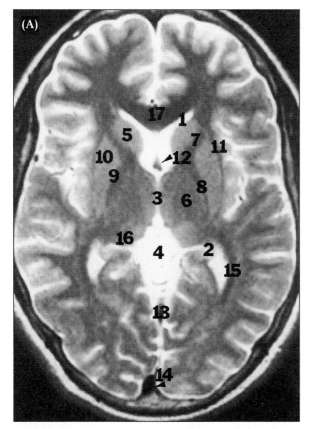 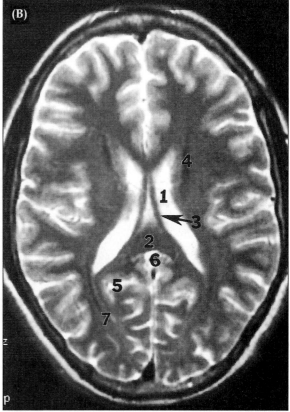

A: Normal axial T2-weighted MRI through the basal ganglia:
 1. frontal horn of lateral ventricle;
 2. choroid plexus of the lateral ventricle;
 3. third ventricle;
 4. quadrigeminal plate cistern;
 5. head of the caudate nucleus;
 6. thalamus;
 7. anterior limb of the internal capsule;
 8. posterior limb of the internal capsule;
 9. globus pallidus;
10. putamen;
11. external capsule;
12. fornix (arrowhead);
13. straight sinus;
14. superior sagittal sinus (arrowhead);
15. optic radiation;
16. pulvinar of the thalamus;
17. genu of the corpus callosum.

Fig. 7.2 Normal axial T2-weighted head MRI.

B: Normal axial T2-weighted MRI through the lateral ventricles:
 1. body of the lateral ventricle;
 2. splenium of corpus callosum;
 3. fornix (arrow);
 4. occipitofrontal fasciculus;
 5. parieto-occipital sulcus;
 6. inferior sagittal sinus;
 7. optic radiation.

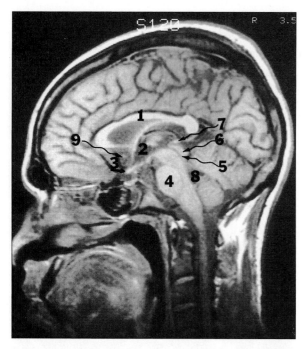

Fig. 7.3 Normal sagittal T1-weighted MRI through the midline:
1. corpus callosum;
2. third ventricle;
3. optic chiasm (straight black arrow);
4. pons;
5. aqueduct of sylvius (wavy black arrow);
6. superior and inferior colliculi (wavy black arrow);
7. pineal gland (wavy black arrow);
8. fourth ventricle;
9. anterior commissure.

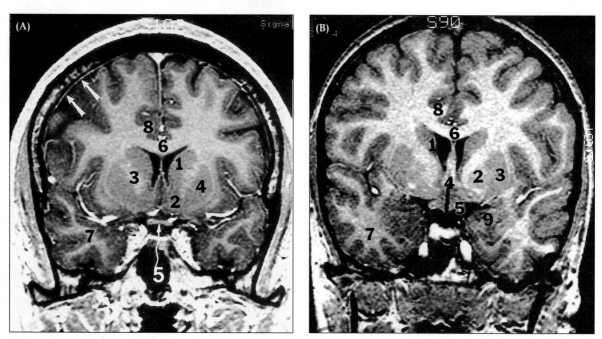

Fig. 7.4 Contrast-enhanced coronal head MRI.

A: Normal coronal gradient-echo (GRE; spoiled gradient recalled acquisition in the steady state, SPGR) T1-weighted MRI with contrast enhancement through the anterior third ventricle: 1, head of the caudate nucleus; 2, nucleus accumbens; 3, anterior limb of the internal capsule; 4, putamen; 5, optic chiasm (wavy white arrow); 6, genu of corpus callosum; 7, anterior temporal lobe; 8, cingulate gyrus; straight white arrows, normal enhancing dura mater (maxim 14).

B: Normal coronal GRE T1-weighted MRI with contrast enhancement through the hippocampal head and amygdala complex: 1, head of the caudate nucleus; 2, globus pallidus; 3, putamen; 4, third ventricle; 5, optic tract; 6, genu of corpus callosum; 7, anterior temporal lobe; 8, cingulate gyrus; 9, hippocampal head and amygdala complex.

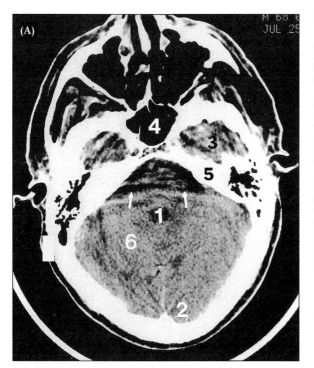

Fig. 7.5 Head CT.

A: Normal noncontrasted axial CT through the posterior fossa: 1, fourth ventricle; 2, occipital lobe (inferior tip); 3, temporal fossa; 4, sphenoid sinus; 5, petrous bone; 6, cerebellum; white arrows, beam-hardening artifacts.

B: Normal noncontrasted axial CT through the suprasellar cistern: 1, suprasellar cistern; 2, interpeduncular cistern (wavy white arrow); 3, superior colliculi; 4, midbrain; 5, sylvian fissure; 6, tip of the basilar artery (wavy white arrow); 7, posterior clinoid (wavy white arrow); 8, ambient cistern (wavy white arrow).

C: Normal noncontrasted axial CT through the frontal horns: 1, frontal horn; 2, third ventricle; 3, pineal gland – calcified (wavy white arrow); 4, choroid plexus – calcified (small straight white arrow); 5, head of the caudate nucleus; 6, thalamus; 7, anterior limb of the internal capsule; 8, posterior limb of the internal capsule; 9, foramen of Monro (wavy white arrow); 10, straight sinus (large straight white arrow).

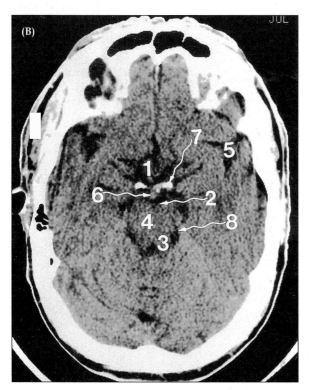

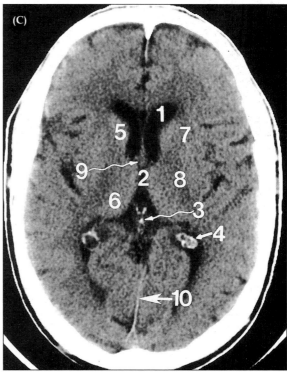

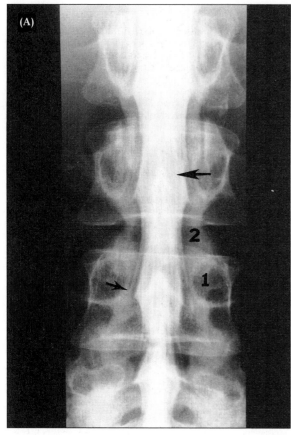

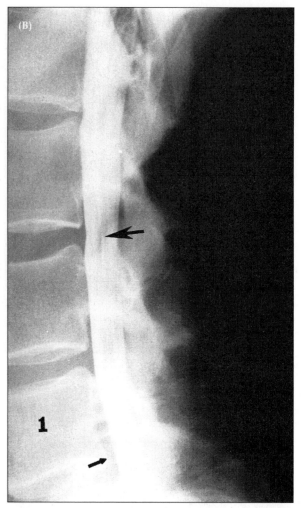

A: Anteroposterior view myelogram: 1, pedicle of L3; 2, L2/L3 disc space; large arrow, cauda equina (nerve roots in thecal sac); small arrow, L3 vertebral body and right L3 nerve root and root sheath.

B: Lateral view myelogram: 1, body of L4; large arrow, cauda equina (nerve roots in the thecal sac); small arrow, epidural space at L4/L5.

Fig. 7.6 Intrathecal contrast myelography. Normal lumbar myelogram and postmyelogram CT.

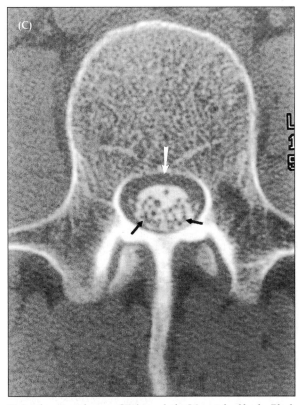

C: Axial postmyelogram CT through the L3 vertebral body. Black arrows, nerve roots in the thecal sac; white arrow, epidural space.

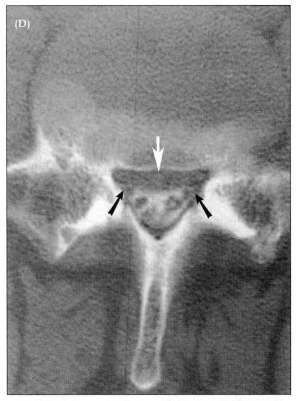

D: Axial postmyelogram CT through the L5 vertebral body. A large disc bulge impinging the dural sac (white arrow) and displacing the L5 nerve roots (black arrows) is present at the L4/L5 disc space. The disc bulge was not apparent on the myelogram.

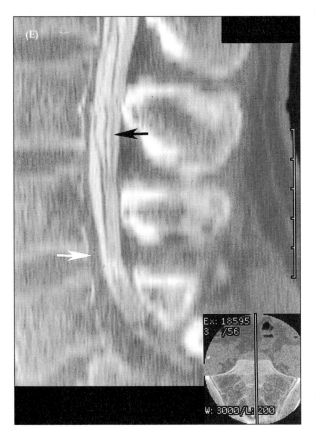

E: Midline sagittal ("lateral view") reconstruction of postmyelogram CT. The nerve roots in the thecal sac (black arrow) and a wide epidural space at L4/L5 (white arrow) due to the disc bulge are discernible. At the L5/S1 disc space, the thecal sac is normally wide.

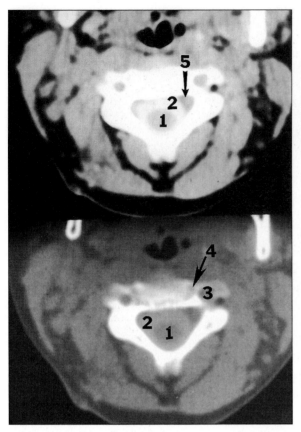

Fig. 7.7 Axial CT through the cervical spine with intrathecal contrast. The upper image was filmed at soft tissue window settings (window width 350, level 40) to optimize visualization of the spinal cord and the surrounding thecal sac. The lower image was filmed at bone window settings (window width 500, level 480). 1, spinal cord; 2, contrast in the thecal sac; 3, foramen transversarium (location of vertebral artery); 4, intervertebral disc space; 5, epidural space.

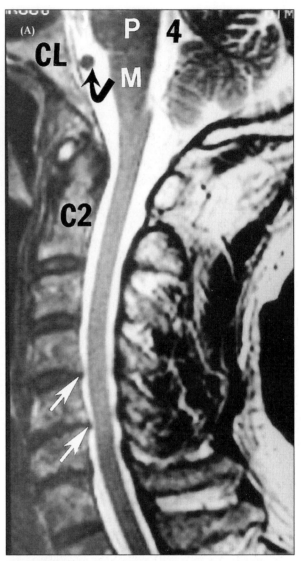

Fig. 7.8 A: Sagittal T2-weighted image of the cervical spine in a 65-year-old woman with mild degenerative changes (osteophytes, straight white arrows): P, pons; M, medulla oblongata; C2, C2 vertebral body; curved black arrow, cross-section of the elongated basilar artery; 4, fourth ventricle; CL, clivus.

Fig. 7.8 MRI of the cervical spine.

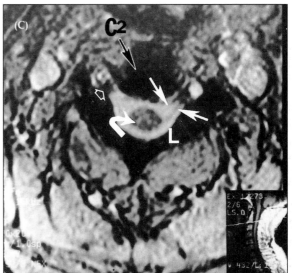

Fig. 7.8 C: Axial T2-weighted gradient-echo image through the C2–3 disc space: C2, C2 vertebral body (black arrow); L, lamina of C2; open white arrow, neural foramen; curved white arrow, spinal cord; straight white arrows, anterior and posterior nerve root.

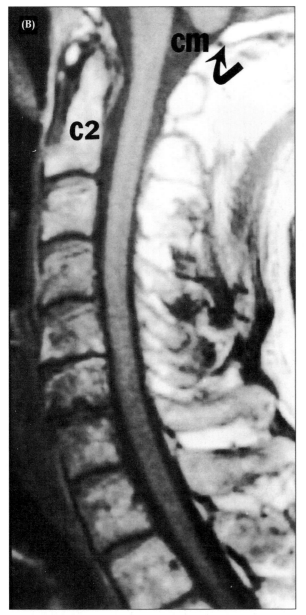

Fig. 7.8 B: Sagittal T1-weighted image through the cervical spine of a 36-year-old man with neck pain and degenerative changes (narrow disc space at C4/C5 and C5/C6): C2, C2 vertebral body; CM, cisterna magna; curved black arrow, posterior lip of the foramen magnum.

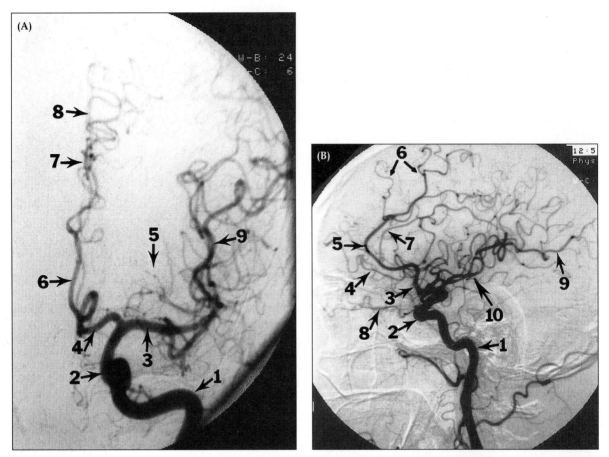

Fig. 7.9 Normal carotid digital subtraction angiography (conventional intra-arterial contrast angiography). **A:** Anteroposterior projection, arterial phase: 1, left internal carotid artery; 2, carotid siphon; 3, M1 segment of the middle cerebral artery; 4, A1 segment of the anterior cerebral artery; 5, lenticulostriate arteries; 6, anterior cerebral artery; 7, pericallosal artery; 8, callosomarginal artery; 9, middle cerebral artery (over insula). **B:** Lateral projection, arterial phase: 1, distal internal carotid artery; 2, carotid siphon; 3, anterior cerebral artery; 4, frontopolar artery; 5, callosomarginal artery; 6, branches off the callosomarginal artery; 7, pericallosal artery; 8, ophthalmic artery; 9, posterior temporal artery; 10, middle cerebral artery group.

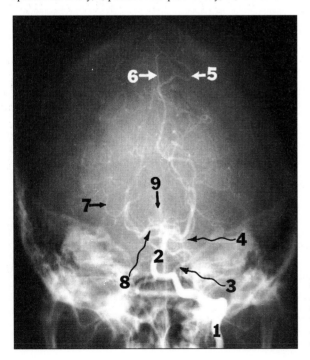

Fig. 7.10 Conventional intra-arterial contrast vertebral angiogram, anteroposterior view, arterial phase: 1, left vertebral artery; 2, basilar artery; 3, posterior inferior cerebellar artery; 4, superior cerebellar artery; 5, parieto-occipital artery; 6, calcarine artery; 7, hemispheric branch of the superior cerebellar artery; 8, posterior cerebral artery; 9, thalamoperforate arteries.

2
Normal Anatomy can Simulate Pathology

8. *Increased PD and T2 signal in the anterior periventricular region is usually ependymitis granularis*

A thin region of increased signal on T2- or PD-weighted images is often present around the frontal horns of the lateral ventricles (Fig. 8.1). This is termed "ependymitis granularis" and is normal, even when asymmetric.[1] Ependymitis granularis probably results from a loose network of axons with low myelin content in the periventricular region with resultant leakage of CSF through the ependyma and into the surrounding white matter.

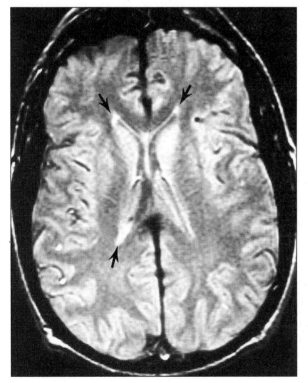

Fig. 8.1 Ependymitis granularis: proton density axial image of a 30-year-old man with headaches. Focal high signal is present around the frontal and occipital horns consistent with ependymitis granularis (arrows). The ventricles are normal in size.

With increasing age, the entire periventricular area can become hyperintense. This more diffuse finding is probably different from ependymitis granularis. Global periventricular hyperintensity ranges from a thin to a thick halo around the ventricles (on T2 and PD images) and may be asymmetric. Hyperintensity of the entire periventricular region has not been correlated to pathology. Care is needed when evaluating this finding as the high signal (especially when asymmetric) can mimic subependymal pathology, such as multiple sclerosis, small vessel disease, or infectious ependymitis.

Reference

1. Sze G, De Armond SJ, Brant-Zawadski M, *et al*. Foci of MR signal (pseudo lesions) anterior to the frontal horns: histologic correlations of a normal finding. *AJR Am J Roentgenol* 1986; **147**: 331–7.

9. *Peritrigonal white-matter T2 hyperintensity is normal in young adults*

The white matter in the dorsal and superior periventricular trigone often has high signal on T2-weighted MRI through young adulthood (Fig. 9.1).[1] This region is the "terminal area of nonmyelination". The white matter in this area largely consists of association fibers that may not myelinate until the second or even third decade of life (especially in people born preterm). The high signal on T2-weighted images should not be confused with demyelinating (multiple sclerosis), dysmyelinating, or other diseases that affect white matter (maxim 19).

Reference

1. Osborn A. *Handbook of neuroradiology*. St Louis: CV Mosby, 1994: 73.

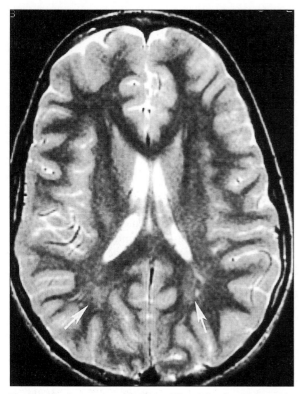

Fig. 9.1 The last area to myelinate. T2-weighted axial image in an 18-year-old male reveals vague high-signal regions posterior to both occipital horns (arrows). This finding is normal through early adulthood. This is the last area of the brain to myelinate.

10. *The pituitary gland undergoes physiologic hypertrophy in pregnancy and puberty*

The pituitary gland normally changes in size and shape during puberty and pregnancy. In children under 12 years of age, the pituitary gland should measure less than 6 mm vertically (from the base of the sella turcica to the location where the pituitary stalk and the gland join). In pubescent girls, the pituitary increases in height, and develops a convex upper surface and sometimes a spherical shape which is normal for this group.[1] A similar change occurs during pregnancy.

In pubescent girls the pituitary gland size and shape normally increase, whereas pubescent boys should have only a transient increase in pituitary size without significant change in overall shape. That is, the pituitary gland is normally larger in adolescent girls (< 10 mm) than in adolescent boys (< 7 mm). A convex upper surface or spherical appearance of the pituitary gland in children, men and boys, and women (except during pregnancy) may indicate a pituitary adenoma and warrants further investigation. In contrast, a similar symmetric change in the shape of the pituitary in a pubescent or pregnant female is an expected finding. Asymmetry of the upper surface always raises the possibility of a pituitary tumor.

In addition to physiologic hypertrophy, MRI reveals small pituitary abnormalities consistent with small adenomas (3–6 mm in greatest diameter) in about 10% of the normal adult population. These appear to remain asymptomatic in the vast majority of patients and do not require treatment.[2]

References

1. Elster AD, Chen MY, Williams DW, *et al*. Pituitary gland MR imaging of physiologic hypertrophy in adolescence. *Radiology* 1990; **174**: 681–5.
2. Hall WA, Luciano MG, Doppman JL, Patronas NJ, Oldfield EH. Pituitary magnetic resonance imaging in normal human volunteers: occult adenomas in the general population. *Ann Intern Med* 1994; **120**: 817–20.

11. *An empty sella is usually an incidental finding*

Recognition of the empty sella syndrome allows differentiation of this benign entity from suprasellar arachnoid cysts and cystic brain tumors. In the empty sella syndrome, the open diaphragma sellae allows the suprasellar cistern to partially herniate into and sometimes enlarge the pituitary fossa.[1] On plain radiographs of the sinuses and skull this condition can simulate a neoplasm in the pituitary area. The empty sella is best visualized on the midline sagittal T1 MR image. The pituitary gland is usually atrophic and displaced posteriorly against the floor of the sella making it difficult to identify. Most patients are asymptomatic, although mild hypopituitarism can occur.[2]

In the empty sella syndrome, the infundibulum is easily identified on MRI (regardless of imaging orientation) as it crosses the prominent CSF space in the suprasellar cistern and pituitary fossa. In contrast, arachnoid cysts cause displacement or nonvisualization of the infundibulum. Contrast enhancement of the border indicates a cystic tumor.

The empty sella syndrome is more common in women than in men, with a peak incidence in the fourth and fifth decades. It is also described in children.

The term "secondarily empty sella" refers to the appearance to the sella turcica after successful treatment of a pituitary lesion. Following pituitary surgery, adhesions can cause retraction of the optic chiasm into the sella with resultant visual problems. The relationship of the optic chiasm to the pituitary fossa is best defined on sagittal and coronal T1 and T2 MR images. Optimal evaluation of this region requires thin sections (3 mm or less), a small or zero interslice gap, and a small field of view (10–18 cm).

References

1. Kaufman B. The "empty" sella turcica: a manifestation of the intrasellar subarachnoid space. *Radiology* 1968; **90**: 931–41.
2. Ekblom M, Ketonen L, Kuuliala I, *et al.* Pituitary function in patients with enlarged sella turcica and primary empty sella syndrome. *Acta Med Scand* 1981; **209**: 31–5.

12. *Large and asymmetric Virchow–Robin spaces are normal; differentiation from gliosis, multiple sclerosis, and infarction is usually possible with MRI*

The perivascular space, also known as the Virchow–Robin space (VR space), consists of CSF surrounding invaginations of the pia as it follows vessels passing from the subarachnoid space into the brain parenchyma. VR spaces exist around the arteries and veins but not the capillaries. Tiny VR spaces are visible in all age groups on brain MRI (especially high field strength) and are normal. Larger, asymmetric VR spaces are common and are also usually normal (Figs. 12.1 and 67.1).[1–3] Because VR spaces are filled with CSF, their signal characteristics follow the pattern of CSF: low signal on T1, isointense to brain on PD, and high signal on T2-weighted images. The vessel contained within the VR space is usually not visualized, although sometimes a flow void is present and appears as a punctate area of low signal in the center of the VR space. Normal VR spaces are not visible on CT.

Typically, VR spaces are present in the inferior third of the basal ganglia (where the lenticulostriate arteries enter the brain creating the anterior perforate substance), the cerebral cortex and centrum semiovale following the course of the larger penetrating cortical vessels. VR spaces may also be present in the pons, midbrain and thalamus.

VR spaces enlarge and become more numerous with increasing age and in hypertensive patients.[4] Meningitis early in life can also result in enlarged VR spaces. Differentiating VR spaces from lacunar infarctions can be difficult (Table 12.1).[5] Lesion location is important. VR spaces are adjacent to the anterior commissure and are bilateral (although sometimes asymmetric). Lacunar infarctions are usually asymmetric or unilateral and are

Table 12.1 Differential diagnosis of common basal ganglion region lesions

Common basal ganglion region lesions	T2	PD	T1
VR space	+	0	–
Gliosis	+	+	0 to –
Small infarction	+	+	–
Infarction with cavitation	+	0 Peripheral +	–
Multiple sclerosis	+	+	0 to –

+ high signal; – low signal; 0 isointense to CSF.

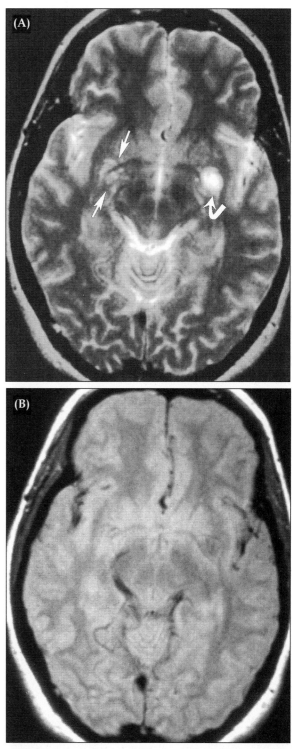

Fig. 12.1 Virchow–Robin (VR) spaces. **A:** T2-weighted axial MRI demonstrates small VR spaces on the right side (straight arrows) and a large VR space on the left side (curved arrow). The VR spaces are due to invaginations of the pia and CSF along the penetrating cerebral arteries. **B:** Proton density image through the same area. Proton density images are needed to distinguish VR spaces from infarctions. Because VR spaces contain CSF, they produce the same signal as the CSF in the ventricles and are essentially isointense to brain on PD images. Old infarctions (gliosis) produce high signal on PD images as in Fig 12.2.

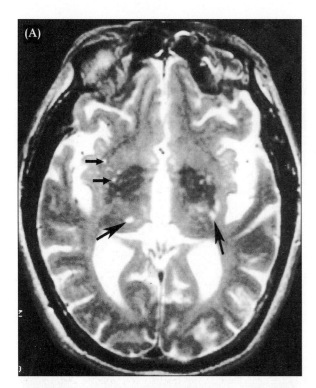

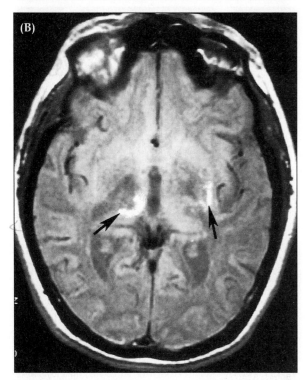

Fig. 12.2 Proton density images are required to differentiate normal Virchow–Robin (VR) spaces (perivascular spaces) from infarcts. **A:** T2-weighted axial image of a 60-year-old demented man with a history of multiple infarcts. Both VR spaces (small arrows) and infarcts (large arrows) generate high signal on this sequence. **B:** Proton density image at the same level. The VR spaces are isointense to the CSF and brain tissue (and are therefore not visualized). The infarcts (gliosis) are easily identified as well-outlined high-signal areas (large arrows).

more rostral, usually in the superior two-thirds of the basal ganglia. VR spaces, lacunae, and gliosis may be indistinguishable on T1- and T2-weighted images but VR spaces "disappear" on PD images, whereas gliosis and lacunae tend to be hyperintense on PD images (Fig. 12.2). In some cases, lacunae have hyperintense edges on PD images but are otherwise isointense to brain tissue.

References

1. Elster AD, Richardson DN. Focal high signal on MR scans of the midbrain caused by enlarged perivascular spaces: MR–pathologic correlation. *AJNR Am J Neuroradiol* 1990; **11**: 1119–22.
2. Cajade-Law AG, Cohen JA, Heier LA. Vascular causes of white matter disease. *Neuroimaging Clin North Am* 1993; **3**: 361–77.
3. Hirabuki N, Fwita N, Fujii K, *et al.* MR appearance of Virchow–Robin spaces along lenticulostriate arteries: spin-echo and two dimensional fast low-angle shot imaging. *AJNR Am J Neuroradiol* 1994; **15**: 277–81.
4. Heier LA, Bauer CJ, Schwartz L, Zimmerman RD, Morgello S, Deck MD. Large Virchow–Robin spaces: MR–clinical correlation. *AJNR Am J Neuroradiol* 1989; **10**: 929–36.
5. Braffman BH, Zimmerman RA, Trojanowski JQ, *et al.* Brain MR: correlation with gross and histopathology: 1. Lacunar infarction and Virchow–Robin spaces. *AJNR Am J Neuroradiol* 1988; **9**: 621–8.

13. *Intracranial calcifications may be normal or pathologic*

Noncontrasted CT is the best modality to detect intracranial calcifications. In the normal adult brain, the choroid plexus, pineal gland, and falx cerebri have dense accumulations of calcium (Fig. 13.1A–C). Calcium may also be present within blood vessel walls, especially in the elderly, due to atherosclerosis. In normal middle-aged adults speckled calcifications may occur in the deep cerebellum and basal ganglia (globus pallidus) (Fig. 13.1A and D).

Visualization of calcium on MRI is suboptimal and can be confusing.[1] Typically low or absent signal is present on all sequences because fat and water are replaced by the calcium and the protons in calcium do not resonate significantly at MRI settings used for imaging. It is difficult to distinguish the signal void produced by calcium from that produced by hemosiderin unless the shape of the region is typical. Often, the signal void is not large enough to be detected in areas with speckled calcification.

Sometimes calcified lesions produce high T1 signal. Particulate calcium (calcium crystals with a large surface area) reduces the T1 relaxation time, presumably due to increased magnetic susceptibility, resulting in higher T1 signal.[2,3] Calcification of the intervertebral disc is a prime example of this phenomenon. Hyperintense discs on T1-weighted images are degenerated and calcified.[4-6] The T2 relaxation time is also reduced, but this just further lowers the T2 signal. Despite this unusual property of particulate calcium, MRI does not reveal calcium that is occult on CT.[7] Gradient-echo sequences and phase imaging can be optimized to detect calcium, but the

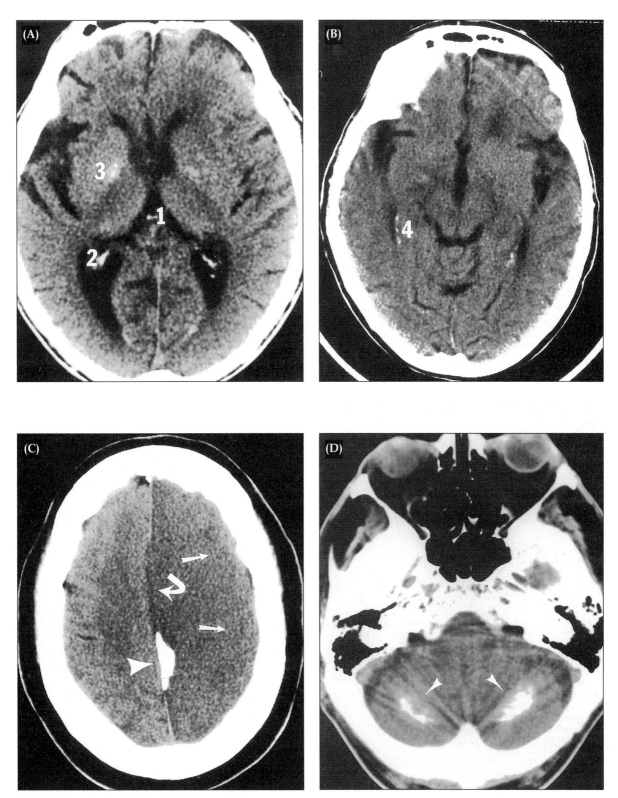

Fig. 13.1 Normal physiologic calcifications on CT.
A: 1, pineal calcification; 2, choroid plexus calcification; 3, idiopathic basal ganglia calcification.
B: 4, calcified choroid plexus in the temporal horn.
C: Normal falx calcification (arrowhead) in an elderly woman with a left hemispheric infarction. Note also the effacement of the sulci (straight arrows) and mild midline shift (curved arrow) (maxim 48).
D: Axial CT through the posterior fossa. Extensive calcifications are present in the deep cerebellum (arrowheads). The patient is normal and has no known metabolic abnormalities.

results are not as reliable as those of CT.[8,9] CT is commonly obtained as a complementary test to MRI to identify calcification.

Certain lesions tend to calcify (maxim 30). These include some tumors (*see below*), vascular malformations (arteriovenous and cavernous malformations), old infectious foci (e.g. cysticercosis), congenital infections (Table 13.1), old hemorrhages, tuberous sclerosis (maxim 26), and certain perinatal conditions (e.g. kernicterus, perinatal hypoxia). Hyperparathyroidism may cause bilateral basal ganglia calcification. Identification of calcium within a suspected tumor narrows the differential diagnosis, as only certain tumors tend to calcify:

- Almost all meningiomas, oligodendrogliomas, craniopharyngiomas, and pineal tumors calcify.
- Some astrocytomas, ependymomas (speckled), choroid plexus papillomas, chordomas, gangliogliomas, dysgerminomas, and any tumor after radiation therapy, calcify.

Table 13.1 Congenital infections (as a group termed TORCH)

T	Toxoplasmosis*
O	Other (syphilis, varicella, HIV, hepatitis B*)
R	Rubella*
C	Cytomegalovirus (CMV)*
H	Herpes

*Congenital infections that result in cerebral calcifications.

Calcium in either the choroid plexus or pineal region in a young child should always raise the possibility of a tumor, as these regions do not physiologically calcify until the second or third decade of life.

Care must be taken not to misinterpret acute (or subacute) hemorrhage as calcium. Examining the CT bone windows before identifying any high-density lesion as calcium should help prevent this error (maxim 61).

References

1. Tsuchiya K, Makita K, Furui S, Nitta K. MRI appearances of calcified regions within intracranial tumours. *Neuroradiology* 1993; **35**: 341–4.
2. Henkelman RM, Watts JF, Kucharczyck W. High signal intensity in MR images of calcified brain tissue. *Radiology* 1991; **179**: 199–206.
3. Boyko OB, Burger PC, Shelburne JD, Ingram P. Non-heme mechanisms for T1 shortening: pathologic, CT, and MR elucidation. *AJNR Am J Neuroradiol* 1992; **13**: 1439–45.
4. Tyrrell PN, Davies AM, Evans N, Jubb RW. Signal changes in the intervertebral discs on MRI of the thoracolumbar spine in ankylosing spondylitis. *Clin Radiol* 1995; **50**: 377–83.
5. Bangert BA, Modic MT, Ross JS, *et al.* Hyperintense disks on T1-weighted MR images: correlation with calcification. *Radiology* 1995; **195**: 437–43.
6. Major NM, Helms CA, Genant HK. Calcification demonstrated as high signal intensity on T1-weighted MR images of the disks of the lumbar spine. *Radiology* 1993; **189**: 494–6.
7. Kucharczyk W, Henkelman RM. Visibility of calcium on MR and CT: can MR show calcium that CT cannot? *AJNR Am J Neuroradiol* 1994; **15**: 1145–8.
8. Henkelman RM, Kucharczyk W. Optimization of gradient-echo MR for calcium detection. *AJNR Am J Neuroradiol* 1994; **15**: 465–72.
9. Gronemeyer SA, Langston JW, Hanna SL, Langston JW Jr. MR imaging detection of calcified intracranial lesions and differentiation from iron-laden lesions. *J Magn Reson Imaging* 1992; **2**: 271–6.

14. *The meninges normally enhance on MRI but not on CT*

Meningeal enhancement on CT is never normal and indicates a pathologic process (maxim 15). However, on MRI, the meninges normally enhance with the extent of normal enhancement dependent on the imaging technique.[1] On 2D SE or FSE images normal meningeal enhancement occurs in discontinuous short segments, whereas on 3D GRE images nearly continuous meningeal enhancement may be present in normal people (Fig. 7.4A).

The meninges are composed of three layers: the dura, arachnoid, and pia mater. The dura consists of two layers that make up the pachymeninges. The outer layer is tightly fused for most of its length to the inner layer. The inner layer invaginates to form the falx cerebri and tentorium cerebelli. Between the inner layer of dura and the arachnoid is the potential subdural space. The subarachnoid space, filled with CSF, lies between the arachnoid and the pia. The pia closely follows the surface of the brain and forms the Virchow–Robin space as it follows blood vessels where they penetrate into the brain parenchyma (maxim 12). Together, the pia and arachnoid make up the leptomeninges.

Normal contrast enhancement on the surface of the brain occurs within the dura mater and the cortical veins. The dural vessels do not have a blood–brain barrier, which probably contributes to enhancement in this layer. In contrast, the arachnoid and pia have a blood–brain barrier and thus do not enhance. However, the cortical veins travelling within these layers do enhance.

Differentiating normal from abnormal meningeal enhancement is mandatory, because many pathologic processes involve the meninges (maxim 15). The major features of normal meningeal enhancement are as follows:

- Thin and smooth
- Follows skull not brain surface (dura)
- Discontinuous on FSE and SE images
- More continuous on 3D GRE images
- Most prominent in the anterior temporal fossa and near the vertex
- Less intense than cavernous sinus enhancement
- Not present on CT.

Reference

1. Farn JW, Mirowitz S. MR imaging of the normal meninges: comparison of contrast enhancement patterns on 3D gradient-echo and spin-echo images. *AJR Am J Roentgenol* 1994; **162**: 131–5.

15. *A thick enhancing dura mater may be due to dural production of CSF in low CSF pressure states (intracranial hypotension)*

Approximately 450 ml of CSF are produced daily in an adult. The choroid plexus generates the vast majority of this CSF. Several other tissues, including the dura mater, produce small amounts of CSF in the normal state. In low CSF pressure conditions, such as with overaggressive shunting or a CSF leak (e.g. after lumbar puncture), the choroid plexus may not be able to produce enough CSF. In these circumstances, production of CSF by the dura may increase. Increased CSF production by the dura is associated with increased blood flow within the dura. In this situation, the dura becomes thickened and can markedly enhance in a slightly inhomogeneous pattern on both CT and MRI. Downward brain displacement may also occur. This condition can reverse with resolution of the intracranial hypotension.[1–3]

Contrast enhancement usually occurs with processes that result in disruption of blood vessel wall integrity, breakdown of the blood–brain barrier or increased blood flow.[4] Meningeal contrast enhancement may result from a variety of causes (Fig. 15.1). Striking meningeal enhancement is present in over three-quarters of patients shortly after craniotomy. Infectious processes are another common cause of abnormally enhancing meninges.[5] However, meningitis should not be diagnosed solely with neuroimaging (maxim 74). The dura can be infiltrated by metastatic carcinoma or sarcoid granulation tissue with resultant pachymeningeal enhancement.[6] Subarachnoid spread of tumor (carcinomatous meningitis) can cause thickening and enhancement of the leptomeninges. Careful examination of the spinal fluid is diagnostic in most of these conditions. The differential diagnosis of abnormal meningeal enhancement is summarized as follows:

1. *Hemorrhage*
   (a) Subacute subarachnoid hemorrhage
   (b) Pial siderosis or fibrosis
2. *Infection or inflammation*
   (a) Bacterial meningitis (including syphilis, Lyme disease)
   (b) Granulomatous meningitis (tuberculosis, fungus, sarcoid)
   (c) Viral meningitis (including AIDS)
   (d) Rheumatoid disease (pachymeningitis)
   (e) Benign leptomeningeal fibrosis:
      • Surgery
      • Shunt
      • Chemotherapy
   (f) Miscellaneous:
      • Histiocytosis X
      • Idiopathic hypertrophic pachymeningitis
      • Vasculitis

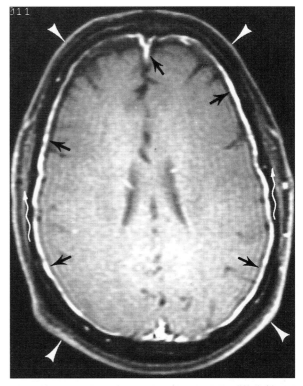

Fig. 15.1 Abnormal pachymeningeal enhancement. An axial contrast enhanced T1-weighted image reveals abnormal bilateral thick dural enhancement (straight black arrows). Note the normal T1-weighted image features of very-high-signal subcutaneous fat (white arrowheads) and modestly high-signal bone marrow within the skull (wavy white arrows). The cortical bone of the skull surrounding the bone marrow is black (very low signal).

3. *Infarction* (overlying a subacute infarction)
4. *Neoplasm*
   (a) Primary meningeal neoplasm: meningioma
   (b) Primary central nervous system tumors with potential for meningeal spread (includes "drop metastases" into spinal canal)
      • Malignant astrocytoma
      • Pineal tumors
      • Medulloblastoma (PNET)
      • Ependymoma
      • Choroid plexus neoplasm
      • Lymphoma
   (c) Orbital tumors with meningeal spread
      • Retinoblastoma
      • Ocular melanoma
   (d) Metastatic extracranial tumors: carcinomatous meningitis (lung, breast, melanoma, and lymphoproliferative malignancies are the most common)
5. *Intracranial hypotension*
   (a) Excessive drainage (shunt)
   (b) CSF leak (trauma, lumbar puncture).

References

1. Pannullo SC, Reich JB, Krol G, Deck MD, Posner JB. MRI changes in intracranial hypotension. *Neurology* 1993; **43**: 919–26.
2. Good DC, Ghobrial M. Pathologic changes associated with intracranial hypotension and meningeal enhancement on MRI. *Neurology* 1993; **43**: 2698–700.
3. Fishman RA, Dillon WP. Dural enhancement and cerebral displacement secondary to intracranial hypotension. *Neurology* 1993; **43**: 609–11.
4. Kilgore DP, Breger RK, Daniels DL, Pojunas KW, Williams AL, Haughton VM. Cranial tissues: normal MR-appearance after intravenous injection of Gd-DTPA. *Radiology* 1986; **160**: 757–61.
5. Sze G, Zimmerman RD. The magnetic resonance imaging of infections and inflammatory diseases. *Radiol Clin North Am* 1988; **26**: 839–59.
6. Phillips ME, Ryals TJ, Kambhu SA, *et al*. Neoplastic *vs.* inflammatory meningeal enhancement with Gd-DTPA. *J Comput Assist Tomogr* 1990; **14**: 536–41.

3
White-Matter Lesions

16. *High-signal T2 white-matter lesions are common: they may be unimportant or indicate serious disease*

Numerous conditions are associated with high-signal T2 and PD white-matter lesions on MRI:

1. *Demyelinating conditions*
   - Multiple sclerosis (MS)
   - Acute disseminated encephalomyelitis (ADEM; postinfectious, postvaccinal)
   - Progressive multifocal leukoencephalopathy (PML; papovavirus)
2. *Metabolic disorders*
   - (a) Leukodystrophies (progressive genetic disorders of myelin metabolism)
     - (i) Diagnosis possible with metabolic studies
       - Metachromatic leukodystrophies (deficiency of arylsulfatase A; autosomal recessive; late infantile, juvenile, and adult onset forms)
       - Krabbe leukodystrophy (globoid cell leukodystrophy; deficiency of galactocerebrosidase; infantile onset)
       - Adrenoleukodystrophy (elevated very long chain fatty acids in serum and skin fibroblasts; X-linked recessive; childhood onset)
       - Canavan disease (spongy degeneration of the nervous system; megalencephaly; deficiency of the enzyme aspartoacylase in skin fibroblasts; autosomal recessive; infantile onset)
     - (ii) Diagnosis presently requires pathologic examination
       - Alexander disease (megalencephaly; Rosenthal fibers on pathologic examination; infantile onset)
       - Pelizaeus–Merzbacher disease (a sudanophilic leukodystrophy; X-linked recessive; infantile onset, survival to adulthood)
       - Cockayne syndrome (microcephaly; autosomal recessive; early childhood onset)
   - (b) Other storage disorders (most are due to a deficient catabolic lysosomal enzyme with accumulation of abnormal material in the lysosomes; most are autosomal recessive, some are X-linked recessive)
     - Lipidoses (group of disorders with excessive storage of lipids; includes gangliosidoses (Tay–Sachs, Sandoff), neuronal ceroid lipofuscinosis, others; onset at all ages)
     - Mucopolysaccharidoses (group of disorders with deficiency of a lysosomal hydrolase resulting in accumulation of mucopolysaccharide; onset at all ages)
     - Mucolipidoses (group of disorders with excessive urinary oligosaccharides or glycopeptides; includes sialidoses; infantile, childhood, and adolescent onset forms)
   - (c) Other metabolic disorders
     - Organic acidurias (includes genetic disorders of amino acid metabolism and fatty acid metabolism; neuroimaging may demonstrate abnormalities in both white matter and basal ganglia)
     - Metabolic acidosis
     - Hypoglycemia
3. *Other conditions*
   - (a) Generally bilateral
     - Hydrocephalus
     - Anoxic injury
     - Chemotherapy
   - (b) Unilateral or bilateral
     - Inflammatory and infectious diseases (HIV, sarcoid, etc.)
     - Tumors (lymphoma, gliomas, etc.)
     - Trauma (shear injury, etc.)
     - Radiation therapy
     - Strokes (embolic or thrombotic)
     - Cerebrovascular dementia
     - Vasculitis
     - Migraine
     - Arteriopathies (nonvasculitic), e.g. CADASIL – cerebral autosomal dominant arteriopathy with subcortical infarctions and leukoencephalopathy.

In addition to these disorders, high-signal lesions, without apparent clinical symptomatology, occur in normal individuals and are increasingly frequent with advancing age.[1-4] High-signal lesions within the white matter are termed "unidentified bright objects" (UBOs) when the pathophysiology or cause is not understood.

When the white-matter changes are prominent, the term leukoaraiosis is sometimes applied.[5] Hopefully, as neuroimaging technology advances and clinical pathologic correlations improve, both of these vague terms will vanish. CT is less sensitive for detecting small white-matter changes, although more widespread white matter changes appear as low densities. The patient's age and clinical history should be used to narrow the radiologic differential diagnosis for white-matter disease.

Numerous inborn errors of metabolism present primarily as white-matter brain disease (*see above*). Most of these are autosomal recessive conditions and present in infancy or childhood. However, some present in adulthood (e.g. adult-onset ceroid lipofuscinosis – Kuf disease). Several of these hereditary metabolic disorders have characteristic patterns on MRI. Demyelinating disease, in particular multiple sclerosis (MS), often has a typical MRI appearance, but its presentation can be variable (maxims 19 and 20).

Vascular disease is probably the most common cause of white-matter lesions. Strokes within the white matter are more numerous in the elderly but also occur frequently in young patients with hypertension, diabetes, or a rarer predisposing condition such as homocystinuria, sickle cell anemia, and hypercoagulable states (e.g. antithrombin III, protein C, and protein S deficiency, resistance to activated protein C, or antiphospholipid antibody syndrome). Extracranial carotid artery disease does not correlate with white-matter disease.[6] Vasculitis and nonvasculitic arteriopathies (e.g. congophilic angiopathy, CADASIL, and possibly migraine) may cause strokes primarily within the white matter. UBOs are commonly present in migraine sufferers.[7,8] The lesions in most of these vascular conditions also involve gray matter in many of the affected patients.

Differentiating pure white-matter infarctions from MS can be difficult. Infarctions tend to be more peripherally located than do MS plaques (maxim 20). Infratentorial lesions are more common in MS than in vascular disease, especially in children and adolescents.[9]

Metastatic cancer usually involves the gray–white matter junction, although metastases can be confined to the white matter and identified as a UBO. Acute strokes, cancer, and abscesses typically have surrounding edema and may have ring enhancement. MS plaques, when active, enhance homogeneously. Occult vascular malformations (cavernomas) minimally enhance and are often calcified. Enhancing lesions are usually not termed UBOs.

Virchow–Robin (VR) spaces should not be mistaken for white-matter disease. Additionally, they should not be classified as a UBO because their nature is understood. A helpful clue is that VR spaces "vanish" on PD images because they are isointense to brain (maxim 12).

References

1. Yetkin FZ, Haughton VM, Papke RA, Fischer ME, Rao SM. Multiple sclerosis: specificity of MR for diagnosis. *Radiology* 1991; **178**: 447–51.
2. Scheltens P, Barkhof F, Leys D, Wolters E, Ravid R, Kamphorst W. Histopathologic correlates of white matter changes on MRI in Alzheimer's disease and normal aging. *Neurology* 1995; **45**: 883–8.
3. Hayman LA. White matter lesions in MR imaging of clinically healthy brains of elderly subjects: possible pathologic basis. *Radiology* 1987; **162**: 509–11.
4. Hendrie HC, Farlow MR, Austrom MG, *et al.* Foci of increased T2 signal intensity on brain MR scans of healthy elderly subjects. *AJNR Am J Neuroradiol* 1989; **10**: 703–7.
5. Hachinci VC, Potter P, Merskey H. Leuko-araiosis. *Arch Neurol* 1987; **44**: 21–3.
6. Streifler JY, Eliasziw M, Benavente OR, Hachinski VC, Fox AJ, Barnett HJM. Lack of relationship between leukoaraiosis and carotid artery disease. *Arch Neurol* 1995; **52**: 21–4.
7. Osborn RE, Alder DC, Mitchell CS. MR imaging of the brain in patients with migraine headaches. *AJNR Am J Neuroradiol* 1991; **12**: 521–4.
8. Soges LJ, Cacayorin ED, Petro GR *et al.* Migraine: evaluation by MR. *AJNR Am J Neuroradiol* 1988; **9**: 425–9.
9. Uhlenbrock D, Seidel D, Gehlen W, *et al.* MR imaging in multiple sclerosis: comparison with clinical, CSF and visual evoked potential findings. *AJNR Am J Neuroradiol* 1988; **9**: 59–67.

17. *Diffuse periventricular T2 and PD hyperintensities are largely inconsequential*

Periventricular T2 and PD hyperintensities of variable thickness often become evident after age 40 years and are present in up to one-quarter of the elderly. These hyperintensities consist of caps over the anterior horns of the lateral ventricles or a diffuse halo (band) surrounding the lateral ventricles. They probably have no impact on cognitive function but indicate an underlying disorder of fluid dynamics. However, they tend to be more prominent in patients with Alzheimer disease (maxim 64).[1]

The pathologic correlate of the periventricular caps and bands is a spongiform zone separated from the ventricular lumen by a rim of subependymal gliosis. There is demyelination in the spongiform zone. Discontinuities within the adjacent ependymal lining are common. These changes are likely the result of chronically elevated quantities of interstitial periventricular fluid. There is no evidence for microscopic infarctions within these lesions and, thus, vascular disease presumably does not play a role in the pathophysiology.[2]

Several possible reasons for increased periventricular fluid have been proposed. Because the brain has no lymphatics, extracellular fluid circulates through the ependyma and into the ventricles (extrachoroidal CSF production). Processes that increase the extracellular fluid volume (such as ischemia or infarction) or impair subependymal fluid resorption (such as increased intracranial pressure associated with hydrocephalus) may result in a backup of fluid in the periventricular space. The tips of the anterior horns have a particularly rich venous plexus, and large veins pass along the ventricular walls making these regions especially vulnerable to venous transudation (as may occur when the central venous pressure is increased in congestive heart failure). Additionally, the discontinuities in the ependymal lining may promote leakage of CSF into the periventricular tissues. The demyelination, gliosis and, to some extent, the increased extracellular fluid result in the T2 hyperintensities surrounding the lateral ventricles.[2,3]

If the periventricular hyperintensities are irregular or extend into the central white matter, a pathologic vascular cause is much more likely (maxim 64). Diffuse periventricular hyperintensities may occur in multiple sclerosis, lymphoma, CSF seeding of intracranial malignant disease, or periventricular infections such as CMV, HIV, or Lyme neuroborreliosis. Contrast enhancement occurs with most of the malignant, infectious, and inflammatory causes, making the differentiation between a disease state and this inconsequential condition fairly easy because age-related hyperintensities do not enhance with contrast.

References

1. Scheltens P, Barkhof F, Leys D, Wolters E, Ravid R, Kamphorst W. Histopathologic correlates of white matter changes on MRI in Alzheimer's disease and normal aging. *Neurology* 1995; **45**: 883–8.
2. Fazekas F, Lieinert R, Offenbacher H, *et al.* Pathologic correlates of incidental MRI white matter signal hyperintensities. *Neurology* 1993; **43**: 1683–9.
3. Boyko OB, Alston SR, Burger PC. Neuropathologic and postmortem MR imaging correlation of confluent periventricular white matter lesions in the aging brain. *Radiology* 1989; **173**: 86 (abstract).

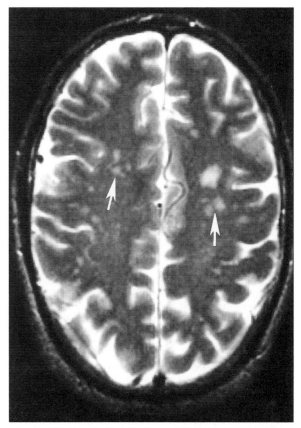

Fig. 18.1 Unidentified bright objects (UBOs). Axial T2-weighted image through the centrum semiovale in a middle-aged woman with headaches. Multiple punctate high-signal lesions are present throughout the white matter (arrows). These UBOs are nonspecific. The patient is neurologically normal.

18. *Subcortical T2 and PD high-signal white-matter lesions are frequently present in the elderly with intact cognition*

Head MRI in "neurologically normal" elderly individuals commonly reveals punctate high-signal T2 and PD lesions ("UBOs" – unidentified bright objects) within the subcortical and periventricular white matter (Fig. 18.1). These hyperintensities are present in 30–80% of people over 40 years old and increase in frequency and size with increasing age, chronic hypertension, other vascular risk factors, past stroke or transient ischemic attack. Although the frequency of scattered punctate UBOs is higher in older age groups, UBOs are also present in up to one-quarter of people under 40 years old.[1] Essentially all individuals 75 years old and older have one or more UBOs.

The exact nature, clinical significance, and pathophysiology of UBOs is not known – hence the name. It is likely that several different causes underlie their presence. They may be the result of an inadequate supply of nutrients necessary to sustain and replace normally catabolized myelin, with the loss of myelin leading to reactive gliosis.[2–5] Atrophic perivascular demyelination appears as high signal on T2 and PD images.

Some, if not most, UBOs are part of a continuum of vascular disease, which in the most extreme form presents as multi-infarct dementia (maxim 64). As the bulk of white-matter disease increases, decreased attention and speed of mental processing is measurable and subjective mental impairment develops.[6,7] However, there are wide variations in the studies to date, and small numbers of UBOs do not appear to be associated with cognitive impairment.

The main issue is differentiating UBOs of "normal aging" from another process (maxim 16). The patient's age and clinical history, and the distribution of the lesions are helpful. Normal perivascular spaces (Virchow–Robin spaces – high signal on T2, isointense to brain on PD) are typically adjacent to the anterior commissure, are present at all ages, and are not UBOs (maxim 12). Migraine sufferers, especially when over 40 years old, have more UBOs than the normal population, usually scattered throughout the white matter (maxim 44). MS lesions are usually situated in the periventricular region. Lacunar infarctions are typically located in the region of the upper basal ganglia. Contrast injection may be warranted to separate nonspecific hyperintensities from small metastases in patients with cancer or to identify small abscesses in patients with immunodeficiency.

References

1. Awad IA, Spetzler RF, Hodak JA, et al. Incidental subcortical lesions identified on magnetic resonance imaging in the elderly. I. Correlation with age and cerebrovascular risk factors. Stroke 1986; 17: 1084–9.
2. Awad IA, Johnson PC, Spetzler RF, et al. Incidental subcortical lesions identified on MRI in the elderly. II. Postmortem pathological correlations. Stroke 1986; 17: 1090–7.
3. Fazekas F, Kleinert R, Offenbacher H, et al. The morphologic correlate of incidental punctate white matter hyperintensities on MR images. AJNR Am J Neuroradiol 1991; 12: 915–21.
4. Braffman BH, Zimmerman RA, Trojanovski JQ, et al. Brain MR: correlation with gross and histopathology. 2. Hyperintense white-matter foci in the elderly. AJNR Am J Neuroradiol 1988; 9: 629–36.
5. Kirkpatrick JB, Hayman A. White matter lesions in MR images of clinically healthy brains of elderly subjects: possible pathologic basis. Radiology 1987; 162: 509–11.
6. Breteler MM, van Swieten JC, Bots ML, et al. Cerebral white matter lesions, vascular risk factors, and cognitive function in a population-based study: the Rotterdam study. Neurology 1994; 44: 1246–52.
7 Ylikoski R, Ylikoski A, Erkinjuntti T, Sulkava R, Raininko R, Tilvis R. White matter changes in healthy elderly persons correlate with attention and speed of mental processing. Arch Neurol 1993; 50: 818–24.

19. *MRI is sensitive but not always specific for the diagnosis of multiple sclerosis*

The diagnosis of multiple sclerosis (MS) is clinical and established when neurologic deficits are disseminated in both time and space. In its fully developed form, demyelinating lesions of differing ages are present in the brain, optic nerves, and spinal cord. Neuroimaging (specifically MRI) can help confirm the diagnosis and is useful for following the course of the disease.[1] The sensitivity of MRI in the diagnosis of MS is over 90%, and MRI is more sensitive than any other standard test including CSF oligoclonal bands, visual-evoked potentials, somatosensory evoked potentials, and CT.[2–4] However, the specificity of MRI is significantly lower than its sensitivity.[5–7] Radiologic differentiation of MS from other diseases that cause white-matter lesions is a major challenge (maxims 16 and 20).

CT has a very limited role in MS, even with double-dose contrast.[8] Mild MS is rarely detectable on CT, and the sensitivity of CT in detecting MS is less than 50%.[9] MS lesions, when visible, have low attenuation on CT. On MRI, multiple isolated or confluent high-signal lesions confined to the white matter on T2 and PD images are typically present (maxim 18). The lesions infrequently involve the gray–white matter junction. Acute and subacute MS plaques can homogeneously enhance with contrast on both CT and MRI. MS plaques do not calcify.

MS can present as a single giant T2 high-signal lesion with or without enhancement and minimal mass effect. In this situation, the diagnosis of MS may be initially overlooked, especially when other typical signs and symptoms are absent. The giant plaque is typically misinterpreted as a tumor or abscess. A large lesion without much mass effect should raise the possibility of MS.[10]

A negative brain MRI does not exclude MS.[8] For instance, a negative brain MRI is present in Devic syndrome (neuromyelitis optica), a form of MS confined to the spinal cord and optic nerves.[11] In patients with myelopathy and a normal appearing spinal cord on MRI, brain lesions may be present. In one MRI series of patients with MS with spinal cord symptomatology, spinal cord lesions with a normal brain MRI were present in 16% and brain lesions with a normal spinal cord MRI were present in 19%.[12]

Isolated optic neuritis, sometimes the first presentation of MS, is associated with two or more white-matter lesions on MRI in over one-quarter of affected individuals. However, MRI rarely identifies another pathologic process in clinically suspected optic neuritis, and the presence or absence of associated white-matter lesions does not clearly predict the development of clinical MS. Thus, MRI of the head is probably not indicated in most cases of isolated optic neuritis.[13]

Several schemes for the diagnosis of MS exist. Initially rigid clinical criteria were used to categorize patients with "definite", "probable", or "possible" MS.[14–16] Recently MRI and other laboratory data were added to these criteria.[6,17]

References

1. Issac C, Genton M, Jardine C, et al. Multiple sclerosis: a serial study using MRI in relapsing patients. Neurology 1988; 38: 1511–15.
2. Yetkin FZ, Haughton VM, Papke RA, Fischer ME, Rao SM. Multiple sclerosis: specificity of MR for diagnosis. Radiology 1991; 178: 447–51.
3. Lee KH, Hashimoto SA, Hooge JP, et al. Magnetic resonance imaging of the head in the diagnosis of multiple sclerosis: a prospective 2-year follow-up with comparison of clinical evaluation, evoked potentials, oligoclonal banding, and CT. Neurology 1991; 41: 657–60.
4. Paty DW, Oger JJ, Kastrukoff LF, et al. MRI in the diagnosis of MS: a prospective study with comparison of clinical evaluation, evoked potentials, oligoclonal banding, and CT. Neurology 1988; 38: 180–5.
5. Uhlenbrock D, Herbe E, Seidel D, Gehlen W. One-year MR imaging follow-up of patients with multiple sclerosis under cortisone therapy. Neuroradiology 1989; 31: 3–7.
6. Poser CM, Paty DW, Scheinberg L, et al. New diagnostic criteria for multiple sclerosis: guidelines for research protocols. Ann Neurol 1983; 13: 227–31.
7. Palo J, Ketonen L, Wikström J. A follow-up study of very low field MRI findings and clinical course in multiple sclerosis. J Neurol Sci 1988; 84: 177–87.
8. Mushlin AI, Detsky AS, Phelps PW, et al. The accuracy of magnetic resonance imaging in patients with suspected multiple sclerosis. JAMA 1993; 269: 3146–51.

9. Runge VM, Price AC, Kirshner HS, Allen JH, Partain CL, James AE Jr. Magnetic resonance imaging of multiple sclerosis: a study of pulse-technique efficacy. *AJR Am J Roentgenol* 1984; **143**: 1015–26.

10. Giang DW, Poduri KR, Eskin TA, *et al*. Multiple sclerosis masquerading as a mass lesion. *Neuroradiology* 1992; **34**: 150–4.

11. Mandler RN, Davis LE, Jeffery DR, Kornfeld M. Devic's neuromyelitis optica: a clinicopathological study of 8 patients. *Ann Neurol* 1993; **34**: 162–8.

12. Ensmond D, DeLaPaz RL, Rubin D. Magnetic resonance of the spine. St Louis: CV Mosby, 1990: 428–30.

13. Beck RW, Arrington J, Murtagh FR, *et al*. Brain magnetic resonance imaging in acute optic neuritis: experience of the optic neuritis study group. *Arch Neurol* 1993; **50**: 841–6.

14. Schumacher GA, Beebe G, Kibler RF, *et al*. Problems of experimental trials of therapy in multiple sclerosis. Report by the panel on the evaluation of experimental trials of therapy in multiple sclerosis. *Ann N Y Acad Sci* 1965; **122**: 552–668.

15. Kurtzke JF. Diagnosis and differential diagnosis of multiple sclerosis. *Acta Neurol Scand* 1970; **46**: 484–92.

16. Matthews WB. Laboratory diagnosis. In: Matthews WB (ed.), *McAlpine's multiple sclerosis*, 2nd edn. New York: Churchill Livingstone, 1991: 190–3.

17. Paty DW, Osborn AK, Herndon RM, *et al*. Use of magnetic resonance imaging in the diagnosis of multiple sclerosis: policy statement. *Neurology* 1986; **36**: 1575–85.

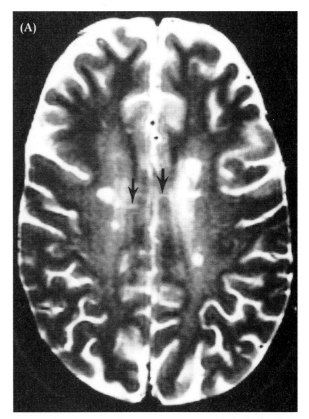

20. *Multiple sclerosis has several characteristic features on MRI*

Multiple sclerosis (MS) lesions often have a characteristic pattern, but many diseases can mimic MS on MRI (maxims 17, 19). MS lesions are most commonly located in the periventricular white matter, adjacent to the body and the trigones of the lateral ventricles. MS plaques are often also present in the corpus callosum, cerebellar hemispheres, middle cerebellar peduncles, and superior and inferior colliculi. The periventricular plaques are characteristically ovoid with the long axis perpendicular to the ventricle surface (Dawson's fingers; Fig. 20.1). This appearance is common and is present in the majority of patients with MS (86% of 59 patients in one study[1]). In chronic MS the periventricular plaques are often confluent (the cumulative effect of multiple attacks). Cortical atrophy and ventricular enlargement are common in chronic MS.

MS lesions are best visualized on PD and heavily T2-weighted MRI. T1-weighted images are less revealing. Fast spin-echo (FSE) can be used to shorten the examination time, because the detection of white-matter lesions with conventional SE and FSE techniques is equal.[2]

A low signal ring encircling a central oval high-signal lesion on a T2-weighted image ("target lesion") is rare, but characteristic of acute MS (Fig. 20.2). The signal characteristics of the ring suggest that it is composed of paramagnetic material (it has a susceptibility effect). The ring likely consists of free radicals within the layer of macrophages that is between the region of demyelination (often surrounding a small vein) and the surrounding edema in an acute plaque.[3]

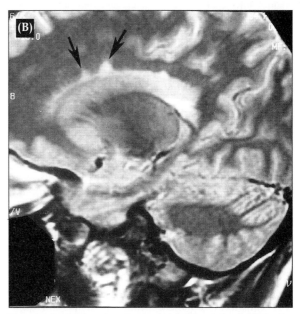

Fig. 20.1 Multiple sclerosis. **A:** Axial T2-weighted MRI of a 40-year-old woman with definite MS. Multiple ovoid, round and oblong lesions are present in the deep periventricular white matter and corpus callosum (arrows). These extend into the centrum semiovale along the deep medullary veins. Periventricular lesions perpendicular to the lateral ventricular surface extending into the deep white matter (Dawson's fingers) are characteristic of MS. Diffuse atrophy is also present. **B:** Parasagittal PD image of a 34-year-old woman with definite MS. The parasagittal image reveals periventricular white-matter MS plaques (arrows).

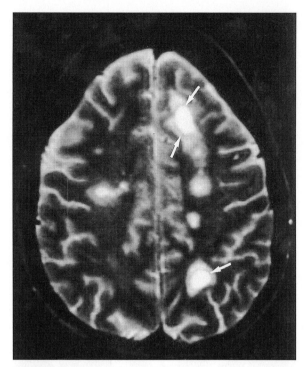

Fig. 20.2 Acute multiple sclerosis ring lesions. Axial T2-weighted image through the centrum semiovale. A low-signal ring is present within two MS lesions (arrows). The signal of the ring is consistent with paramagnetic material. The ring may be the result of free radicals in macrophages at the margins of acute plaques. Edema surrounds the ring. Numerous other plaques are apparent.

Contrast does not significantly help with the differential diagnosis but can define the age of the plaque. Acute plaques may enhance with contrast whereas chronic plaques do not enhance. Disruption of vessel wall integrity (indicated by contrast enhancement) is a marker of an acute MS attack and correlates with histologic activity. Enhancing lesions have moderate-to-marked macrophage infiltration.[4] The enhancement is a transient phenomenon, usually lasting less than 5 weeks,[5] although it may persist for up to 5 months.[4] The enhancement can be homogeneous, patchy, or ring-like (which may "fill in" when imaging is delayed). Acute plaques progressively enlarge, achieving a maximum diameter at approximately 4 weeks after the onset of an attack, and then shrink to some degree.[6–8]

MS lesions characteristically are isointense or rarely hypointense on T1 and hyperintense on T2 and PD weighted images (Fig. 20.1). They contain regions of demyelination and gliotic scarring and, in the acute state, edema. Proton density imaging is ideal for screening because MS plaques adjacent to CSF spaces are readily identified. On heavily T2-weighted images, the high-signal plaque may blend with the normal high-signal CSF within the ventricle, hindering visualization of the lesions.

The best MRI orientations for MS lesion detection are axial and sagittal.[9] Use of thin sagittal T2 sections (4 mm or less) improves detection of tiny low-contrast lesions

within the corpus callosum (Fig. 20.3). Partial volume effects are minimized with this technique. Axial orientation is superior for detecting lesions within the brain stem. Careful evaluation for midline corpus callosum lesions increases the specificity of MRI (to 98%) because vascular lesions are uncommon in the corpus callosum due to its rich blood supply.[10] The sagittal orientation is also useful for identifying corpus callosum atrophy, which is often present in patients with chronic MS.[11,12]

Detailed examination of the corpus callosum is essential because high-signal lesions within the corpus callosum are highly specific for MS and predict dementia. Both clinical and autopsy studies demonstrate that MS patients with intellectual impairment have severe atrophy of the corpus callosum.[13,14] In contrast, neither the number of lesions, the distribution of lesions, nor the extent of diffuse brain atrophy is significantly greater in demented than in nondemented patients with MS. Corpus callosum atrophy correlates better with dementia than with physical disability.

Approximately 90% of adults with MS have premature iron deposition in the putamen and the thalamic pulvinar region (Fig. 20.4). This most likely represents nonheme iron. This nonspecific finding also occurs in other systemic and chronic diseases.[15,16] An increase of nonheme brain iron parallels the decrease in oxygen utilization in aging.[17] Liberation of iron probably plays an important role in free radical propagation and subsequent tissue damage.[18] The abnormal iron accumulation may be due to interruption of iron transport or direct cell injury. The accumulated iron produces a magnetic susceptibility effect that is easily identified with

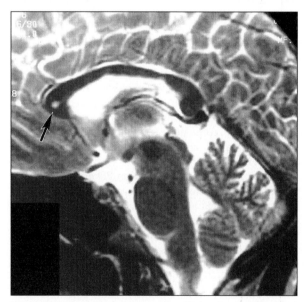

Fig. 20.3 Corpus callosum lesion in multiple sclerosis. Magnified sagittal T2-weighted image through the corpus callosum demonstrating a small MS plaque in the genu (arrow). The location of the plaque, within the corpus callosum, is highly specific for MS. Also note the generalized and corpus callosal atrophy.

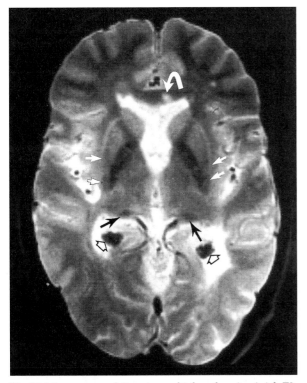

Fig. 20.4 Iron accumulation in multiple sclerosis. Axial T2-weighted image at the level of the basal ganglia in a 32-year-old patient with MS. Abnormal low signal is seen in the pulvinar of the thalami (black arrows) and the posterior putamen (straight white arrows) due to premature accumulation of iron, a finding which is often present in patients with MS. An MS plaque is apparent within the corpus callosum (curved white arrow). Note that calcium in the choroid plexus produces very little signal and is black (signal void; open black arrows).

conventional SE images. The FSE technique is less sensitive to magnetic susceptibility effects and thus it is more difficult to detect premature iron deposition using the FSE sequence.[2]

The use of MRI in MS can be summarized as follows:

- MRI is far more sensitive than CT in detecting MS lesions.
- A negative head MRI does not exclude MS.
- MRI studies need to include PD and T2 axial and sagittal "thin" sections to be adequate.
- Contrast material is not necessary for the diagnosis of MS but can assist in determining the age of an MS lesion.
- Although MRI is sensitive for MS, it is not specific. Other white-matter diseases may mimic MS (maxim 16).

References

1. Horowitz AL, Kaplan RD, Grewe G, *et al.* The ovoid lesion: a new MR observation in patients with multiple sclerosis. *AJNR Am J Neuroradiol* 1989; **10**: 303–5.
2. Norbash AM, Glover GH, Enzmann DR. Intracerebral lesion contrast with spin-echo and fast spin-echo pulse sequences. *Radiology* 1992; **185**: 661–5.
3. Powell T, Sussman JG, Davies-Jones GAB. MR imaging in acute multiple sclerosis: ring-like appearance in plaques suggesting the presence of paramagnetic free radicals. *AJNR Am J Neuroradiol* 1992; **13**: 1544–6.
4. Nesbit GM, Forbes GS, Scheithaujer AW, Okazakitz H, Rodiquez M. Multiple sclerosis: histopathology and MRI and/or CT correlation in 37 cases at biopsy and 3 cases at autopsy. *Radiology* 1991; **180**: 467–74.
5. Miller DH, Rudge P, Johnson G, *et al.* Serial gadolinium enhanced magnetic resonance imaging in multiple sclerosis. *Brain* 1988; **111**: 927–39.
6. Willoughby EW, Grochowski E, Li DK, *et al.* Serial magnetic resonance scanning in multiple sclerosis: a second prospective study in relapsing patients. *Ann Neurol* 1989; **25**: 43–9.
7. Grossman RI, Gonzales-Scarano F, Atlas SW, Galetta S, Silberberg DH. Multiple sclerosis: gadolinium enhancement in MR imaging. *Radiology* 1986; **161**: 721–5.
8. Koopmans RA, Li DK, Oger JJ, Mayo J, Paty DW. The lesion of multiple sclerosis: imaging of acute and chronic stages. *Neurology* 1989; **39**: 959–63.
9. Wilms G, Marchal G, Kersschot E, *et al.* Axial *vs* sagittal T2-weighted brain MR images in the evaluation of multiple sclerosis. *J Comput Assist Tomogr* 1991; **15**: 359–64.
10. Gean-Marton AD, Vezina LG, Marton KI, *et al.* Abnormal corpus callosum: a sensitive and specific indicator of multiple sclerosis. *Radiology* 1991; **180**: 215–21.
11. Simon JH, Holtas SL, Schiffer RB, *et al.* Corpus callosum and subcallosal-periventricular lesions in multiple sclerosis: detection with MR. *Radiology* 1986; **160**: 363–7.
12. Rao SM, Bernardin L, Leo JG, Ellington L, Ryan SB, Burg LS. Cerebral disconnection in multiple sclerosis: relationship to atrophy of the corpus callosum. *Arch Neurol* 1989; **46**: 918–20.
13. Barnard RO, Triggs M. Corpus callosum in multiple sclerosis. *J Neurol Neurosurg Psychiatr* 1974; **37**: 1259–64.
14. Huber SJ, Paulson GW, Shuttleworth EC, *et al.* Magnetic resonance imaging correlates of dementia in multiple sclerosis. *Arch Neurol* 1987; **44**: 732–6.
15. Drayer BP, Burger P, Hurwitz B, Dawson D, Cain J. Reduced signal intensity on MR images of thalamus and putamen in multiple sclerosis: increased iron content? *AJNR Am J Neurodiol* 1987; **8**: 413–19.
16. Milton WJ, Atlas SW, Lexa FJ, Mozley PD, Gur RE. Deep gray matter hypointensity pattern with aging and healthy adults: MR imaging at 1.5 T. *Radiology* 1991; **181**: 715–19.
17. Hallgren B, Sourander P. The effect of age on the non-haemin iron in the human brain. *J Neurochem* 1958; **3**: 41–51.
18. Komara J, Nayini N, Bialick H, *et al.* Brain iron delocalization and lipid peroxidation following cardiac arrest. *Ann Emerg Med* 1986; **15**: 384–9.

21. *Osmotic myelinolysis may involve many regions of the brain (central pontine and extrapontine myelinolysis)*

Central pontine myelinolysis (CPM) was first described in chronic alcoholics in 1959.[1] This demyelinating condition occurs mainly in patients with hyponatremia that is corrected too rapidly.[2] Most affected patients have associated severe medical illnesses including malignancies, sepsis, extensive burns, and advanced liver disease. CPM is rare in patients with the daily cycling hyponatremia of psychogenic polydipsia and the stable chronic hyponatremia of SIADH. It typically occurs in patients with hyponatremia present for several days or longer who are treated to the extent that the sodium corrects at a rate greater than 15 mEq/L per day during the first 1–2 days.

The clinical presentation may be biphasic. Initial improvement of mental status may occur. However, within several days, rapid deterioration occurs with varying degrees of altered mental status (ranging from a behavioral disorder to coma), ophthalmoparesis, bulbar and pseudobulbar palsy, hyperreflexia, and quadriparesis. A locked-in state may result. The long-term outcome extends from full recovery (rare) to death.

Demyelination is classically present within the center of the pons, sparing the periphery.[3] Extrapontine involvement occurs in at least 10% identified by postmortem microscopic examination. Most commonly, the putamen, thalami, caudate nuclei, and central white matter are affected in addition to the pons.[4–6] The relatively myelin-poor medial thalami and pulvinar regions are spared. A better term to describe this condition is osmotic myelinolysis.

MRI performed in the acute phase may be normal. In one case low signal was present on T2-weighted MRI during the first day.[7] Several days or weeks after onset, T2-weighted images reveal high signal within the center of the pons (Fig. 21.1). CT may be normal or demonstrate low density in the same region. Mass effect is minimal or absent and contrast enhancement does not occur. Extrapontine lesions may be present in the areas described above.

The differential diagnosis of osmotic myelinolysis includes stroke, primary and metastatic neoplasm, multiple sclerosis, and encephalitis.[8] If the lesion enhances with contrast one of these other entities is most likely. When thalamic and basal ganglia regions are affected more than the pons, the diagnosis of osmotic myelinolysis may be difficult. In this situation, the medical history and patient's age usually help differentiate osmotic myelinolysis from other white-matter diseases such as Leigh disease, Wilson disease, and Krabbe disease (maxim 16).

References

1. Adams RD, Victor M, Mancall EL. Central pontine myelinolysis: a hitherto undescribed disease occurring in alcoholic and malnourished patients. *Arch Neurol Psychiatr* 1959; **81**: 154–72.
2. Sterns RH, Riggs JE, Schochett SS Jr. Osmotic demyelination syndrome following correction of hyponatremia. *N Eng J Med* 1986; **314**: 1535–42.
3. Thompson AJ, Brown MM, Swash M, *et al.* Autopsy validation of MRI in central pontine myelinolysis. *Neuroradiology* 1988; **30**: 175–7.
4. Laureno R, Karp BI. Pontine and extrapontine myelinolysis following rapid correction of hyponatraemia. *Lancet* 1988; **i**: 1439–41.
5. Wright D, Laureno R, Victor M. Pontine and extrapontine myelinolysis. *Brain* 1979; **102**: 361–85.
6. Koci TM, Chiang F, Chowp, J. Thalamic extra pontine lesions in central pontine myelinolysis. *AJR Am J Roentgenol* 1990; **11**: 1229–33.
7. Rouanet F, Tison F, Dousset V, Corand V, Orgogozo JM. Early T2 hypointense signal abnormality preceding clinical manifestations of central pontine myelinolysis. *Neurology* 1994; **44**: 979–80.
8. Miller GM, Baker HL, Okazaki H, Whisnant JP. Central pontine myelinolysis and its imitators: MRI findings. *Radiology* 1988; **168**: 795–802.

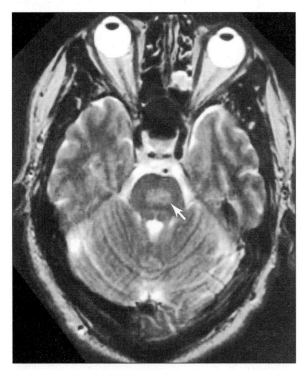

Fig. 21.1 Central pontine myelinolysis (CPM) (osmotic myelinolysis). Axial T2-weighted MRI following rapid correction of hyponatremia. The typical imaging finding of osmotic myelinolysis is present in the pons. The central pons is hyperintense (arrow). Normal signal is present in the periphery of the pons.

4
Epilepsy

22. *Most patients with seizures should be evaluated with neuroimaging*

In general, patients with a new-onset seizure should undergo neuroimaging to search for a lesion that requires early treatment. MRI is the best modality, but CT is adequate to rule out pathology that requires immediate attention. The exception are those patients, predominantly children and adolescents, who present with primary generalized epilepsy, benign rolandic epilepsy or typical febrile convulsions. Important causes for seizures are:[1]

- Meningitis/encephalitis, e.g. bacterial, fungal, viral (herpes)
- Head trauma, e.g. subdural hematoma, cerebral contusion or injury
- Vascular, e.g. arteriovenous malformation, embolic stroke, bland infarction, subdural hematoma
- Metabolic derangements, e.g. uremia, hyponatremia or hypernatremia, hypocalcemia or hypercalcemia, hypomagnesemia, hypoglycemia or hyperglycemia
- Alcohol withdrawal
- Other drugs or chemical toxins, e.g. cocaine, organophosphates
- Neoplasia (primary or metastatic)
- Hereditary, e.g. primary generalized epilepsies, benign rolandic epilepsy, lower seizure threshold, phakomatoses
- Congenital, e.g. brain malformation, perinatal stroke, inborn errors of metabolism
- Anoxia/hypoxia/hypoperfusion, e.g. syncopal seizure
- Degenerative conditions, e.g. Alzheimer dementia, Creutzfeldt–Jakob disease, Rett syndrome.

Approximately one in ten people will have a seizure at some time during their life. About half of these are seizures associated with a febrile illness before the age of 5 years (febrile convulsions). Only one-tenth of patients with a seizure, or 1% overall, develop epilepsy. Epilepsy is a general term that encompasses many different syndromes with recurrent epileptic seizures. The epilepsies can be divided into three large categories: primary generalized epilepsy, localization-related epilepsy, and secondary generalized epilepsy. To optimize management, the epilepsy syndrome should be determined for all patients with epilepsy.

The primary generalized epilepsies are genetically determined with the trait inherited in an autosomal dominant pattern. Childhood absence epilepsy (CAE) and juvenile myoclonic epilepsy (JME) are the major phenotypes. It is likely that the classification and treatment of these disorders will change when the genes responsible are identified. Patients with primary generalized epilepsy have absence seizures, myoclonic seizures, and primarily generalized tonic–clonic seizures. They never have an aura, although they may develop a foggy state from repetitive absence seizures before a major convulsion (major motor seizure; generalized tonic–clonic seizure). The EEG pattern is relatively specific with 3–4 Hz generalized spike and wave discharges. In CAE the EEG discharges and simultaneous absence seizures are precipitated by hyperventilation, and in JME the discharges with absence or myoclonic seizures are precipitated by photic stimulation (photosensitivity). The clinical history and EEG patterns are highly specific for a primary generalized epilepsy and usually no further evaluation is required. Neuroimaging results are normal in this syndrome and, as a rule, need not be performed.

Seizures that arise from a focal region of the cortex are termed focal, partial, or localization-related seizures. Patients with recurrent partial seizures have localization-related epilepsy. Most patients with localization-related epilepsy have intellectual function in the normal range. The initial diagnosis of this condition is often psychosocially devastating. Partial seizures most frequently arise from the temporal lobe. Seizures with no impairment of responsiveness (usually an internal thought or sensation) are termed simple partial seizures (auras). Complex partial seizures involve impairment of responsivity and usually consciousness due to ictal involvement of the limbic system. Automatic behaviors like lip smacking or picking (automatisms) often occur during complex partial seizures, although they may also occur during an absence seizure in a patient with primary generalized epilepsy. Major motor seizures that follow a focal seizure are termed secondary generalized convulsive (tonic–clonic) seizures. Partial frontal lobe seizures often occur out of sleep and may consist of complicated involuntary behaviors (yelling, bicycling, fencing postures) while the patient has preserved awareness – a mortifying experience.

Patients with localization epilepsy who do not become seizure free with the first appropriate antiseizure medication often continue to have seizures for the remainder of their life. Neuroimaging frequently reveals a lesion, either a vascular malformation, infarction, encephalomalacia, or tumor (usually benign or low grade). In patients with temporal lobe epilepsy, mesial temporal sclerosis may be identified on MRI with thin sections through the hippocampus (maxim 24). These patients, especially those with radiologically identifiable lesions, are candidates for epilepsy surgery evaluation. Cerebellar and sometimes generalized atrophy is often present in patients who have taken antiseizure medications (especially phenytoin) for many years (Fig. 22.1). Diffuse cerebral atrophy may be present in patients who have had episodes of status epilepticus.

Benign rolandic epilepsy (BRE) and the other benign idiopathic partial epilepsies have a characteristic presentation, EEG and natural history. The seizures begin during childhood and resolve during adolescence. Neuroimaging is normal and is not necessary in patients with typical BRE. BRE is familial, and once the gene or genes are identified this form of epilepsy is likely to be reclassified with other genetic (developmental) epilepsies.

Secondary generalized epilepsies typically occur in children with developmental delay, although they may begin in otherwise normal children. Most patients with secondary generalized epilepsy have some degree of mental retardation and diffuse cerebral dysfunction. Tonic, atonic, myoclonic, atypical absence, and generalized convulsive seizures predominate in these syndromes but any seizure type may occur. The Lennox–Gastaut syndrome is the best known secondary generalized epilepsy and often develops in children who had infantile spasms. Neuroimaging usually reveals diffuse cerebral atrophy. Evidence for delayed myelination may be present in young children (maxim 77). Cerebral dysplasias or other congenital malformations are sometimes present (maxim 78). Surgical treatment of the secondary generalized epilepsies is limited to palliation with corpus callosotomy in those who have frequent drop seizures.

Three-quarters of patients with epilepsy achieve complete or nearly complete seizure control with antiseizure medications. Some epilepsies are self-limited (e.g. BRE and CAE), and after several years seizures stop. Seizures may persist despite the use of antiseizure medications for a number of reasons:

- Antiseizure medication noncompliance
- Inadequate dosage of antiseizure medication
- Inappropriate antiseizure medication, e.g. using carbamazepine to treat primary generalized epilepsy
- Incorrect diagnosis, e.g. psychogenic seizures, syncopal seizures
- Physical stresses, e.g. sleep deprivation, febrile illness
- Emotional stresses
- Alcohol
- True medication intractability
- Change in lesion, e.g. growing tumor, arteriovenous malformation with hemorrhage.

These patients should be referred for evaluation in an epilepsy center. A substantial portion of patients with medically refractory (intractable) seizures are effectively treated and sometimes cured by surgical removal of the epilepsy focus. Epilepsy surgery is usually reserved for patients with seizure onset in noneloquent cortex (e.g. the anterior temporal lobe). An extensive evaluation including video–EEG monitoring with surface and sometimes intracranial electrodes, neuropsychologic testing, and an intracarotid amobarbital procedure (Wada test) to lateralize language and confirm adequate contralateral memory function is necessary before surgical treatment.[2] Neuroimaging with MRI is invaluable in this evaluation. It often reveals structural lesions that are not otherwise apparent (maxim 24). Occasionally functional imaging with interictal PET (Fig. 22.2) or SPECT and ictal imaging with SPECT (Fig. 22.3) reveal a seizure focus that is not apparent on structural imaging.[3] Relative hypometabolism within a temporal lobe on PET is highly correlated with side of seizure onset, but PET is often not necessary in the surgical evaluation when MRI and EEG give concordant results. CT is used to identify calcification but has low overall sensitivity in patients with epilepsy. MR spectroscopy and functional MRI will likely have a role in the future for identifying seizure foci and in surgical planning.[4]

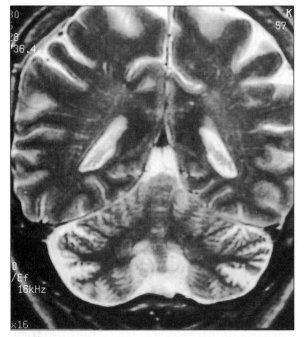

Fig. 22.1 Cerebellar atrophy in epilepsy. Coronal T2-weighted image in a patient with chronic epilepsy and long-term use of phenytoin. Marked cerebellar and modest cerebral atrophy are present. Note the normal linear Virchow–Robin spaces coursing through the white matter.

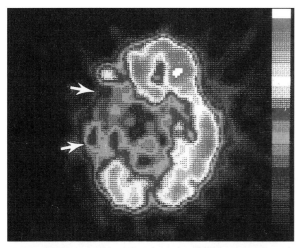

Fig. 22.2 PET in epilepsy. Axial interictal PET image demonstrating decreased left frontal and temporal metabolic function in a patient with medically intractable focal seizures arising in the left hemisphere (arrows). Note that, although not labeled (Fig. 1.1), the left side of the image is the left side of the patient (the opposite of the standard CT and MRI convention).

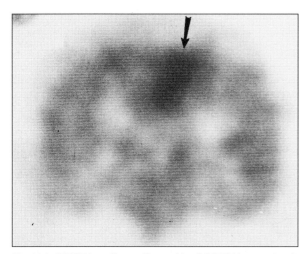

Fig. 22.3 SPECT in epilepsy. Coronal ictal SPECT image revealing an area of markedly increased metabolic activity in the right parasagittal frontal cortex (arrow). Note that, although not labeled (Fig 1.1), the right side of the image is the right side of the patient (the opposite of the standard CT and MRI convention). The interictal study showed normal activity in this region and throughout the brain.

References

1. Porter RJ. *Epilepsy: 100 elementary principles*, 2nd edn. London: WB Saunders, 1989: 9–15.
2. Jeffery PJ, Monsein LH, Szabo Z, *et al*. Mapping the distribution of amobarbital sodium in the intracarotid Wada test by use of Tc-99m HMPAO with SPECT. *Radiology* 1991; **178**: 847–50.
3. Chugani HT. The role of PET in childhood epilepsy. *J Child Neurol* 1994; **9**(Suppl 1): S82–8.
4. Zigun JR, Frank JA, Barrios FA, *et al*. Measurement of brain activity with bolus administration of contrast agent and gradient-echo imaging. *Radiology* 1993; **186**: 353–6.

23. *The hippocampus and its connections are important in memory formation and epilepsy, and are readily discernible on MRI*

The hippocampus is located in the mesial temporal lobe adjacent to the parahippocampal gyrus (Fig. 23.1). These structures have a highly organized architecture with recurrent circuits which are likely related to both the important role they have in forming new memories and their susceptibility to producing seizures.

The hippocampus has an enlarged anterior portion, the pes or head, which is bordered on its superior and anterior margin by the amygdala. The hippocampus extends posteriorly as the body and then tail and ends at the splenium of the corpus callosum. The temporal horn of the lateral ventricle resides just lateral to the hippocampus.[1,2]

The hippocampus consists of two C-shaped gray-matter structures that are hooked together, the cornu ammonis (hippocampus proper) and the dentate gyrus. They are separated by the hippocampal sulcus, a potential space that is normally largely obliterated, although small CSF-filled vesicles often persist. The cornu ammonis is anatomically divided into four fields (in the coronal cross-sectional plane) based on the morphology of the pyramidal neurons: CA1, CA2, CA3, and CA4. CA4 is bordered on three sides by the dentate granule cells of the dentate gyrus. CA2 is sometimes termed the resistant zone because it is usually not affected by conditions that damage the other hippocampal subfields, such as epilepsy and hypoxia. The cells of CA1 are continuous with the subiculum of the parahippocampal gyrus.

The major afferent input to the hippocampus is from the entorhinal cortical cells which reside in clumps in layer two of the anterior parahippocampal gyrus. Excitatory (glutaminergic) axons from the entorhinal cells travel in the angular bundle superiorly and posteriorly, perforating through the subiculum (the perforant path) to innervate the dentate granule cells and, to a lesser extent, the pyramidal cells in CA1. The most prominent hippocampal circuit continues with excitatory (glutaminergic) dentate granule cells innervating pyramidal cells in CA4 and CA3. CA4 and CA3 project into the alveus but CA3 also innervates CA1 via the Schaffer collaterals. CA1 projects back to the subiculum, which then supplies the majority of the alveus.

The alveus, the white matter "coating" of the hippocampus, forms into the fimbriae on the superior mesial surface of the hippocampus. The fimbriae form the fornix at the posterior margin of the hippocampal tail, inferior to the corpus callosum. The fornix travels anteriorly, forming the roof of the third ventricle and eventually terminates in the septal nuclei and mammillary bodies. (One circuit then continues as the mammillothalamic tract into the anterior nucleus of the thalamus. The anterior nucleus connects to the cingulate gyrus, which supplies the entorhinal cortex. This circuit is sometimes referred to as the Papez circuit.)

Most of the structures of the hippocampus can be visualized on high-quality, meticulously performed MRI (Fig. 23.1). The best gray–white matter differentiation is achieved with thin (1.5–3.0 mm) coronal sections obtained with a small field of view (18 cm or less) using a three-dimensional gradient-echo protocol producing T1-weighted images (spoiled gradient recalled acquisition in the steady state (SPGR) or fast low-angle shot (FLASH)). The superb anatomical details improve the ability to detect hippocampal atrophy visually and quantitatively (maxim 24). Interpretation of the MRI requires knowledge of the complex hippocampal anatomy to ensure that abnormal signals are within the hippocampus and do not represent nonanatomical artifacts.[3] T2-weighted spin-echo (long TR) coronal images are useful for evaluating the intrinsic hippocampal signal and other possible pathologies (maxim 24).

The normal hippocampus on coronal MRI is an oval or nearly oval shaped structure that is isointense (or with slightly higher T2 signal) to the temporal neocortical gray matter (the right hippocampus is normal in Fig. 23.1). The subiculum, which occupies the superior curvature of the parahippocampal gyrus, is inferior and medial to the hippocampus. The hippocampus, which consists of primitive cortex (allocortex), is separated from the temporal neocortex by a transitional zone of neurons. The alveus, a compact white-matter tract of efferent axons located between the hippocampus and temporal horn, is often discernible on MRI (open arrows in Fig. 23.1). The fimbriae form the free edge of the alveus on the mesial and superior surface of the hippocampus (wavy white arrow). The fimbriae ultimately form the fornix in the region of the hippocampal tail. The head of the hippocampus is typically identified at the level of the basilar artery by its digitations (infoldings of the various layers of the hippocampus proper; straight black arrows). The amygdala is anterior and superior to the hippocampal head (curved black arrows). The boundary between amygdala and hippocampus is not always well resolved on MRI. Medially the parahippocampal gyrus continues into the uncus. The flatter hippocampal tail forms an arch posteriorly extending to the level of the quadrigeminal plate.

The abnormal left hippocampus in Fig. 23.1 is atrophic with low signal on the T1 and high signal on the T2 weighted image. Most cases of hippocampal sclerosis are not as severe as in this case.

References

1. Duvernoy HM. *The human hippocampus: an atlas of applied anatomy.* Munich: JD Berman, 1988: 1–45.
2. Amaral DG, Insausti R. Hippocampal formation. In: Paxinos G (ed.), *The human nervous system.* San Diego: Academic Press, 1990: 711–56.
3. Hahn FJ, Chu WK, Coleman PE, *et al.* Artifacts and diagnostic pitfalls on magnetic resonance imaging: a clinical review. *Radiol Clin North Am* 1988; **26**: 717–35.

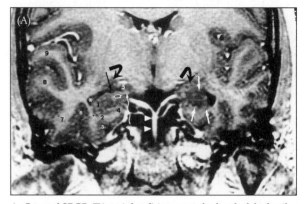

A: Coronal SPGR (T1-weighted) image at the level of the basilar artery (white arrowheads). The right hippocampus is normal and the left hippocampus is atrophic and low signal. 1, hippocampus; 2, parahippocampal gyrus white matter; 3, collateral sulcus; 4, uncus (location of the gray matter of the subiculum); 5, choroidal fissure; 6, occipitotemporal sulcus; 7, inferior temporal gyrus; 8, middle temporal gyrus; 9, superior temporal gyrus; *, temporal horn of the lateral ventricle; straight black arrow, digitations of the hippocampal head; straight white arrows, abnormal left hippocampus; open white arrow, alveus (high signal white matter); wavy white arrow, fimbriae; curved black arrows, amygdala.

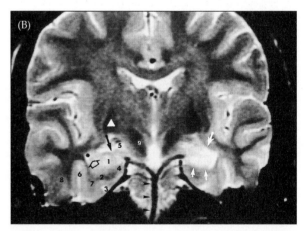

B: Coronal T2-weighted image at the level of the basilar artery (black arrowheads). 1, hippocampus; 2, parahippocampal gyrus white matter; 3, collateral sulcus; 4, uncus (region of the subiculum); 5, choroidal fissure; 6, occipitotemporal sulcus; 7, fusiform gyrus; 8, inferior temporal gyrus; 9, substantia nigra (low signal due to high iron content – maxim 31); white triangle, globus pallidus (low signal due to high iron content – maxim 30); *, temporal horn of the lateral ventricle; black arrow, digitations of the hippocampal head; white arrows, abnormal left hippocampus (high signal); open white arrow, alveus (low signal, white matter).

Fig. 23.1 Normal right hippocampus and abnormal left hippocampus.

24. *The hippocampus is usually small and sclerotic in temporal lobe epilepsy*

Temporal lobe epilepsy is a major cause of seizures that are refractory to antiseizure medications. In this condition, a structural abnormality is often present in the mesial temporal lobe on MRI. Although the relationship between the mesial temporal lesions and the presence of seizures is not fully understood, surgical removal of the anterior and mesial temporal lobe cures or eliminates most medically refactory seizures in over three-quarters of patients with unilateral temporal lobe epilepsy.[1–3] Hippocampal sclerosis (sometimes loosely termed mesial temporal sclerosis) is the single most common lesion associated with medically intractable temporal lobe epilepsy.[4,5] Because surgical removal of a unilateral sclerotic hippocampus is associated with a good outcome, preoperative diagnosis of hippocampal sclerosis is important and may obviate the need for potentially dangerous invasive EEG monitoring.[6,7]

Hippocampal sclerosis is characterized on MRI by a small hippocampus and high T2 and PD signal within the hippocampus (Fig. 23.1). The size of the hippocampus is best evaluated on coronal thin section GRE T1-weighted images (maxim 23). Typically the atrophy and signal change are asymmetric and most prominent in the body and head of the hippocampus. These macroscopic MRI findings correspond to neuronal cell loss and gliosis on pathologic examination.[8–10] The findings on MRI may be quite subtle, and careful review is important. Direct measurement of the hippocampus provides slightly higher sensitivity for detecting hippocampal atrophy, but visual assessment by a trained interpreter is usually adequate.[2,11] The T2 signal can also be quantified using a special sequence to determine the T2 relaxometry of a specific region.[12] The macroscopic internal structure of the hippocampus is disrupted when hippocampal sclerosis is present.

In addition to atrophy and increased T2 and PD signal, the ipsilateral amygdala is sometimes similarly affected.[10] The ipsilateral mesial temporal lobe white matter anterior to the head of the hippocampus produces an abnormal T2 signal isointense to gray matter, especially within the parahippocampal gyrus, in up to two-thirds of patients with medically intractable temporal lobe epilepsy.

It is important not to mistake CSF within the adjacent temporal horn or residual hippocampal fissure vesicles as hippocampal sclerosis. The PD image helps make this differentiation because CSF is isointense to brain and sclerosis is hyperintense. The temporal horns are often asymmetric and suprisingly, sometimes larger on the normal side.

MRI usually clearly delineates lesions other than hippocampal sclerosis that may cause seizures, such as a tumor, cortical dysplasia, or vascular malformation. Invading tumors can enlarge the hippocampus without causing significant signal change. Tumors or vascular malformations immediately adjacent to the hippocampus (e.g. in the parahippocampal gyrus) may be associated with hippocampal sclerosis (dual pathology), but the hippocampus is often normal if the lesion is further away (e.g. in the fusiform gyrus). Intravenous contrast is not helpful in the routine examination of the hippocampus unless a tumor or vascular malformation is present.[13,14] CT and angiography are generally not useful for evaluating the hippocampus.

References

1. Spencer SS. Surgical options for uncontrolled epilepsy. *Neurol Clin* 1986; **4**: 669–95.
2. Jack CR, Sharbrough FW, Twomey CK, *et al.* Temporal lobe seizures: lateralization with MR volume measurements of the hippocampal formation. *Radiology* 1990; **175**: 423–9.
3. Levesque MF, Nakasato N, Vinters HV, Babb TL. Surgical treatment of limbic epilepsy associated with extrahippocampal lesions: the problem of dual pathology. *J Neurosurg* 1991; **75**: 364–70.
4. Falconer MA, Serafetinides EA, Corsellis JA. Etiology and pathogenesis of temporal lobe epilepsy. *Arch Neurol* 1964; **10**: 233–48.
5. Babb TL, Brown WJ. Pathological findings in epilepsy. In: Engel J Jr (ed.), *Surgical treatment of the epilepsies*. New York: Raven Press, 1987: 511–40.
6. Burton CJ. *The neuropathology of temporal lobe epilepsy*. Oxford: Oxford University Press, 1988.
7. VanBuren JM. Complications of surgical procedures in the diagnosis and treatment of epilepsy, in Engel JJ (ed.), *Surgical treatment of the epilepsies*. New York: Raven Press, 1987: 465–75.
8. Bronen RA, Cheung G, Charles JT, *et al.* Imaging findings in hippocampal sclerosis: correlation with pathology. *AJNR Am J Neuroradiol* 1991; **12**: 933–40.
9. Ashtari M, Barr WB, Schaul N, Bogerts B. Three-dimensional fast low-angle shot imaging and computerized volume measurement of the hippocampus in patients with chronic epilepsy of the temporal lobe. *AJNR Am J Neuroradiol* 1991; **12**: 941–7.
10. Berkovic SF, Andermann F, Olivier A, *et al.* Hippocampal sclerosis in temporal lobe epilepsy demonstrated by magnetic resonance imaging. *Ann Neurol* 1991; **29**: 175–82.
11. Jack CR, Bentley MD, Twomey CK, Zinsmeister AR. MR imaging-based volume measurements of the hippocampal formation and anterior temporal lobe: validation studies. *Radiology* 1990a; **176**: 205–9.
12. Jackson GD, Connelly A, Duncan JS, Grunewald RA, Gadian DG. Detection of hippocampal pathology in intractable partial epilepsy: increased sensitivity with quantitative magnetic resonance T2 relaxometry. *Neurology* 1993; **43**: 1793–9.
13. Cascino GD, Hirschhorn KA, Jack CR, Sharbrough FW. Gadolinium-DTPA-enhanced magnetic resonance imaging in intractable partial epilepsy. *Neurology* 1989; **39**: 1115–18.
14. Elster AD, Mirza W. MR imaging in chronic partial epilepsy: role of contrast enhancement. *AJNR Am J Neuroradiol* 1991; **12**: 165–70.

25. *Agyria (lissencephaly) forms a spectrum with pachygyria and laminar heterotopia. These and other neuronal migration abnormalities are well visualized on MRI and often cause epilepsy*

The structure of the human brain forms during the latter portion of the first trimester and second trimester of gestation.[1,2] Neuronal migration is complete by the end of the second trimester. During embryogenesis, neuroblasts reside in the germinal matrix adjacent to the ventricles. Most neuronal migration occurs between 7 and 16 weeks. The neurons migrate along radial glial fibers, guided in part by cell adhesion molecules (CAMs). The layers of the neocortex are formed in an inside-out fashion, with the deepest layer developing first. Differentiation of a neuron follows completion of migration but the morphologic class of each neuron is determined by the order of generation, not the final position. Once neuronal migration is completed, the radial glia disappear. Some radial glia may transform into astrocytes, ependyma, and oligodendroglia.

Agyria, pachygyria, heterotopia, and polymicrogyria are the primary forms of cerebral dysgenesis. Agyria (lissencephaly) refers to a smooth brain without gyri (Fig. 25.1). It is due to arrested migration of the neuroblasts and is the result of an insult or migration error between the 11th and 13th week of gestation. Agyria may be associated with other malformations. Classically a thick four-layered cortex is present. Layer 1 consists of a molecular layer, layer 2 of neurons with normal cortex layer III, IV, and V morphology, layer 3 is cell sparse with tangential myelinated fibers, and layer 4 has heterotopic neurons. Infants with agyria are microcephalic and severely retarded. Usually seizures and other developmental anomalies are present.

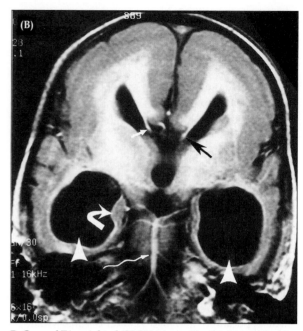

B: Coronal T1-weighted (SPGR) image. The third ventricle rides high and is continuous with the interhemispheric fissure. The lateral ventricles are wing shaped. The fornix is located medial to the lateral ventricles (white and black straight arrows). The Probst bundles (uncrossed callosal fibers) are just superior to the fornix. The temporal horns are large (white arrowheads), and the hippocampi are hypoplastic and vertically oriented (curved arrow). The cortex is thick and smooth. The white matter is not arborized, resulting in a smooth gray–white matter interface. Note the clarity of the vertebral-basilar arterial system in this noncontrasted SPGR image (wavy white arrow).

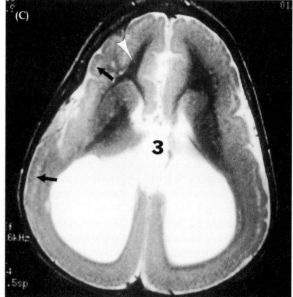

C: Axial T2-weighted image demonstrates colpocephaly (small frontal horns and large occipital horns, not separated by a corpus callosum) and a high riding third ventricle (3) due to the absent corpus callosum. Only a small amount of myelinated white matter is present in the frontal lobes (white arrowhead). The posterior brain is nearly agyric. The thickened cortex is striated, with a layer of high signal dividing gray-matter regions (black arrows).

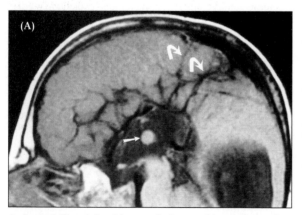

A: Sagittal T1-weighted image of a 3-year-old revealing only a few medial hemispheric sulci (curved arrows) that abnormally radiate into the third ventricle because of the absence of the cingulate gyrus and corpus callosum. The massa intermedia is prominent (straight arrow).

Fig. 25.1 Agyria (lissencephaly) and agenesis of the corpus callosum (maxim 78).

Death typically occurs before 2 years of age. Most cases of agyria are sporadic, but familial forms exist (e.g. Miller–Dieker syndrome due to chromosome 17p deletion).

Brains with a paucity of gyri that are coarse, broadened, and flattened have pachygyria. Pachygyria forms a spectrum with agyria and laminar heterotopia. The insult typically occurs near or after the 13th week. Focal pachygyria is common and often associated with focal heterotopias (cortical dysplasias).

Heterotopia may be nodular or laminar. Nodular heterotopias consist of masses of gray matter in close proximity to the ventricular walls (subependymal). They may be focal (Fig. 25.2) or diffuse (Fig. 25.3). Nodular heterotopias are the most common cortical dysplasia and have multiple causes. Sometimes they are associated with polymicrogyria.

Laminar heterotopias are divided into two types. The first consist of islands of isolated gray matter that sometimes span from the ventricular surface to the cortex. The terminology is not consistent, and this form of laminar heterotopia is sometimes classified as a form of nodular heterotopia or termed a focal cortical dysplasia. Many small focal cortical dysplasias are probably not detectable with any current neuroimaging modality.

The second form of laminar heterotopia is the band heterotopias (double cortex) (Fig. 25.4). Most patients with band heterotopias are female. The bands are likely to be due to an abnormality on the X chromosome. Probably most males with this genetic abnormality die *in utero*; those that survive to birth have lissencephaly. Presumably, band heterotopias are a layer of neurons arrested in mid-migration. They are usually bilateral symmetric ribbons (sheets) of gray matter located within the centrum semiovale that partially follow the gyral convolutions. They are usually fairly symmetric and diffuse, although they may be largely unilateral. They may be thick or thin. Several layers may be present (Fig. 25.4B). The adjacent cortex may be normal or have four layers with associated pachygyria. Band heterotopias were first reported in 1893 in a patient with epilepsy and slight psychomotor retardation. By 1930 twenty cases were present in the literature. With the advent of MRI, band heterotopias of varying degrees are fairly commonly recognized.[3–7] The affected individual may be otherwise normal or may have any degree of impairment up to a severe seizure disorder and mental retardation.

Polymicrogyria (multiple, small, shallow gyri) may be present in small patches, bordering porencephalic cysts, or over the entire cortical surface. Porencephalic cysts are usually the result of an intrauterine infarction and are termed schizencephalic clefts. Typically an abnormal four-layer cortex is present within the polymicrogyric cortex. Polymicrogyria is caused by an insult between 20 and 28 weeks of gestation.

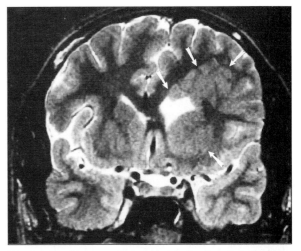

Fig. 25.2 Focal heterotopic gray matter (focal cortical dysplasia; focal laminar heterotopia). Coronal T2-weighted MRI of a 15-year-old patient with intractable seizures. Masses of heterotopic gray matter (arrows) deform the left lateral ventricle which is enlarged, perhaps due to decreased white matter connections.

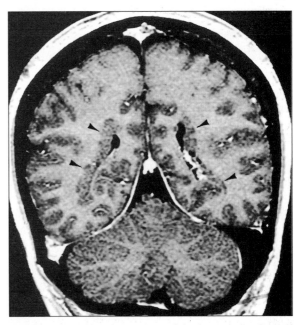

Fig. 25.3 Diffuse periventricular nodular heterotopia. Coronal T1-weighted GRE in a patient with mild multifocal epilepsy and borderline intellectual function. Surrounding the lateral and inferior surface of both lateral ventricles is a rim of nodular gray matter (arrowheads).

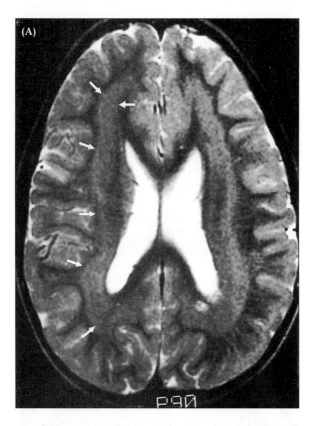

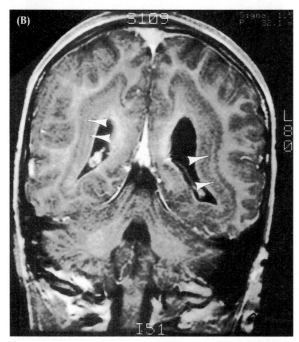

Fig. 25.4 Band heterotopia (double cortex). **A:** T2-weighted axial image reveals a continuous layer of gray matter (arrows) separated from the cortex by a layer of myelinated white matter. **B:** Coronal SPGR (T1-weighted) image showing that the band consists of several layers of heterotopic neurons separated by thin bands of white matter (arrowheads). The surface of the heterotopic band partially follows the enfoldings of the cortical gyri.

References

1. Sarnat HB. *Cerebral dysgenesis: embryology and clinical expression.* New York: Oxford University Press, 1992.
2. Friede RL. *Developmental neuropathology*, 2nd edn. Berlin: Springer-Verlag, 1989.
3. Ricci S, Cusmai R, Fariello G, Fusco L, Vigevano F. Double cortex: a neuronal migration anomaly as a possible cause of Lennox–Gastaut syndrome. *Arch Neurol* 1992; **49**: 61–4.
4. Livingston JH, Aicardi J. Unusual MRI appearance of diffuse subcortical heterotopia or "double cortex" in two children. *J Neurol Neurosurg Psychiatr* 1990; **53**: 617–20.
5. Palmini A, Andermann F, Aicardi J, *et al.* Diffuse cortical dysplasia, or the "double cortex" syndrome: the clinical and epileptic spectrum in 10 patients. *Neurology* 1991; **41**: 1656–62.
6. Barkovich AJ, Guerrini R, Battaglia G, *et al.* Band heterotopia: correlation of outcome with magnetic resonance imaging parameters. *Ann Neurol* 1994; **36**: 609–17.
7. Ketonen L, Roddy S, Lannan M. Band heterotopia. *J Child Neurol* 1994; **9**: 384–5.

26. *CT is warranted for visualization of calcifications in tuberous sclerosis*

Tuberous sclerosis (Bourneville disease), a neurocutaneous disorder (phakomatosis), is inherited as an autosomal dominant trait with an incidence of approximately 1 per 100 000.[1] However, many cases are sporadic. It is characterized by multiple hamartomas of the brain, retina, skin, and viscera. The classic clinical triad is epilepsy, mental retardation, and facial adenoma sebaceum.

Intracranial abnormalities occur in up to 90%.[2] The most common radiologic manifestation, present in 50%, is calcified intracerebral tubers (glial hamartomas), which increase in number with age. CT is required to visualize the parenchymal or subependymal calcifications accurately within the tubers. These calcifications may go undetected on MRI (maxim 13) (Fig. 26.1). On CT, the tubers appear as single or multiple hyperdense (calcified), nonenhancing masses typically located in the subependymal region. In contrast, noncalcified tubers are best visualized with MRI.[3] However, MRI may fail to demonstrate pathologically confirmed cortical tubers.

The radiologic diagnosis of tuberous sclerosis may be difficult during infancy because the tubers are often not yet calcified (the major feature on CT) and the white matter is not fully myelinated, obscuring the differentiation of a tuber from normal tissue on MRI (maxim 76). In infancy, tuberous sclerosis presents with infantile spasms; the only radiologic finding at this age may be ventricular dilatation.

The radiologic differential diagnosis for parenchymal calcifications includes congenital infections such as toxoplasmosis and cytomegalovirus (maxim 13). Both CT and MRI of the head are often warranted in suspected tuberous sclerosis. CT identifies the calcifications and MRI defines the tubers, heterotopias, degree of myelination, and myelination disorders.

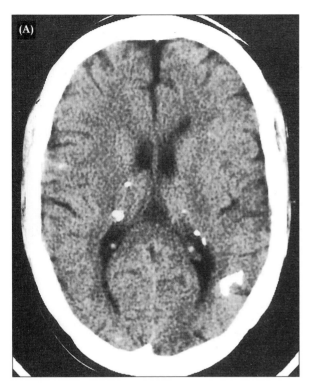

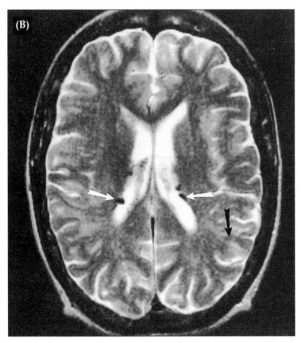

Fig. 26.1 Tuberous sclerosis. **A:** Axial non-contrasted CT of a patient with tuberous sclerosis. Both subependymal and parenchymal calcifications are present. **B:** Corresponding T2-weighted MRI on same patient demonstrates small subependymal calcified nodules (white arrows). The large parenchymal calcification easily visualized on CT is barely visible on MRI (black arrow).

Tubers located in the subependymal region can undergo neoplastic degeneration into giant cell astrocytomas. Subependymal hamartomas and giant cell astrocytomas are opposite extremes of a spectrum. The cortical tubers, unlike the subependymal tubers, rarely, if ever, undergo neoplastic degeneration. Giant cell astrocytomas develop in approximately 15% of patients with tuberous sclerosis. The most common location is near the foramen of Monro. Because tubers usually do not enhance, any abnormal periventricular enhancement is highly suspicious for neoplasm.

Giant cell astrocytomas characteristically appear isodense or hyperdense to brain tissue on CT and are high signal on T1 and T2. They enhance with intravenous contrast.[4] When located near the foramen of Monro, they can cause hydrocephalus.[5] Serial MRI is the best method for following the evolution of the tubers. High-resolution MRI is the primary radiologic modality for planning surgical resection of a subependymal tuber that may be malignant or a cortical tuber causing intractable seizures.

The clinical examination in tuberous sclerosis may be revealing. Hypopigmented macules (ash leaf spots), best seen with ultraviolet light (Wood's lamp), appear in infancy and, when associated with infantile spasms, strongly suggest tuberous sclerosis. The facial adenoma sebaceum (actually hamartomatous fibromas and not adenomas of the sebaceous glands) do not appear until childhood, but are the most characteristic cutaneous lesion. Shagreen patches (a connective tissue hamartoma consisting of subepidermal fibrosis) are coarse, yellow-brown elevated plaques located over the lumbosacral spine and do not become evident until the second decade. Periungual and subungual fibromas also occur during adolescence. Retinal hamartomas (phakomas), renal (often bilateral) angiomyolipomas, and renal cysts occur frequently. Less commonly present are depigmented patches in the iris, cardiac rhabdomyomas, pulmonary lymphangiomyomas and other visceral hamartomas. The wide spectrum of associated pathologic entities requires that all clinical symptoms be carefully considered and evaluated appropriately. Occasional laboratory screening (including urine analysis) and cardiac and renal ultrasonography are warranted.

In infancy tuberous sclerosis may present as West syndrome, which consists of infantile spasms, hypsarrhythmia and mental retardation (or developmental delay). Infantile spasms are brief flexor, extensor, or mixed tonic "spasms", usually occurring in clusters many times per day. Hypsarrhythmia is a characteristic EEG pattern consisting of a high-voltage, slow disorganized background with multifocal epileptiform spikes. West syndrome, which is not specific for tuberous sclerosis, occurs in a wide variety of developmental and acquired abnormalities of the brain. It often develops into the Lennox–Gastaut syndrome (a secondary generalized epilepsy characterized by tonic/atonic, myoclonic, atypical absence, generalized tonic–clonic, and sometimes partial seizures, with an interictal slow spike and wave EEG pattern and associated mental retardation).

References

1. Barkovich AJ, Gressens P, Evrard P. Formation, maturation and disorders of brain neocortex. *AJNR Am J Neuroradiol* 1992; **13**: 423–46.
2. Osborn AG. *Handbook of neuroradiology.* St Louis: Mosby, 1991.
3. Braffman BH, Bilaunik LT, Naidich TP, *et al.* MR imaging of tuberous sclerosis: pathogenesis of this phacomatosis, use of gadopentate dimeglumine, and literature review. *Radiology* 1992; **183**: 227–38.
4. Martin N, Debussche C, De Broucker T, *et al.* Gadolinium-DTPA enhanced MR imaging in tuberous sclerosis. *Neuroradiology* 1990; **31**: 492–7.
5. Gardeur D, Palmieri A, Mashaly R. Cranial computed tomography in phacomatoses. *Neuroradiology* 1983; **25**: 293–304.

27. *Neurofibromatosis type 1 (NF-1) is peripheral neurofibromatosis (von Recklinghausen disease)*

The term neurofibromatosis (NF) encompasses several distinct neurocutaneous syndromes (phakomatoses) with the two most prominent being NF-1 and NF-2. Both NF-1 and NF-2 are autosomal dominant and share the clinical features of multiple cutaneous hyperpigmented lesions (cafe-au-lait spots) and multiple neurofibromas. Neurofibromatosis was first described by von Recklinghausen in 1882. Until recently these separate conditions were lumped together. Thus the literature, especially before the late 1980s, is confusing.

Neurofibromatosis type 1

NF-1 is the most common neurocutaneous disorder, with a frequency of 1 per 3000.[1] The abnormal gene is located on the long arm of chromosome 17. There is nearly 100% penetrance but extremely variable expression. Although it is inherited as an autosomal dominant trait, up to 50% of the cases are due to a spontaneous mutation. The clinical diagnosis is made by identification of two or more of the following features:[2]

- Six or more cafe-au-lait spots with a diameter over 5 mm before puberty and 15 mm after puberty
- Axillary or inguinal freckling
- Two or more neurofibromas or one plexiform neurofibroma
- NF-1 in a first-degree relative
- Two or more iris hamartomas (Lisch nodules)
- Bone lesions – sphenoid wing dysplasia or thinning of the long bone cortex with or without pseudo-arthrosis.

Brain pathology in NF-1

An assortment of CNS lesions can occur in patients with NF-1:

1. *Intracranial lesions*
   (a) Neoplasia
      - Optic nerve glioma
      - Glioma (astrocytoma)
      - Plexiform neurofibroma
      - Neurofibrosarcoma (malignant nerve sheath tumor)
   (b) Non-neoplastic hamartomatous lesions
      - White matter
      - Basal ganglia
2. *Meningeal and skull dysplasias*
   - Hypoplastic sphenoid wing
   - Suture defect
   - Dural ectasia
   - Scoliosis
3. *Vascular lesions*
   - Cerebral arterial occlusions
   - Aneurysm
   - Vascular ectasia
   - Arteriovenous fistula.

MRI is better than CT for detecting neoplasms, unidentified bright objects, and dysplasias of the meninges. CT is better for defining dysplasias of the skull.

In nearly 80% of patients with NF-1, the T2-weighted MRI demonstrates high-signal lesions within the basal ganglia, thalami, optic radiations, brain stem, midbrain, cerebellar peduncles, deep cerebellar white matter, dentate nucleus, or the supratentorial white matter (Fig. 27.1). These lesions are isointense on T1-weighted images. The T2 white-matter hyperintensities are probably malformations (foci of abnormal myelination, hamartomas, or heterotopias) and not neoplastic, although pathologic studies are lacking. The differential diagnosis includes low-grade gliomas and infarctions. When there is no associated mass effect, contrast enhancement, hemorrhage, or vasogenic edema, these lesions are usually benign. The basal ganglia lesions may have mild mass effect with low signal on T1-weighted images and thus represent a different histopathology from the typical white-matter lesions.[3]

White-matter lesions typically increase in size and number in early childhood and then diminish with age and are rare in adults.[3] The risk for neoplastic transformation is low in these transient lesions. They are sometimes termed "UNOs" (unidentified neurofibromatosis objects).[4]

Optic nerve gliomas (ONGs) are the most common intracranial neoplasm in NF-1, occurring in 5–15% (Fig. 27.2). They may involve one or both optic nerves and often extend into the chiasm. Posterior involvement of the optic tracts, lateral geniculate body, and optic radiations may occur. Most ONGs associated with NF-1 are histologically benign (low-grade astrocytomas, pilocytic type) and enlarge slowly. However, in about

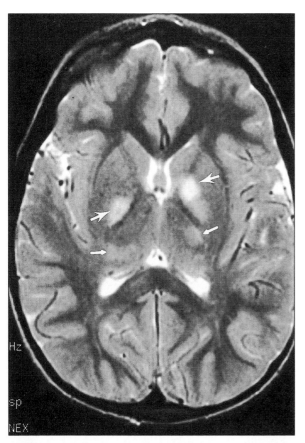

Fig. 27.1 Neurofibromatosis type 1 (NF-1): brain lesions. Axial T2-weighted image through the basal ganglia showing hyperintense foci without mass effect in a 12-year-old with NF-1. Several high signal foci are located in the basal ganglia and thalami bilaterally (arrows). These lesions are most probably caused by foci of abnormal myelination, hamartomatous changes, or heterotopia, and usually disappear by adulthood.

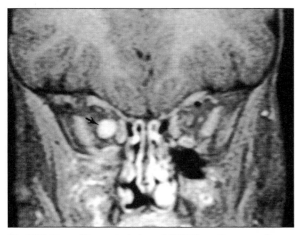

Fig. 27.2 Neurofibromatosis type 1 (NF-1): optic nerve glioma. Coronal T1-weighted image with contrast enhancement and fat suppression in a patient with NF-1 reveals an optic nerve glioma (arrow).

20%, they are aggressive. Yearly MRI is the primary modality for determining which ONGs are aggressive.

Nonoptic gliomas (usually low grade) arise in the brain stem, tectum, and periaqueductal regions. They may result in aqueduct obstruction with hydrocephalus. However, patients with NF-1 may develop aqueduct obstruction without detectable mass lesions.

Plexiform neurofibromas, a hallmark of NF-1, frequently arise in the first (ophthalmic) division of the trigeminal nerve. They do not metastasize but are locally invasive. They are often associated with sphenoid wing dysplasia (empty orbit sign) and may extend posteriorly to involve the cavernous sinus but do not extend beyond Meckel's cave. The "empty orbit" is usually the result of mesodermal dysplasia and is rarely caused by direct erosion from a neurofibroma. When present, the temporal lobe may herniate through the bony defect into the orbit, producing pulsatile exophthalmos. CT with rendered three-dimensional views is a superior method to define the bony pathology.

Spinal pathology in NF-1

Spinal column abnormalities vary from mild scoliosis to severe acute angle cervical kyphosis. Occult spinal tumors are present in over half of asymptomatic patients.[5] These occult tumors include neurofibromas, schwannomas, and meningiomas. Dural ectasia, vertebral body scalloping, and erosion of the adjacent bone with enlargement of the neural foramen may occur. Lateral thoracic meningoceles are common.

References

1. Gardeur D, Palmieri A, Mashaly R. Cranial computed tomography in the phakomatoses. *Neuroradiology* 1983; **25**: 293–304.
2. NIH consensus. Neurofibromatosis. *Arch Neurol* 1988; **45**: 575–8.
3. Sevick RJ, Barkovich AJ, Edwards MS, Koch T, Berg B, Lempert T. Evolution of white matter lesions in neurofibromatosis type 1: MRI findings. *AJR Am J Roentgenol* 1992; **159**: 171–5.
4. Pont MS, Elster AD. Lesions of skin and brain: modern imaging of the neurocutaneous syndromes. *AJR Am J Roentgenol* 1992; **158**: 1193–203.
5. Egelhoff JC, Bates DJ, Ross JS, Rothner AD, Cohen BH. Spinal MR findings in neurofibromatosis types 1 and 2. *AJNR Am J Neuroradiol* 1992; **13**: 1071–7.

28. *Neurofibromatosis type 2 (NF-2) is central neurofibromatosis (bilateral acoustic neuroma syndrome)*

NF-2 is much less common than NF-1. The genetic defect is on the long arm of chromosome 22. Like NF-1, NF-2 has autosomal dominant inheritance and nearly complete penetrance with marked variable expression.[1] Acoustic nerve schwannomas are the hallmark of NF-2. Some authors recommend renaming NF-2 "schwannomatosis" or MISME (multiple inherited schwannomas, meningiomas and ependymomas). This acronym more precisely describes the pathology in NF-2. The clinical diagnosis may be made by identifying any of the following features:[2]

- Bilateral eighth nerve tumors
- Unilateral eighth nerve tumor and a first-degree relative with NF-2
- A first-degree relative with NF-2 and two of the following: neurofibroma, meningioma, schwannoma, glioma, or juvenile posterior subcapsular lenticular opacity.

Brain abnormalities in NF-2

CNS tumors in NF-2 arise from the coverings of the brain and nerves (Schwann cells and meninges). The most common neoplasm arises in the vestibular nerve and appears, on MRI, as a rounded, contrast-enhancing mass within the cerebellopontine angle centered at the external auditory canal (Fig. 28.1A). The trigeminal nerve is the next most frequently involved nerve. Schwannomas may occur in other cranial nerves (with the exception of the olfactory and optic nerves, which are brain tracts).

Intracranial meningiomas are relatively common in NF-2 (Fig. 28.1B). They are often multiple. Prominent choroid plexus calcifications, cerebellar cortex calcifications, and occasionally cerebral cortex calcifications are also features of NF-2 and are best visualized with CT. MRI may miss even dense calcifications (maxim 13).

Spinal abnormalities in NF-2

Intramedullary ependymomas are common in NF-2, as are multiple intradural extramedullary masses (meningiomas or schwannomas) at multiple levels. Bone changes, commonly present in the spine of patients with NF-2, are usually secondary to the spinal cord or nerve root tumors and are not due to dysplasia.

References

1. Sainio M, Strachan T, Blomstedt G, et al. Presymptomatic DNA and MRI diagnosis of neurofibromatosis 2 with mild clinical course in an extended pedigree. *Neurology* 1995; **45**: 1314–22.
2. NIH consensus. Neurofibromatosis. *Arch Neurol* 1988; **45**: 575–8.

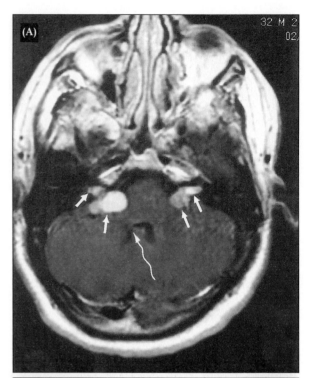

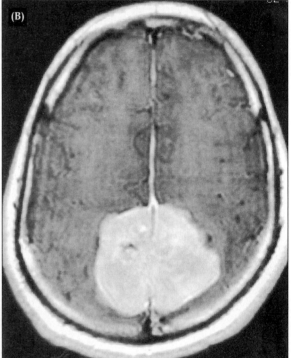

Fig. 28.1 Neurofibromatosis type 2 (NF-2) or MISME (multiple inherited schwannomas, meningiomas and ependymomas). **A:** Axial T1-weighted image with contrast enhancement in a patient with NF-2 reveals rounded contrast-enhancing mass lesions within the cerebellopontine angles centered at the orifice of the internal auditory canal and extending into the canal (straight arrows). These are bilateral acoustic schwannomas. Note the asymmetric compression of the fourth ventricle (wavy arrow). **B:** T1-weighted contrast-enhanced axial image of the same patient at the level of the centrum semiovale. A large enhancing falx meningioma is present.

29. MRI with contrast reveals the leptomeningeal angiomatosis of Sturge–Weber syndrome

Encephalotrigeminal angiomatosis, commonly referred to as Sturge–Weber or Sturge–Weber–Dimitri syndrome, is a neurocutaneous syndrome (phakomatosis) with characteristic radiographic findings. The syndrome consists of an ipsilateral facial port-wine stain, leptomeningeal angiomatosis, focal brain atrophy with neurologic deficits, seizures, mental retardation, and glaucoma. The leptomeningeal angiomatosis typically occurs in the parieto-occipital region (although the temporal and frontal lobes may also be involved) and dramatically enhances on contrasted MRI (Fig. 29.1A).[1] Enlarged deep veins and ipsilateral choroid plexus angiomas are often present. On plain skull radiographs, classic "railroad track" or "tram track" calcifications are

usually observed. These calcifications, which are located in the outer layers of the cortex, are also readily apparent on CT (Fig. 29.1B). Contrasted MRI with GRE sequences is the best single test for diagnosis,[2,3] but CT is complementary and both should be performed.[4-6] Before age 2 years, the calcifications may not be present, and MRI may be the only revealing neuroimaging study.[7]

Most cases of the syndrome are sporadic. A genetic origin is postulated, but no causative gene abnormality has yet been identified. Approximately half of affected patients are mentally retarded; 40–70% develop hemiparesis. Homonymous hemianopsia is usually present when the angiomatosis covers the occipital lobe. When cerebral disease is present, the port-wine stain (nevus flammeus), almost without exception, involves the cutaneous distribution of the ophthalmic division of the trigeminal nerve, including the eyelid. Rarely, there is no cutaneous involvement. The size of the cutaneous lesion, which may extend to involve any region of the trigeminal nerve distribution, does not correlate with the degree of intracranial involvement. Glaucoma and

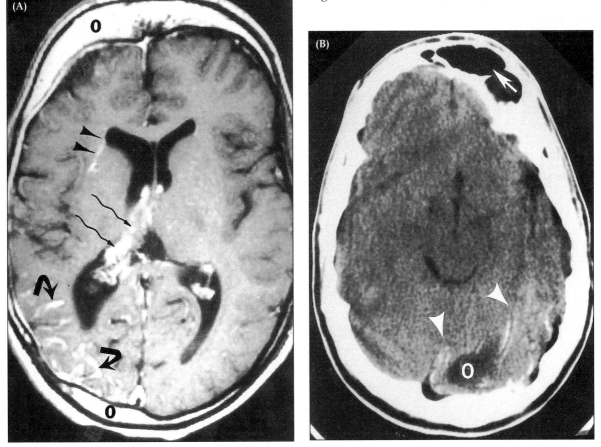

Fig. 29.1 Sturge–Weber syndrome. **A:** Axial T1-weighted image with contrast enhancement in a patient with Sturge–Weber syndrome. The pia enhances over the entire right hemisphere (curved arrows) and the right choroid plexus is enlarged (wavy arrows). An enlarged parenchymal vein enhances (arrowheads). The right hemisphere is smaller with thicker bone marrow (0s) (Dyke–Davidoff–Mason syndrome). **B:** Noncontrasted axial CT of a different patient demonstrates the classic cortical calcification in the region affected by the angioma (arrowheads). The calcifications consist of paired curving lines that follow the gyral topography. They are primarily located in the cortex but are also present in the subcortical white matter. Focal atrophy is present in the same area (prominent sulci and a dilated occipital horn (0)). The entire left hemisphere is smaller, with a thickened skull and dilated frontal sinus (straight arrow).

buphthalmos (enlargement and distention of the fibrous coats of the eye) occur in approximately one-third of patients with Sturge–Weber syndrome. The cerebral and ocular abnormalities are invariably unilateral to the cutaneous lesion. Up to 20% of patients have bilateral cerebral involvement, usually with more severe disease.[8,9] Some 75–90% develop epilepsy, with the majority having seizure onset within the first 2 years of life. The clinical course is highly variable and is sometimes benign.[10]

Episodes of recurrent seizures or status epilepticus can be associated with temporary or permanent focal neurologic deficits. The leptomeningeal malformation results in impaired venous drainage, which is inadequate during the high metabolic activity that occurs during recurrent seizures. MRA, at settings for venous flow, may demonstrate prominent deep venous collaterals and a paucity of superficial cortical veins.[11] This anatomy predisposes to venous congestion occurring during periods of high metabolic demand. Cortical damage can result from the venous congestion, leading to calcifications, focal atrophy, neurologic deficits, and worsening of the seizure disorder.[10,12] If the cortical damage occurs early in life, compensatory skull thickening may occur (Dyke–Davidoff–Mason syndrome).

Surgical treatment, usually involving lobectomy or partial hemispherectomy, is considered when refractory seizures or relapsing or permanent neurologic deficits are present. PET can demonstrate decreased glucose utilization,[13] and SPECT decreased regional cerebral blood flow.[14] PET and SPECT may be used to follow the course of the disease and in evaluation for surgery.

References

1. Benedikt RA, Brown DC, Walker R, Ghaed VN, Mitchell M, Geyer CA. Sturge–Weber syndrome: cranial MR imaging with Gd-DTPA. *AJNR Am J Neuroradiol* 1993; **14**: 409–15.

2. Marti-Bonmati L, Menor F, Poyatos C, Cortina H. Diagnosis of Sturge–Weber syndrome: comparison of the efficacy of CT and MR imaging in 14 cases. *AJR Am J Roentgenol* 1992; **158**: 867–71.

3. Marti-Bonmati L, Menor F, Mulas F. The Sturge–Weber syndrome: correlation between the clinical status and radiological CT and MRI findings. *Childs Nerv Syst* 1993; **9**: 107–9.

4. Elster AD, Chen MY. MR imaging of Sturge–Weber syndrome: role of gadopentate dimeglumine and gradient-echo techniques. *AJNR Am J Neuroradiol* 1990; **11**: 685–9.

5. Chamberlain MC, Press GA, Hesselink JR. MR imaging and CT in three cases of Sturge–Weber syndrome: prospective comparison. *AJNR Am J Neuroradiol* 1989; **10**: 491–6.

6. Wasenko JJ, Rosenbloom SA, Duchesneau PM, Lanzieri CF, Weinstein MA. The Sturge–Weber syndrome: comparison of MR and CT characteristics. *AJNR Am J Neuroradiol* 1990; **1**: 131–4.

7. Sperner J, Schmauser I, Bittner R, *et al.* MR-imaging findings in children with Sturge–Weber syndrome. *Neuropediatrics* 1990; **21**: 146–52.

8. Martina Bebin E, Gomez MR. Prognosis in Sturge–Weber disease: comparison of unihemispheric and bihemispheric involvement. *J Child Neurol* 1988; **3**: 181–4.

9. Pascual-Castroviejo I, Diaz-Gonzalez C, Garcia-Melian RM, Gonzalez-Casado I, Munoz-Hiraldo E. Sturge–Weber syndrome: study of 40 patients. *Pediatr Neurol* 1993; **9**: 283–8.

10. Erba G, Cavazzuti V. Sturge–Weber syndrome: natural history and indications for surgery. *J Epilepsy* 1990; **3**(suppl): 287–91.

11. Vogl TJ, Stemmler J, Bergman C, Pfluger T, Egger E, Lissner J. MR and MR angiography of Sturge–Weber syndrome. *AJNR Am J Neuroradiol* 1993; **14**: 417–25.

12. Wohlwill FJ, Yakovlev PI. Histopathology of meningo-facial angiomatosis. *J Neuropathol Exp Neurol* 1957; **16**: 341–64.

13. Chugani HT, Mazziotta JC, Phelps ME. Sturge–Weber syndrome: a study of cerebral glucose utilization with positron emission tomography. *J Pediatr* 1989; **114**: 244–53.

14. Chiron C, Raynaud C, Dulac O, Tzourio N, Plouin P, Tran-Dinh S. Study of the cerebral blood flow in partial epilepsy of childhood using the SPECT method. *J Neuroradiol* 1989; **16**: 317–24.

5
Movement Disorders

30. *The basal ganglia are susceptible to damage and heavy metal accumulation in a variety of conditions. Iron is normally present in an adult's substantia nigra, red nucleus, and globus pallidus, and in an elder's putamen*

The basal ganglia, which are part of the telencephalon and form the extrapyramidal motor system, consist of the neostriatum (caudate, putamen, and nucleus accumbens) and pallidostriatum (globus pallidus). The putamen and globus pallidus are often grouped as the lentiform nucleus because these two structures together have a lens-like shape. The head of the caudate and the lentiform nucleus are separated anteriorly by the anterior limb of the internal capsule (which contains white matter connecting the frontal lobe and the thalamus). The posterior limb of the internal capsule (which contains motor fibers, sensory fibers, and a portion of the optic radiations) is located between the posterior half of the lentiform nucleus and the thalamus. The thalamus, subthalamus, epithalamus, and hypothalamus constitute the diencephalon. The thalamus can be thought of as the main switching station of the brain because the motor system, the limbic system, and all sensory modalities except olfaction have prominent connections within its many nuclei. The subthalamus is involved in motor control, and lesions of this structure result in contralateral hemiballismus (involuntary large-amplitude, ballistic-like movements).

These extrapyramidal structures are clearly distinguished on MRI (Figs. 7.1–7.5). Iron, which normally and abnormally deposits within several of these structures, is readily apparent on T2-weighted images because its magnetic susceptibility effect results in low signal (Fig. 23.1B; maxim 4).[1,2] Calcium, which deposits in normal and abnormal conditions within some of these structures usually requires CT for identification. GRE images accentuate the appearance of iron and calcium on MRI compared with spin-echo images.

At birth and during childhood, MRI evidence of iron is not present. During the second decade, gradually increasing hypointensity develops within the substantia nigra, red nucleus, and globus pallidus on T2-weighted images.[3] These areas are normally hyperintense with respect to white matter on T2-weighted images at 2 years, isointense to about 15 years, and hypointense by 25 years. The globus pallidus is the first structure to demonstrate this normal T2 shortening (resulting in lower signal), followed by the red nucleus and substantia nigra. These patterns parallel the finding of increasing deposits of iron with increasing age within these structures on pathologic examination.[4] In adults (30 years and older) the globus pallidus has the highest iron content. Iron deposits (normal or abnormal) are easier to detect on images acquired with the conventional spin-echo sequence than with the more widely used fast spin-echo sequence.[5]

The caudate nucleus and thalamus do not normally accumulate iron. Healthy adults younger than 60 years old also have no discernible hypointensity in the putamen.[6] However, in the normal elderly, iron deposition occurs within the putamen.

The basal ganglia are uniquely prone to damage from a variety of rare conditions (Table 30.1; Fig. 30.1). Various therapies can successfully reverse damage to the

Table 30.1 Basal ganglia damage

Acute process	Chronic process
Globus pallidus (high signal on T2-weighted images and low density on CT)	
Carbon monoxide	Wilson disease
Hypoxia	
Cyanide	
Hydrogen sulphate	
Ethylene glycol (antifreeze)	
Globus pallidus (high signal on T1-weighted images)	
	Liver failure (cirrhosis)
	Manganese
Putamen (high signal on T2-weighted images and low density on CT)	
Methanol (Fig. 30.1)	Wilson disease
Basal ganglia (high signal on T2-weighted images)	
Hypoxic/ischemic encephalopathy (including perinatal hypoxia)	Leigh syndrome
	Mitochondrial encephalopathy (MELAS, Kearns-Sayre syndrome)
Severe hypoglycemia	
Osmotic myelinolysis	Wilson's disease
Internal cerebral vein thrombosis	Methylmalonic acidemia

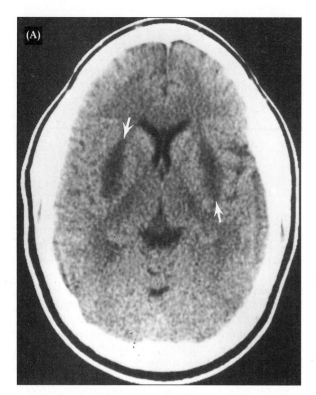

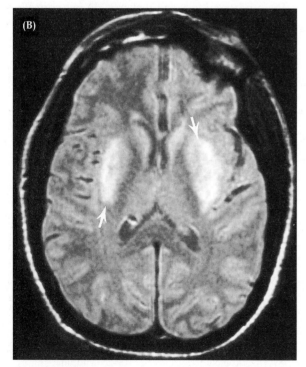

Fig. 30.1 Putamen damage from methanol. **A:** Axial CT of a patient who unsuccessfully attempted to commit suicide by consuming methanol. The putamen is symmetrically low density presumably due to necrosis. **B:** Axial PD-weighted image in the same patient. The putamen produces high signal bilaterally (arrows). The T2-weighted image displayed a similar signal pattern.

basal ganglia in some of the conditions listed. Several disorders are associated with dystrophic calcification within the basal ganglia as follows:

1. *Idiopathic*
2. *Endocrine*
   - Hypoparathyroidism
   - Pseudohypoparathyroidism
   - Pseudopseudohypoparathyroidism
   - Hyperparathyroidism
   - Hypothyroidism
3. *Metabolic*
   - Fahr disease (familial cerebrovascular ferro-calcinosis)
   - Mitochondrial disease
   - Hallervorden–Spatz syndrome (dystrophic calcification in addition to iron deposition)
   - Hypoxia
4. *Postinflammatory*
   - Tuberculosis
   - Toxoplasmosis
   - Rubella
   - Cytomegalovirus
   - Systemic lupus erythematosus
   - Epstein–Barr virus (congenital chicken pox)
   - Human immunodeficiency virus
5. *Toxic*
   - Lead poisoning
   - Radiation therapy
   - Methotrexate therapy
   - Carbon monoxide intoxication.

References

1. Drayer BP, Burger P, Darwin R, Riederer S, Herfkens R, Johnson GA. Magnetic resonance imaging of brain iron. *AJNR Am J Neuroradiol* 1986; **7**: 373–80.
2. Drayer BP. Basal ganglia: significance of signal hypointensity in T2-weighted MR images. *Radiology* 1989; **173**: 311–12.
3. Aoki S, Okada Y, Nishinura K, *et al*. Normal deposition of brain iron in childhood and adolescence: MR imaging at 1.5 T. *Radiology* 1989; **172**: 381–5.
4. Hallgren B, Sourander P. The effect of age on the non-heme iron in the human brain. *J Neurochem* 1958; **3**: 41–51.
5. Norbash AM, Glover GH, Enzmann DR. Intracerebral lesion contrast with spin-echo and fast spin-echo pulse sequences. *Radiology* 1992; **185**: 661–5.
6. Milton WJ, Atlas SW, Lexa FJ, *et al*. Deep gray matter hypointensity patterns with aging in healthy adults: MR imaging at 1.5 T. *Radiology* 1991; **181**: 715–19.

31. *The substantia nigra has increased iron content in idiopathic Parkinson disease, but this is not consistently discernible on standard MRI*

Idiopathic Parkinson disease is characterized by progressively worsening rigidity, resting tremor, bradykinesia, freezing, gait instability with a flexed posture, and loss of postural reflexes. Symptoms usually first appear asymmetrically between the ages of 40 and 70 years. The major pathologic finding is loss of dopaminergic cells in the substantia nigra pars compacta. Increased iron is present in the substantia nigra as well.[1,2] However, increased iron deposition may also be present in progressive supranuclear palsy and multiple system atrophy.[3] Both genetic and environmental factors probably play a role in the pathophysiology of this disease. Patients with Parkinson disease respond to dopamine replacement therapy.

MRI may reveal decreased width of the pars compacta of the substantial nigra on T2-weighted images in advanced cases.[4-6] Generalized cerebral atrophy is often present. However, the degree of atrophy overlaps with the amount present in normal aging. Periventricular and deep white matter T2 and PD hyperintensities are more common in patients with Parkinson disease than in age-matched controls. Those with periventricular hyperintensities may have a more rapidly progressive course.[7]

Attempts to image the increased iron within the substantia nigra with conventional MRI (including GRE, which is more sensitive to iron; maxim 30) have had variable success. Special techniques using ultrahigh (3.0 T) MRI have demonstrated that the amount of excessive iron within this region correlates with a decline in motor performance.[8] In advanced Parkinson disease decreased amounts of iron may be present within the lentiform nucleus, suggesting a complicated process of iron metabolism.[9] This observation contrasts with the Parkinson-plus syndromes, where increased iron may be present in the putamen (maxim 34).

Many other causes for a parkinsonian state exist. Typically these conditions respond poorly to dopamine replacement therapy. Neuroimaging, specifically MRI, differentiates several of these conditions. Multiple lacunae (maxim 12) suggests symptomatic vascular parkinsonism. Increased iron content (T2 hypointensity) within the striatum suggests multiple system atrophy (maxim 34). Midbrain atrophy is most consistent with progressive supranuclear palsy (maxim 33). Hallervorden–Spatz disease is characterized by a "tiger's eye" lesion within the pallidum.[10,11]

References

1. Sofic E, Paulus W, Jellinger K, Riederer P, Youdim MB. Selective increase of iron in substantia nigra zona compacta of parkinsonian brains. *J Neurochem* 1991; **56**: 978–82.
2. Chan JC, Hardy PA, Kucharczyk W, *et al*. MR of human post-mortem brain tissue: correlative study between T2 assays of iron and ferritin and in Parkinson and Huntington disease. *AJNR Am J Neuroradiol* 1993; **14**: 275–81.
3. Dexter DT, Sian J, Jenner P, Marsden CD. Implications of alterations in trace element levels in brain in Parkinson's disease and other neurological disorders affecting the basal ganglia. *Adv Neurol* 1993; **60**: 273–81.
4. Huber SJ, Chakeres DW, Paulson GW, *et al*. Magnetic resonance imaging in Parkinson's disease. *Arch Neurol* 1990; **47**: 735–7.
5. Duguid JR, DeLaPaz R, DeGroot J. Magnetic resonance imaging of the midbrain in Parkinson's disease. *Ann Neurol* 1986; **20**: 744–7.
6. Braffman BH, Grossman RI, Goldberg HI, *et al*. MR imaging of Parkinson disease with spin-echo and gradient-echo sequences. *AJR Am J Roentgenol* 1989; **152**: 159–65.
7. Piccini P, Pavese N, Canapicchi R, *et al*. White matter hyperintensities in Parkinson's disease. *Arch Neurol* 1995; **52**: 191–4.
8. Gorell JM, Ordidge RJ, Brown GG, Deniau JC, Buderer NM, Helpern JA. Increased iron-related MRI contrast in the substantia nigra in Parkinson's disease. *Neurology* 1995; **45**: 1138–43.
9. Ryvlin P, Broussolle E, Piollet H, *et al*. Magnetic resonance imaging evidence of decreased putamenal iron content in idiopathic Parkinson's disease. *Arch Neurol* 1995; **52**: 583–8.
10. Angelini L, Nardocci N, Rumi V, *et al*. Hallervorden–Spatz disease: clinical and MRI study of 11 cases diagnosed in life. *J Neurol* 1992; **239**: 417–25.
11. Savoiardo M, Halliday WC, Nardocci N, *et al*. Hallervorden–Spatz disease: MR and pathologic findings. *AJNR Am J Neuroradiol* 1993; **14**: 155–62.

32. *The classic sign of Huntington disease is atrophy of the head of the caudate nucleus*

Huntington disease is an autosomal dominant inherited neurodegenerative disorder linked to an abnormal trinucleotide repeat on chromosome 4. It is clinically characterized by a movement disorder initially consisting of chorea (hyperkinetic form) and later rigidity and dystonia, personality disorder (depression, antisocial behavior, and psychosis) and cognitive decline (subcortical dementia with apathy and impairment of memory and intellectual capacity). The clinically apparent onset of symptoms is usually in middle age. However, approximately 10% of patients present in childhood (most with inheritance from their father) with prominent rigidity and sometimes seizures (akinetic–rigid form). The diagnosis is usually made from the clinical presentation and family history. Huntington disease is always fatal and no treatment is known to slow the underlying process.

The classic radiologic and pathologic finding is atrophy of the head of the caudate nucleus, best visualized by axial CT and MRI. Compensatory, fairly symmetric enlargement of the anterior horns of the lateral ventricles is invariably present. The lateral walls of the anterior horns lose their normal concave curvature and flatten or become convex. An increased bicaudate diameter, decreased ratio of the maximum width of the anterior horns to the intercaudate distance (<1.6), and increased ratio of the intercaudate distance to the distance between the outer skull tables (>0.2; bicaudate index) all correlate with the diagnosis of Huntington disease.[1-4] However, area and volume measurements of the caudate and putamen provide better evidence of brain dysfunction in Huntington disease.[5] A small putamen volume is a more sensitive indicator of mild Huntington disease than a small head of the caudate volume.[6] Care must be taken not to confuse obstructive hydrocephalus with caudate atrophy.[7]

Cortical, subcortical, cerebellar, and brain stem atrophy may be present, especially in advanced disease.[8] T2 hyperintensity within the striatum is more common in younger patients with more advanced Huntington disease and in those with the akinetic–rigid form.[9,10] hypometabolism of the caudate and putamen is present on fluorodeoxyglucose PET before caudate atrophy is apparent on structural imaging studies.[11–16] Hypometabolism of the frontal cortex also occurs in early Huntington disease.[17]

References

1. Starkstein SE, Folstein SE, Brandt J, Pearlson GD, McDonnell A, Folstein M. Brain atrophy in Huntington's disease: a CT-scan study. *Neuroradiology* 1989; **31**: 156–9.
2. Bamford KA, Caine ED, Kido DK, Plassche WM, Shoulson I. Clinical–pathologic correlation in Huntington's disease: a neuropsychological and computed tomography study. *Neurology* 1989; **39**: 796–801.
3. Barr AN, Heinze WJ, Dobben GD, Valvassori GE, Sugar O. Bicaudate index in computerized tomography of Huntington disease and cerebral atrophy. *Neurology* 1978; **28**: 1196–200.
4. Stober T, Wussow W, Schimrigk K. Bicaudate diameter: the most specific and simple CT parameter in the diagnosis of Huntington's disease. *Neuroradiology* 1984; **26**: 25–8.
5. Wardlaw JM, Sellar RJ, Abernethy LJ. Measurement of caudate nucleus area: a more accurate measurement for Huntington's disease? *Neuroradiology* 1991; **33**: 316–19.
6. Harris GJ, Pearlson GD, Peyser CE. Putamen volume reduction on magnetic resonance imaging exceeds caudate changes in mild Huntington's disease. *Ann Neurol* 1992; **31**: 69–75.
7. Lang C. Is direct CT caudatometry superior to indirect parameters in confirming Huntington's disease? *Neuroradiology* 1985; **27**: 161–3.
8. Simmons JT, Pastakia B, Chase TN, Shults CW. Magnetic resonance imaging in Huntington disease. *AJNR Am J Neuroradiol* 1986; **7**: 25–8.
9. Oliva D, Carella F, Savoiardo M, *et al*. Clinical and magnetic resonance features of the classic and akinetic–rigid variants of Huntington's disease. *Arch Neurol* 1993; **50**: 17–19.
10. Savoiardo M, Strada L, Oliva D, Girotti F, D'Incerti L. Abnormal MRI signal in the rigid form of Huntington's disease. *J Neurol Neurosurg Psychiatr* 1991; **54**: 888–91.
11. Grafton ST, Mazziotta JC, Pahl JJ, *et al*. A comparison of neurological, metabolic, structural, and genetic evaluations in persons at risk for Huntington's disease. *Ann Neurol* 1990; **28**: 614–21.
12. Grafton ST, Mazziotta JC, Pahl JJ, *et al*. Serial changes of cerebral glucose metabolism and caudate size in persons at risk for Huntington's disease. *Arch Neurol* 1992; **49**: 1161–7.
13. Young AB, Penney JB, Starosta-Rubinstein S, *et al*. PET scan investigations of Huntington's disease: cerebral metabolic correlates of neurological features and functional decline. *Ann Neurol* 1986; **20**: 296–303.
14 Hayden MR, Martin WR, Stoessl AJ, *et al*. Positron emission tomography in the early diagnosis of Huntington's disease. *Neurology* 1986; **36**: 888–94.
15. Kuhl DE, Phelps ME, Markham CH, Metter EJ, Riege WH, Winter J. Cerebral metabolism and atrophy in Huntington's disease determined by 18FDG and computed tomographic scan. *Ann Neurol* 1982; **12**: 425–34.
16. Kuwert T, Ganslandt T, Jansen P, *et al*. Influence of size of regions of interest on PET evaluation of caudate glucose consumption. *J Comput Assist Tomogr* 1992; **16**: 789–94.
17. Martin WR, Clark C, Ammann W, Stoessl AJ, Shtybel W, Hayden MR. Cortical glucose metabolism in Huntington's disease. *Neurology* 1992; **42**: 223–9.

33. *Midbrain atrophy occurs in patients with progressive supranuclear palsy*

Progressive supranuclear palsy (PSP, Steele–Richardson–Olszewski syndrome) is a sporadic neurodegenerative disease characterized clinically by progressively worsening axial rigidity, supranuclear gaze palsy, and dementia. The supranuclear gaze palsy initially involves saccades and pursuit followed by impairment of vertical, predominantly downward, gaze. It is one of the Parkinson-plus disorders and may initially be misdiagnosed as Parkinson disease. However, absence of tremor and poor response to dopaminergic agents is typical. Dysphagia, dysarthria, and bradyphrenia develop as the disease progresses. Death usually occurs within a decade of onset.

Atrophy of the midbrain and pons is present in most patients with progressive supranuclear palsy on MRI and CT.[1] The anterior to posterior midbrain diameter may decrease to less than 1.5 cm. Generalized atrophy is present in advanced disease. PET studies reveal hypometabolism and hypoperfusion in the caudate, putamen, upper midbrain and frontal cortex.[2–8]

Differentiating progressive supranuclear palsy from corticobasal ganglionic degeneration is sometimes difficult. Both conditions may present with rigidity, bradykinesia, supranuclear gaze abnormalities, and dementia. The presence of prominent asymmetry in the akinetic–rigid syndrome or the cortical signs of hemineglect, hemisensory loss, cortical reflex myoclonus, limb apraxia, or the alien limb phenomenon favors the diagnosis of corticobasal ganglionic degeneration. Neuroimaging is useful for making this distinction because asymmetric atrophy in the posterior frontal and parietal regions is usually present in corticobasal ganglionic degeneration, and midbrain atrophy is often present in progressive supranuclear palsy.[9]

References

1. Savoiardo M, Strada L, Girotti F, *et al*. MR imaging in progressive supranuclear palsy and Shy–Drager syndrome. *J Comput Assist Tomogr* 1989; **13**: 555–60.
2. Karbe H, Grond M, Huber M, Herholz K, Kessler J, Heiss WD. Subcortical damage and cortical dysfunction in progressive supranuclear palsy demonstrated by positron emission tomography. *J Neurology* 1992; **239**: 98–102.
3. Bhatt MH, Snow BJ, Martin WR, Peppard R, Calne DB. Positron emission tomography in progressive supranuclear palsy. *Arch Neurol* 1991; **48**: 389–91.
4. Blin J, Baron JC, Dubois B, *et al*. Positron emission tomography study in progressive supranuclear palsy: brain hypometabolic pattern and clinicometabolic correlations. *Arch Neurol* 1990; **47**: 747–52.
5. Otsuka M, Ichiya Y, *et al*. Cerebral blood flow, oxygen and glucose metabolism with PET in progressive supranuclear palsy. *Ann Nucl Med* 1989; **3**: 111–18.
6. Goffinet AM, De Volder AG, Gillain C, *et al*. Positron tomography demonstrates frontal lobe hypometabolism in progressive supranuclear palsy. *Ann Neurol* 1989; **25**: 131–9.
7. Leenders KL, Frackowiak RS, Lees AJ. Steele–Richardson–Olszewski syndrome: brain energy metabolism, blood

flow and fluorodopa uptake measured by positron emission tomography. *Brain* 1988; **111**: 615–30.

8. Foster NL, Gilman S, Berent S, Morin EM, Brown MB, Koeppe RA. Cerebral hypometabolism in progressive supranuclear palsy studied with positron emission tomography. *Ann Neurol* 1988; **24**: 399–406.
9. Gimenez-Roldan S, Mateo D, Benito C, Grandas F, Perez-Gilabert Y. Progressive supranuclear palsy and corticobasal ganglionic degeneration: differentiation by clinical features and neuroimaging techniques. *J Neural Transmission* 1994; **42** (suppl): 79–90.

34. *Striatonigral degeneration, olivopontocerebellar atrophy (OPCA, sporadic form only), and Shy–Drager syndrome are variants of a single entity: multiple system atrophy (MSA)*

Several of the Parkinson-plus syndromes, namely striatonigral degeneration, olivopontocerebellar atrophy (OPCA, sporadic form only), and Shy–Drager syndrome, are probably different extremes of the same disease: multiple system atrophy (MSA).[1–3] These conditions present clinically with various combinations of parkinsonism and pyramidal, cerebellar, and autonomic dysfunction. Pathologic examination reveals common features of neuronal cell loss, gliosis, and characteristic glial cytoplasmic inclusions in the putamen, substantia nigra, pons, cerebellar cortex, and inferior olives. MSA may be fairly common, present in up to 10% of patients with parkinsonism. It is progressive and responds poorly to treatment.

In patients with MSA, an abnormally low T2 signal is present within the putamen, indicating abnormal accumulation of iron (or another paramagnetic substance).[4] Patients diagnosed with idiopathic Parkinson disease who have low signal within the putamen on T2-weighted images may have MSA (maxim 31). Atrophy of the brain stem (especially the pons) and the cerebellum is typically present. Generalized atrophy may also occur.

PET studies demonstrate reduced cerebral blood flow and glucose metabolism in the cerebellum.[5] Proton MR spectroscopy reveals changes in the putamen consistent with cell loss and may be useful for differentiating MSA from idiopathic Parkinson disease.[6]

MSA appears to be distinct from the primarily hereditary neurodegenerative conditions classified pathologically as cerebellar cortical atrophy (CCA), OPCA (hereditary form), and spinal atrophy. The most common spinal atrophy syndrome is Friedreich's ataxia, an autosomal recessive disorder consisting of progressive ataxia, areflexia, proprioceptive loss, and extensor plantar responses, associated with degeneration of the posterior columns, spinocerebellar tracts, and corticospinal tracts on pathologic examination. MRI reveals a small cervical spinal cord in this condition.[7]

References

1. Rinne JO, Burn DJ, Mathias CJ, *et al.* Positron emission tomography studies on the dopaminergic system and striatal opioid binding in the olivopontocerebellar atrophy variant of multiple system atrophy. *Ann Neurol* 1995; **37**: 568–73.
2. Penney JB. Multiple systems atrophy and nonfamilial olivopontocerebellar atrophy are the same disease. *Ann Neurol* 1995; **37**: 553–4. (editorial).
3. Perani D, Bressi S, Testa D, *et al.* Clinical/metabolic correlations in multiple system atrophy. *Arch Neurol* 1995; **52**: 179–85.
4. Drayer BP, Olanow W, Burger P, *et al.* Parkinson plus syndrome: diagnosis using high field imaging of brain iron. *Radiology* 1986; **159**: 493–8.
5. Gilman S, St Laurent RT, Koeppe RA, Junck L, Kluin KJ, Lohman M. A comparison of cerebral blood flow and glucose metabolism in olivopontocerebellar atrophy using PET. *Neurology* 1995; **45**: 1345–52.
6. Davie CA, Wenning GK, Barker GJ, *et al.* Differentiation of multiple system atrophy from idiopathic Parkinson's disease using proton magnetic resonance spectroscopy. *Ann Neurol* 1995; **37**: 204–10.
7. Wullner U, Klockgether T, Petersen D, Naegele T, Dichgans J. Magnetic resonance imaging in hereditary and idiopathic ataxia. *Neurology* 1993; **43**: 318–25.

6
Brain Tumors

35. *Neither tumor type nor pathologic grade can be determined with certainty from the radiographic appearance*

There are approximately equal numbers of primary brain tumors and metastatic brain tumors in adults. Gliomas, metastases, meningiomas, pituitary adenomas, and acoustic neuromas together account for 95% of brain tumors. Although some tumors have a characteristic neuroimaging appearance (e.g. meningiomas, neurocytomas – Fig. 35.1), it is unwise to make an unequivocal diagnosis based on neuroimaging alone. Radiologists should speak in terms of a differential diagnosis when discussing lesions that may be tumors (or infections). Pathologic analysis is required for a pathologic diagnosis. A more explicit title for this maxim is "radiology is not pathology".

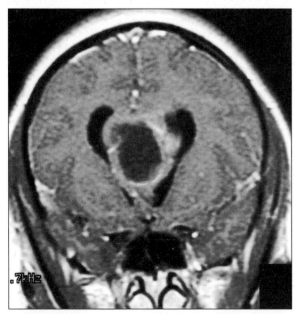

Fig. 35.1 Neurocytoma. T1-weighted contrast-enhanced coronal image of a young woman. A ring-enhancing lesion is present, situated between the ventricles. This is the typical location for a neurocytoma (intraventricular, adjacent to the foramen of Monro). As in this case, neurocytomas usually mildly enhance. Biopsy confirmed the radiologic suspicion of a neurocytoma.

All three of the major diagnostic neuroimaging methods – MRI, CT, and angiography – are used in brain tumor diagnosis. MRI has the greatest sensitivity for tumor detection. The original development of MRI was largely motivated by the desire to distinguish neoplasms from normal tissue.[1] CT enables detection of calcification and bony destruction and is especially useful in tumors involving the skull base. Angiography is sometimes performed before surgery to determine the degree of vascularity and to permit embolization of arterial feeders. Tumor volume (or, less ideally, maximal cross-sectional area or diameter) may be measured and followed on MRI or CT for assessment of lesion growth or regression (therapeutic response). Normally, these studies are complementary rather than redundant.

Although very few neoplastic lesions have a pathognomonic radiographic appearance, the combination of lesion location, patient age, noncontrast CT and MRI appearance, and contrast-enhancement characteristics give helpful hints about the nature of the lesion.[2] Lesions occupying certain anatomical locations have a specific differential diagnosis:

1. *Sellar/suprasellar*
   - Pituitary adenoma (microadenoma and macroadenoma)
   - Craniopharyngioma
   - Meningioma
   - Aneurysm
   - Arachnoid cyst
   - Abscess
   - Ectopic neurohypophysis
   - Hamartoma (of the tuber cinereum)
   - Histiocytosis X
   - Metastasis
   - Rathke's cleft cyst
   - Sarcoid
2. *Pineal*
   - Germ cell tumors (germinoma, teratoma, endodermal sinus tumor)
   - Pineal parenchymal tumors (pinealoma, pinealoblastoma)
   - Pineal cyst
   - Metastasis
   - Vascular malformation (with or without vein of Galen enlargement)

- Lipoma
- Glioma
- Meningioma

3. *Cerebellopontine angle*
- Acoustic neuroma (schwannoma)
- Meningioma
- Epidermoid tumor
- Arterial loop (vascular ectasia)
- Metastasis
- Arachnoid cyst
- Dermoid

4. *Foramen magnum*
- Chiari I malformation
- Meningioma
- Glioma
- Syringohydromyelia
- Metastasis
- Vertebrobasilar dolichoectasia
- Aneurysm (PICA)
- Epidermoid
- Ependymoma
- Medulloblastoma
- Chordoma
- Congenital anomalies

5. *Intraventricular*
- Cavum septum pellucidum and vergae
- Colloid cyst
- Neuroepithelial cyst
- Astrocytoma
- Giant cell astrocytoma
- Choroid plexus papilloma
- Choroid plexus cyst (xanthogranuloma)
- Enlarged choroid plexus (neurofibromatosis type 2, Sturge–Weber syndrome)
- Medulloblastoma
- Ependymoblastoma
- Ependymoma
- Meningioma
- Neurocytoma
- Subependymoma

6. *Skull base and cavernous sinus*
- Metastasis
- Invasive nasopharyngeal tumor
- Chordoma
- Juvenile nasopharyngeal angiofibroma
- Aggressive pituitary adenoma
- Schwannoma
- Meningioma
- Aneurysm
- Carotid–cavernous fistula
- Fibrous dysplasia
- Paget's disease

7. *Meninges (focal)*
- Meningioma
- Metastasis
- Postoperative
- Sarcoid
- Lymphoma
- Leukemia

8. *Calvarium (focal)*
- Hyperostosis frontalis interna
- Paget's disease
- Fibrous dysplasia
- Meningioma
- Osteoma
- Metastasis

9. *Calvarium (diffuse)*
- Normal variant
- Chronic use of phenytoin
- Microcephaly
- Acromegaly
- Hematologic disorders

Intravenous contrast should generally be given with MRI (or CT) because most (but not all) tumors at least partially enhance. The enhancement may help to clarify the diagnosis and better define the tumor size. Defects in the blood–brain barrier and vascular proliferation are responsible for the enhancement. The degree of tumor enhancement tends to correlate with the degree of malignancy for astroglial tumors but not for other tumor types. Nevertheless, occasionally low-grade astroglial tumors enhance and high-grade ones do not (maxim 36). Malignant tumor cells are typically present within the area that enhances. Malignant cells are also often present beyond the enhancing region, an important consideration at surgery.

In general, tumors result in prolongation of both the T1 and T2 times due to increased water content. Thus, tumors usually generate high signal on T2-weighted images and low signal on T1-weighted images. The surrounding edema is easily visualized on MRI but may be difficult to differentiate from the tumor. The contrast enhancement pattern and the T1 and T2 relaxation times are of little use in diagnosing the specific tumor type.

Tumors may be hypodense, isodense, hyperdense, or of mixed density on CT. A hypodense appearance is due to increased water content, cystic components, necrosis, or fat. Astrocytomas usually appear hypodense. A hyperdense appearance is due to hemorrhage, calcification, or a high nuclear to cytoplasmic ratio. The following tumors are hyperdense (without calcification) on non-enhanced CT:

1. *In children*
   (a) Medulloblastoma
   (b) Hamartoma
2. *In adults*
   (a) Meningioma
   (b) Metastases (hemorrhage common)
   - Colon
   - Small-cell lung
   - Renal cell
   - Melanoma
   - Choriocarcinoma
   - Osteogenic sarcoma
   - Lymphoma

It is common for tumors to have mixed density, a result of a combination of hemorrhage, cystic components, calcification, and necrosis.

Certain tumors tend to calcify (maxim 13). Calcium is easy to detect with CT even when it is punctate. However, calcium is often difficult or impossible to discern on MRI. Even large amounts of calcium may be invisible on MRI. Calcium usually appears as a void, but sometimes it is high signal on T1-weighted images (maxim 13). Gradient-echo scans are more sensitive to calcium than are spin-echo scans.

Some general rules apply for brain tumors. Multiple lesions are typically metastases, although primary CNS lymphomas and glioblastomas may be multicentric. The most common solid tumors that metastasize to the brain are breast, lung, renal cell, melanoma, and colon (Fig. 35.2).

Carcinomatous meningitis is usually caused by breast cancer or lymphoma (lymphomatous meningitis). However, many other tumors may also invade the meninges. Classically, hyporeflexia and multiple cranial and segmental nerve palsies develop. The MRI may reveal leptomeningeal, dural, subependymal, or cranial nerve enhancement. Superficial cerebral lesions and communicating hydrocephalus also occur.[3]

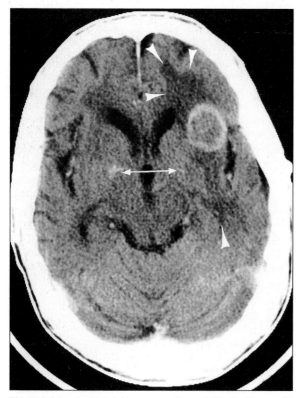

Fig. 35.2 Metastasis. Axial contrast-enhanced CT of a 66-year-old man demonstrating a classic ring-enhancing lesion (see maxim 71). The patient had known metastatic renal cell carcinoma, and multiple metastatic foci were present at different levels. This left frontotemporal lesion has ring enhancement, with minimal core enhancement, and substantial surrounding edema (arrowheads). A metastasis usually has more associated edema than does a primary brain tumor of the same size. Normal idiopathic calcifications of the basal ganglia are present bilaterally (straight double arrow) (maxim 13).

Primary CNS lymphoma deserves special attention. Most CNS lymphomas enhance, but occasionally they present as nonenhancing lesions, a phenomenon that may delay diagnosis and affect treatment.[4]

References

1. Kleinfield S. *A machine called indomitable.* New York: Times Books (Random House), 1985. (Read the whole book, it's a great story!)
2. Osborn A. *Diagnostic neuroradiology.* St Louis: Mosby, 1994: 399–670.
3. Freilich RJ, Krol G, DeAngelis LM. Neuroimaging and cerebrospinal fluid cytology in the diagnosis of leptomeningeal metastasis. *Ann Neurol* 1995; **38**: 51–7.
4. DeAngelis LM. Cerebral lymphoma presenting as a nonenhancing lesion on computed tomographic/magnetic resonance scan. *Ann Neurol* 1993; **33**: 308–11.

36. *Glial neoplasms account for most primary brain tumors*

Glial tumors, which account for over half of the primary brain tumors, include astrocytomas, oligodendrogliomas, and ependymomas. Choroid plexus tumors are sometimes included as an additional group, because they contain modified ependymal cells.

The grade of a glioma is assigned according to the most malignant portion of the tumor. Glial tumors are notoriously heterogeneous. Grading errors occur when a biopsy sample is obtained from a lower grade portion of the lesion. Additionally, the tumor grade may change with time: a benign lesion may transform into a more malignant one.

The pathologic grade is determined by ascertaining the number and quality of the mitotic figures, degree of vascular proliferation, and presence of necrosis. In general, the higher the grade the more neovascularity and the greater the contrast enhancement. Low-grade gliomas characteristically have minimal or no enhancement. The glial tumors typically arise within the white matter of the cerebral hemispheres or the deep nuclei.[1]

Astrocytomas

Grading of astrocytomas according to histopathology is standard but somewhat arbitrary, as the pathology has a continuous spectrum, not discrete divisions. The traditional (Kernohan) astrocytoma grading system used four divisions, with types III and IV considered malignant. The current World Health Organization (WHO) classification uses three divisions: low grade or benign (although the clinical course may not be benign), anaplastic, and glioblastoma multiforme.[2–4] The prognosis for malignant astrocytomas (anaplastic astrocytomas and glioblastomas) is grim, and the response to treatment is poor.

In adults, low-grade astrocytomas are solid tumors with little or no edema. In children, most astrocytomas are low-grade cystic tumors located in the cerebellum. They are poorly visualized on CT (low density when evident) and can be missed completely. Calcification is present in 20% and mild to moderate inhomogeneous enhancement is present in 40%. On MRI, low-grade astrocytomas are isointense or hypointense to brain on T1-weighted images and hyperintense on T2-weighted images. Nonenhancing, infiltrating low-grade astrocytomas may look like edema and be misidentified as a stroke.

Pathologically, low-grade astrocytomas are areas of slight hypercellularity and pleomorphism without vascular proliferation or necrosis. Although they have the best prognosis of the astrocytomas, they are still typically progressive. Transformation from a low to a higher grade of astrocytoma is a common cause of death in patients with low-grade astrocytomas. An exception is childhood cystic astrocytoma, which is usually cured by surgical removal.

Anaplastic astrocytomas have moderate hypercellularity, pleomorphism, and vascular proliferation but no necrosis on pathologic examination. They infiltrate into the brain tissue, resulting in an ill-defined boundary between the neoplasm and brain on neuroimaging, and on surgical and pathologic examination. Like glioblastomas, they may spread along the ependyma or leptomeninges and behave like carcinomatous meningitis, a condition termed leptomeningeal gliomatosis.

Anaplastic astrocytomas are usually heterogeneous without contrast, and enhance with contrast on MRI and CT.[5] They are commonly hyperintense on T2-weighted images. Significant mass effect and surrounding edema are typical. Hemorrhagic components may be present. Enhancement patterns vary widely and include a thin to a thick ring to irregular or dense diffuse involvement. Any enhancement, or a small blush on conventional angiography, within a suspected astrocytoma raises the possibility that the lesion is anaplastic. However, some anaplastic astrocytomas appear as cystic lesions with minimal or no enhancement and no surrounding edema (Fig. 36.1). An astrocytoma adjacent to the cortex may mimic an extra-axial mass lesion (e.g. meningioma) or cystic lesion (e.g. arachnoid cyst). Because the neuroimaging presentation of anaplastic astrocytomas is highly variable, radiologists should be careful about characterizing the degree of malignancy of a glioma.

Glioblastoma multiforme is the most malignant form of astrocytoma, and is characterized by necrosis on pathologic examination. Glioblastomas may be multicentric and radiologically indistinguishable from metastatic disease. A "butterfly glioma" has a specific pattern that involves both cerebral hemispheres with continuity through the corpus callosum. Lymphomas can also present with a butterfly pattern.

Glioblastomas are mainly supratentorial. CT and MRI demonstrate significant heterogeneity. Considerable mass effect and edema are characteristic. Marked peripheral contrast enhancement surrounding a necrotic

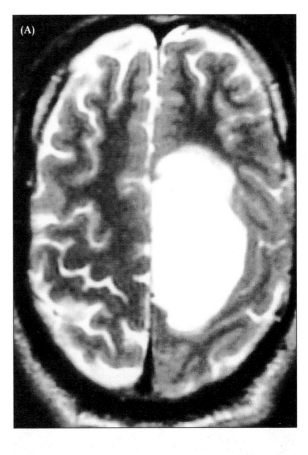

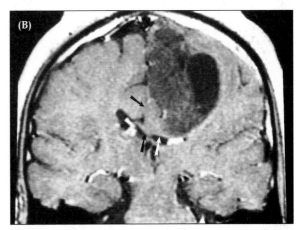

Fig. 36.1 Anaplastic astrocytoma. **A:** T2-weighted axial image through the centrum semiovale of a 34-year-old man with a mild right hemiparesis. A large intraparenchymal anaplastic astrocytoma is present in the left centrum semiovale. There is no surrounding edema. The lesion displaces but does not invade the white matter tracts. Cystic and solid components were present at another level. **B:** T1-weighted contrast-enhanced coronal image with cystic and solid components. Only minimal enhancement is present. Note the mass effect on the body of the lateral ventricle and mild subfalcial herniation (arrows) (maxim 42). The diagnosis of anaplastic astrocytoma was made on pathologic examination. This lesion lacks enhancement, invasion of the parenchyma, and surrounding edema, several of the classic neuroimaging features expected of an anaplastic astrocytoma.

center (ring lesion) is also typical (Fig. 36.2). The striking neovascularity is readily apparent on angiography. The mass effect, edema, and contrast enhancement may diminish with steroid therapy (maxim 38).

Gliomatosis cerebri, a rare presentation of malignant glioma, is characterized by diffuse glial overgrowth involving large portions of the cerebral hemispheres and sometimes extending into the cerebellum and brain stem. CT may be normal or contain only minor mass effect and minimal enhancement. Usually MRI demonstrates heterogeneous hyperintensity on T2-weighted images with some mass effect. There is variability in the degree of enhancement.

Oligodendrogliomas

Oligodendrogliomas, because of their tendency to calcify, were one of the few tumor types reliably diagnosed with plain radiographs before the availability of CT. CT reveals lesions with mixed densities, including high-density regions of calcification. Sometimes hemorrhage is present. Approximately one-half of oligodendrogliomas enhance. A cystic component is commonly present. Edema is usually mild or completely absent.

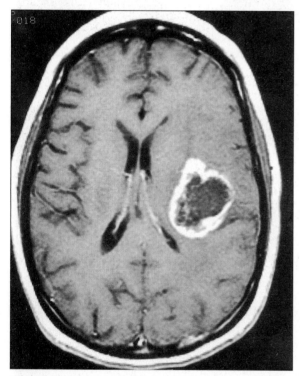

Fig. 36.2 Glioblastoma multiforme. T1-weighted axial image with contrast enhancement in a 22-year-old woman with headaches and a right hemiparesis. An irregular densely enhancing ring lesion is present in the left insular region. The lateral wall, closest to the brain's outer surface, is thinner than the other walls, which is consistent with the diagnosis of tumor (maxim 71). The center of the lesion is necrotic, with only minimal enhancement. The diagnosis on pathologic examination was glioblastoma multiforme. This example demonstrates the marked imaging heterogeneity of these lesions. Most glioblastomas have more mass effect and edema.

MRI demonstrates mixed hypointense and hyperintense signal on T2-weighted images because of the calcified and cystic components. Chronic and subacute blood products may be present. As on CT, focal edema is uncommon.

Most oligodendrogliomas are indolent (although malignant forms occur) and seizures are the most frequent clinical manifestation. The malignant variety may have astrocytic components and may be responsive to chemotherapy. Oligodendrogliomas tend to infiltrate into the brain substance.

Ependymomas

Ependymomas are gliomas arising from the ependymal cells usually within the ventricles or the central canal of the spinal cord. Cranial ependymomas usually develop in childhood, whereas ependymomas of the spinal canal are more frequent in adults. Intracranial ependymomas usually occur during the first 5 years of life. Generally ependymomas have a poor prognosis, especially when present in early childhood.

Most intracranial ependymomas are infratentorial (70%), usually within the fourth ventricle. Fourth ventricle ependymomas may extend through the foramen of Luschka or foramen of Magendie into the cerebellar vallecula, foramen magnum, or cerebellopontine angle. Supratentorial ependymomas are often parenchymal in location rather than intraventricular, the result of nests of ependymal neuroglia left within the white matter during development.

Ependymomas vary from hypodense to isodense to hyperdense on CT. Mixed-density lesions are common. Speckled calcification is present in approximately 50%. Patchy enhancement typically occurs.

These tumors usually have heterogeneous signal on T2-weighted images owing to the mixture of calcification, hemorrhage, and small cystic areas. On T1-weighted images the lesion is commonly hypointense or isointense to brain signal. As on CT, heterogeneous enhancement occurs.

Gangliogliomas

The ganglion cell neoplasms are classified according to their stage of differentiation and relative proportion of glial and neural (ganglion cell) components. The most differentiated is the gangliocytoma, followed by the ganglioglioma, ganglioneuroblastoma, anaplastic ganglioglioma, and neuroblastoma. Gangliogliomas and gangliocytomas (also called ganglioneuromas) are the most common ganglion cell tumors and represent two ends of a spectrum. They should be considered as a single entity, as different nervous tissue components may predominate in different parts of a single lesion. A related condition, dysplastic gangliocytoma of the cerebellum (Lhermitte–Duclos disease), is probably not a true neoplasm but a congenital malformation with thickened cerebellar folia.

Gangliogliomas are most commonly located within the temporal or frontal lobe but may be found in the parietal or occipital lobe, the cerebellum, or the spinal cord.[6–9] When the third ventricle or hypothalamus is involved, signs of hypothalamic dysfunction may occur, and it may be difficult to differentiate a ganglioglioma from a hypothalamic astrocytoma. Gangliogliomas within the cerebral hemispheres commonly present with focal seizures. Neurosurgical resection usually cures the epilepsy.

Gangliogliomas are well-circumscribed lesions that are typically located in the brain periphery, sometimes eroding the inner table of the skull. Mass effect is otherwise not present. Solid parts of the tumor may be isodense or hypodense on CT. Calcification is visible in one-third of cases (Fig. 36.3A). The appearance of gangliogliomas on MRI is not very specific (Fig. 36.3B and C). They have variable (mostly high) signal on T2-weighted images and mixed signal on T1-weighted images. Cystic components are common. A small amount of surrounding edema may be present. Contrast enhancement is variable. The location of the tumor in the periphery of the cerebral hemisphere with minimal associated mass effect is more helpful in making the diagnosis than are the radiographic characteristics. Erosion of the adjacent inner table of the skull, when present, narrows the differential diagnosis.

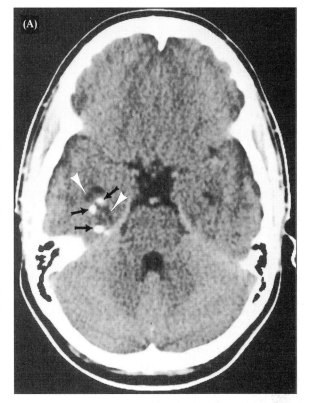

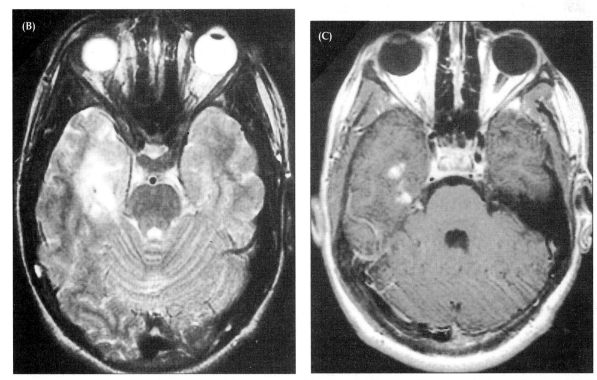

Fig. 36.3 Ganglioglioma. **A:** Axial noncontrasted CT in a 15-year-old with medically intractable seizures for 5 years. Several calcified nodules (black arrows) intermixed with patchy low-density (cystic) regions (white arrowheads) are present in the right temporal lobe. There is no significant mass effect. **B:** Axial T2-weighted image in the same patient demonstrating high signal, predominantly in the white matter of the right temporal lobe. **C:** Axial contrast-enhanced T1-weighted image in the same patient. Patchy contrast enhancement is present in the right temporal lobe lesion. Right temporal lobectomy and lesionectomy resulted in seizure freedom. The pathologic examination demonstrated a ganglioglioma.

References

1. Atlas SW. Adult supratentorial tumors. *Semin Roentgenol* 1990; **25**: 130–54.
2. Kleihues P, Burger PC, Scheithauer BW. The new WHO classification of brain tumours. *Brain Pathology* 1993; **3**: 255–68.
3. Castillo M, Scatliff JH, Boulden TW, *et al*. Radiologic–pathologic correlation: intracranial astrocytoma. *AJNR Am J Neuroradiol* 1992; **13**: 1609–16.
4. Watanabe M, Tanaka R, Takeda N. Magnetic resonance imaging and histopathology of cerebral gliomas. *Neuroradiology* 1992; **35**: 463–9.
5. Dean BL, Drayer BP, Bird CR, *et al*. Gliomas, classification with MR imaging. *Radiology* 1990; **174**: 411–15.
6. Castello M, Davis PC, Takei Y, *et al*. Intracranial ganglioglioma. *AJNR Am J Neuroradiol* 1990; **11**: 109–15.
7. Celli P, Scarpinati M, Nardacci B, Cervoni L, Cantore GP. Gangliogliomas of the cerebral hemispheres: report of 14 cases with long-term follow-up and review of the literature. *Acta Neurochir* 1993; **125**: 52–7.
8. Krouwer HG, Davis RL, McDermott MW, Hoshino T, Prados MD. Gangliogliomas: a clinicopathological study of 25 cases and review of the literature. *J Neuroncol* 1993; **17**: 139–54.
9. Park SH, Chi JG, Cho BK, Wang KC. Spinal cord ganglioglioma in childhood. *Pathol Res Pract* 1993; **189**: 189–96.

37. *A calcified, enhancing globular mass abutting the brain's surface is usually a meningioma*

Meningiomas are common intracranial tumors. Although most are histologically benign and encapsulated, they tend to progressively enlarge and cause neurologic symptoms due to mass effect.[1,2] Meningiomas arise from the arachnoid mater at locations where it folds into the brain. In decreasing order of frequency, intracranial meningiomas are found at the following locations:[1–3]

1. Parasagittal (sagittal sinus or falx cerebri)
2. Cerebral convexity
3. Sphenoid ridge
4. Olfactory groove
5. Parasellar
6. Tentorium cerebelli
7. Posterior fossa (both foramen magnum and CP angle)
8. Temporal fossa
9. Intraventricular
10. Optic nerve sheath.

They adhere firmly to the dura and may invade the adjacent bone, resulting in hyperostosis. They rarely invade the brain,[3] but may have associated extensive cerebral edema. The degree of edema and the size of the meningioma do not always correlate.

Presentation is uncommon before middle age. Meningiomas are more frequent in women. Multiple meningiomas occur in 1–2% of cases and are more common in patients with neurofibromatosis type 2 (maxim 28). Almost all meningiomas are slow growing, although rapid growth is reported, especially in pregnancy. Meningiomas are typically very vascular and are usually supplied by dural, not cerebral, vessels.

Meningiomas are frequently calcified, making identification straightforward on CT. In contrast, they are often isointense to gray matter on MRI (calcium is poorly visualized on MRI; maxim 13) and may be overlooked if significant mass effect is absent (Fig. 37.1). Meningiomas characteristically have pronounced, nearly homogeneous contrast enhancement on both CT and MRI (Figs. 37.1B and 28.1B), although heavily calcified tumors may not enhance.

On CT, a meningioma is typically a homogeneous hyperdense mass. Adjacent edema is variably present. Detection may be difficult if a meningioma is located near the skull base or in the high parasagittal region. Calvarial hyperostosis is best visualized with CT.

MRI is a complementary study to CT. Although the tumor is usually not well defined on noncontrasted MRI, with contrast it is possible to distinguish the broad-based dural margin and confirm an extra-axial location.[4] MRI allows identification of bone marrow invasion and a dural tail (Fig. 37.1B). Dural tails also occur with gliomas, acoustic neuromas, and metastases.[5,6,7] Phase-contrast MRA should be performed if the tumor is located near a venous sinus to determine whether venous sinus thrombosis is present.

Meningiomas may grow as a diffuse sheet of tumor over the brain or the base of the skull. These are termed "*en plaque* meningiomas". They may be difficult to discern on CT but are usually visible on MRI.

Differentiating a meningioma from a peripherally located lymphoma, neurosarcoidosis (maxim 75), or other lesion that has prominent contrast enhancement may be difficult.[8] Blastic calvarial metastasis (e.g. breast and prostate cancer) can mimic a meningioma, especially when an epidural component is present. Occasionally, a calcified giant aneurysm is mistaken for a meningioma on CT (and vice versa); MRI and MRA easily differentiate these two lesions.

Suprasellar meningiomas may be incorrectly identified as pituitary macroadenomas. Meningiomas arise from the diaphragma or tuberculum sellae and are located above the pituitary gland, whereas pituitary adenomas are usually intrasellar and enlarge the pituitary fossa with secondary extension into the suprasellar cistern.

The treatment is usually surgical resection. Its success depends on the location, extent of normal tissue involved, and degree of vascularity. Conventional angiography is sometimes used preoperatively to identify the blood supply (often derived from the extracranial circulation) and to observe the capillary blush (which extends into the late venous phase) to further outline the extent of the tumor. Embolization of feeding vessels is done before resection in some cases.

Complete surgical resection, resulting in cure, is

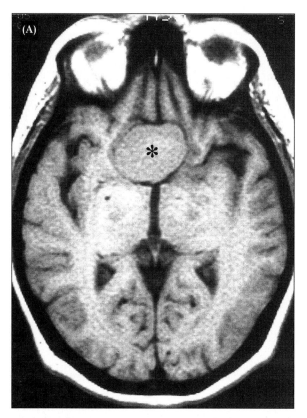

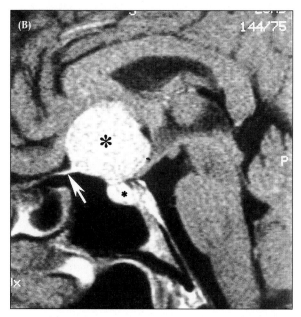

Fig. 37.1 Meningioma. **A:** Axial T1-weighted image of a planum sphenoidale meningioma. The mass (✱) is isointense to gray matter. **B:** Contrast-enhanced sagittal T1 image. The strongly enhancing meningioma (✱) is anterior and superior to the pituitary gland (∗). A dural tail sign (arrow) is present. The dural tail sign is highly suggestive but not specific for meningioma. The contrast-enhanced CT (not shown) revealed this mass located in the suprasellar region; giving a differential diagnosis that included pituitary adenoma, giant aneurysm, meningioma, and craniopharyngioma. The MRI minimized the likelihood of these other possibilities.

usually possible. When vital structures such as the major intracerebral vessels are involved, operative morbidity and mortality rate may be high. In these situations, only debulking may be possible. Even so, surgery is almost always palliative because tumor regrowth is slow.

References

1. Zee CS, Chin T, Segul HD, *et al.* Magnetic resonance imaging of meningiomas. *Semin Ultrasound CT MR* 1992; **13**: 154–69.
2. Rohringer M, Sutherland GR, Louw DF, *et al.* Incidence and clinicopathological features of meningioma. *J Neurosurg* 1989; **71**: 665–72.
3. Maier H, Offer D, Hittmmair A, *et al.* Classic, atypical, and anaplastic meningioma: three histopathologic subtypes of clinical relevance. *J Neurosurg* 1992; **77**: 616–23.
4. Kaplan RD, Coon S, Drayer BP, *et al.* MR characteristics of meningioma subtypes at 1.5 T. *J Comput Assist Tomogr* 1992; **16**: 366–71.
5. Tien RD, Yang PJ, Chu PK. "Dural tail sign": a specific MR sign for meningioma? *J Comput Assist Tomogr* 1991; **15**: 64–6.
6. Goldsher D, Litt AW, Pinto RS, *et al.* Dural "tail" associated with meningiomas on Gd-DTPA-enhanced MR images: characteristics, differential diagnostic value, and possibly implications on treatment. *Radiology* 1990; **176**: 447–50.
7. Wilms G, Lammeus M, Marchal G, *et al.* Prominent dural enhancement adjacent to non-meningiomatous malignant lesions on contrast-enhanced MR images. *AJNR Am J Neuroradiol* 1991; **12**: 761–4.
8. Odake G. Cystic meningiomas: report of three patients. *Neurosurgery* 1992; **30**: 935–40.

38. *Steroids can dramatically change the appearance of a tumor: inform the radiologist when steroids are used*

Malignant neoplasms (gliomas, lymphomas, and metastases) are often very responsive, although only transiently, to treatment with glucocorticoid steroids. This effect is clinically useful, especially when a tumor with associated edema is producing a life-threatening mass effect. However, this property may seriously complicate the diagnostic evaluation, as the tumor (especially a lymphoma) may temporarily disappear.[1–3] In patients with a suspected malignant brain tumor, all initial neuroimaging (often MRI and CT with contrast) should be performed before the administration of steroids. Additionally, steroids should be delayed, if possible, until after biopsy.

Steroids decrease edema, mass effect, and degree of contrast enhancement of malignant tumors (Fig. 38.1). This effect occurs within 72 hours of steroid administration and is presumably due to stabilization of the blood–brain barrier. CNS lymphomas can temporarily vanish after treatment with steroids. When one or multiple lesions disappear after steroid treatment is initiated, lymphoma should be suspected.

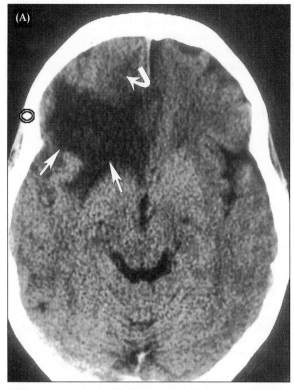

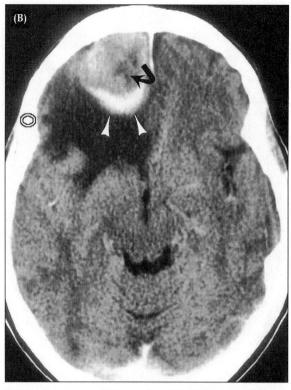

A: Noncontrasted axial CT demonstrates a large low-attenuation region in the right frontal lobe consistent with edema (arrows). A partially isodense tumor is located anterior to the edema (curved arrow).

B: Contrast-enhanced CT reveals striking lesion enhancement. The most prominent enhancement is in the posterior aspect of the tumor (white arrowheads). A cystic/necrotic region is present in the center of the lesion (black curved arrow).

C: Contrast-enhanced CT 48 hours later, after treatment with steroids. The degree of enhancement is dramatically decreased within the tumor (white arrowheads) despite increased enhancement of the vasculature (short straight double-headed arrow) and tentorium cerebelli (long straight double-headed arrow). The amount of brain edema is essentially unchanged in this example.

The response to the specific antitumor treatment (chemotherapy or radiation therapy) may be difficult to assess if concomitant steroids are administered. If pulse steroid treatment is used, follow-up neuroimaging should be performed as remote from the previous pulse as possible. The confusing effect of steroids on neuro-imaging is minimized when a chronic constant steroid dose is used.

In contrast to malignant tumors, benign neoplasms, such as meningiomas and acoustic neuromas, rarely respond to steroid treatment.

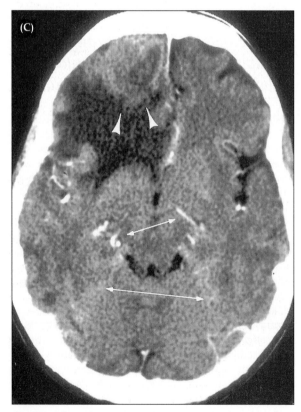

Fig. 38.1 Effect of steroids on a brain tumor.

References

1. Cairncross J, McDonald DR, Pexman W, Ives FJ. Steroid-induced CT changes in patients with recurrent malignant glioma. *Neurology* 1988; **38**: 724–6.
2. Watling CJ, Lee DH, McDonald DR, *et al.* Steroid-induced magnetic resonance imaging changes in patients with recurrent malignant glioma. *J Clin Oncol* 1994; **12**: 1886–9.
3. Hatam A, Bergstrom M, Noren G. Effects of dexamethasone treatment on acoustic neuromas: evaluation by computed tomography. *J Comput Assist Tomogr* 1985; **9**: 857–60.

39. *Differentiating between tumor recurrence and late radiation injury is difficult*

The effect of radiation on normal brain tissue is related to cumulative dose, dose fraction size, duration of treatment, and volume of tissue irradiated. Normal brain can acutely tolerate fractionated regimens of 6000 cGy. In general, "acute" postradiation complications are due to swelling, "early-delayed" injury to demyelination, and "late-delayed" injury to necrosis. Acute and early-delayed postradiation effects are relatively infrequent and usually transient. The most common radiation injury to the central nervous system is late-delayed and takes two forms: focal and diffuse.[1] The focal form often occurs in the location of the pre-existing tumor and can resemble a tumor on neuroimaging. Late-delayed radiation injury develops in the same timeframe as tumor recurrence (months to years after treatment). Differentiation between these two entities is often impossible using MRI and CT.

PET metabolic studies may distinguish tumor recurrence from radiation injury in some cases.[2] Recurrent tumor usually exhibits hypermetabolism, whereas areas of radiation necrosis are hypometabolic. However, hypometabolic (well-differentiated) gliomas may be mistaken for radiation injury. The correlation between the PET result and type of lesion is poor if intensive radiotherapy has been administered.[3]

In focal, late-delayed radiation injury, CT and MRI may reveal a lesion with edema, mass effect, and contrast enhancement (ring or irregular).[4] The appearance mimics that of a glioma or metastasis. The enhancement correlates with the pathologic finding of white-matter coagulation necrosis and fibrinoid necrosis of blood vessel walls. Although the radiation-induced lesion usually develops near the site of the pre-existing tumor, remote radiation necrosis may occur, especially after whole brain irradiation. Sometimes the lesion develops contralateral to the pre-existing tumor. Radiation damage affects white matter more than gray matter.

Diffuse, late-delayed radiation injury manifests as enlarged ventricles, cerebral atrophy, and extensive white-matter abnormalities. More pronounced degrees of brain injury are associated with larger exposed brain volume or whole brain irradiation. Impairment of cognitive function, personality changes, confusion, and, especially in children, learning disabilities may occur. Most people with mild damage are clinically asymptomatic.

The white-matter abnormality in late-delayed radiation injury is usually the most remarkable feature. It varies from small foci at the corners of the frontal and occipital horns of the lateral ventricles to confluent regions extending from the ventricular surface to the gray–white matter junction. The subcortical U-fibers are typically spared. The affected white matter appears hyperintense on PD and T2-weighted images and hypodense on CT secondary to increased water content, gliosis, and demyelination. Diffuse, late-delayed radiation white-matter injury is often symmetric and indistinguishable from the white-matter changes that occur in some normal elderly and people with risk factors for vascular disease. The frequency of late white-matter changes following high-dose brain irradiation is high. Almost all patients have mild changes and a third have severe changes.[5] The occurrence of severe white-matter lesions increases with age, volume of brain irradiated, radiation dose, and time between treatment and neuroimaging.[6]

Dystrophic calcification can occur as a late consequence of radiation. Calcium is easily identified on CT and often invisible on MRI (maxim 13).

Radiation exposure may produce a new neoplasm. The most common radiation-induced brain tumors are meningiomas and gliomas.[7]

References

1. Valk PE, Dillon WP. Radiation injury to the brain. *AJNR Am J Neuroradiol* 1991; **12**: 45–62.
2. Lilja A, Lundqvist H, Olsson Y, *et al*. Positron emission tomography and computed tomography in differential diagnosis between recurrent or residual glioma and treatment-induced brain lesions. *Acta Radiol* 1989; **30**: 121–8.
3. Janus TJ, Kim EE, Tilbury R, Bruner JM, Yung WKA. Use of [^{18}F]fluorodeoxyglucose positron emission tomography in patients with primary malignant brain tumors. *Ann Neurol* 1993; **33**: 540–8.
4. Brismar J, Roberson GH, Davis KR. Radiation necrosis of the brain: neuroradiological considerations with computed tomography. *Neuroradiology* 1976; **12**: 109–13.
5. Constine LS, Konski A, Ekholm S, McDonald S, Rubin P. Adverse effects of brain irradiation correlated with MR and CT imaging. *Int J Radiat Oncol Biol Phys* 1988; **15**: 319–30.
6. Tsuruda JS, Kortman KE, Bradley WG, *et al*. Radiation effects on cerebral white matter: MR evaluation. *AJR Am J Roentgenol* 1987; **149**: 165–71.
7. Kumar PP, Good RR, Skultety FM, *et al*. Radiation induced neoplasms of the brain. *Cancer* 1987; **59**: 1274–82.

40. *Acoustic neuromas can be detected very early with contrasted MRI*

Acoustic neuromas (schwannomas, neurinomas, and neurilemomas) account for 80% of cerebellopontine angle (CP angle) tumors in adults. Meningiomas (10%), epidermoid tumors (4%), metastases, trigeminal neuromas, chordomas, and arachnoid cysts comprise most of the remainder of the CP angle lesions.[1,2]

Intermittent and then progressive hearing loss, disequilibrium, and tinnitus are the cardinal clinical features. Larger lesions may result in headaches, facial numbness and weakness, nystagmus, nausea, and otalgia. Surgical resection often improves the disturbing symptoms of tinnitus and disequilibrium but may result in ipsilateral deafness.

Neuromas arise from the schwann cells within the cranial and peripheral nerves. Sensory nerves are involved more frequently than are motor nerves. Most acoustic neuromas originate in the vestibular branch of the eighth cranial nerve. They are slow-growing benign neoplasms that almost never undergo malignant transformation. Regrowth following incomplete removal is common.

Contrast-enhanced MRI with thin sections in axial and coronal planes is the current gold standard for the diagnosis of acoustic neuromas.[3] Lesions as small as 2 mm can be detected. Most neuromas are slightly hyperintense on T2-weighted images. Contrast enhancement is pronounced and fairly homogeneous (Fig. 40.1). Larger lesions sometimes have a mixed signal pattern and inhomogeneous enhancement.

A CP angle meningioma is the principal lesion that must be differentiated from an acoustic neuroma. Both neuromas and meningiomas may be isointense on T2-weighted images and enhance with contrast. Both may become cystic. Meningiomas in this location are typically broad-based and form an obtuse angle with the adjacent dura. When a dural tail is present, meningioma is very likely. Acoustic neuromas do not generally cause edema in the adjacent pons. In contrast to acoustic neuromas, meningiomas cause narrowing of the internal auditory canal due to hyperostosis. The differential diagnosis also includes other enhancing posterior fossa lesions such as metastasis, hemangioma, postoperative fibrosis, and inflammatory disease.

When MRI is contraindicated, CT cisternogram with air contrast is suitable for diagnosis of acoustic neuromas. Conventional angiography is not typically used for this, although it is sometimes warranted to identify the position of regional arteries. CT, even with contrast enhancement, may not reveal acoustic neuromas until they grow enough to enlarge the internal auditory canal, although with higher resolution CT small acoustic neuromas may be visualized.

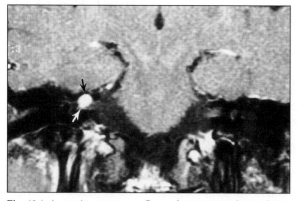

Fig. 40.1 Acoustic neuroma. Coronal contrast-enhanced T1-weighted image of a 47-year-old man with right sensorineural hearing loss. A typical enhancing intracanalicular vestibulocochlear schwannoma (acoustic neuroma) is present (arrows). The tumor measures 6 mm in diameter.

References

1. Hasso AN, Fahmy JL, Hindshaw DB. Tumors of the posterior fossa. In: Stark DD, Bradley WG (eds), *Magnetic resonance imaging*. St Louis: CV Mosby, 1988.
2. Mark AS. The vestibulocochlear system. *Neuroimaging Clin North Am* 1993; **3**: 153–70.
3. Brogan M, Chakeres DW. Gd-DTPA enhanced MRI of cochlear schwannoma. *AJNR Am J Neuroradiol* 1990; **11**: 407–8.

41. *Primitive neuroectodermal tumors often have leptomeningeal metastases*

Medulloblastoma, primary intracranial neuroblastoma, ependymoblastoma, and pinealoblastoma are primitive neuroectodermal tumors (PNETs). They are highly aggressive and carry a poor prognosis. Most occur in children under 10 years of age. The remainder occur in young adults. Leptomeningeal dissemination is present in most, if not all, cases.

The most common PNET is the medulloblastoma, which is one of the most common posterior fossa tumors in children.[1] Boys are affected four times more frequently than girls. Posterior fossa tumors usually present with hydrocephalus once they have grown large enough to obstruct CSF flow. The typical history is one of several days or weeks of progressively worsening headaches with the development of nausea, vomiting, and altered mental status. Surgery with radiation or chemotherapy is palliative but rarely results in cure.

Medulloblastomas have a characteristic appearance on CT. Without contrast they are well-defined hyperdense masses, classically within the midline of the cerebellum in children. The high density is due to a high nuclear-to-cytoplasmic volume ratio. Profound, nearly homogeneous, enhancement occurs with contrast. The appearance of medulloblastomas on MRI, on the other hand, is variable and less specific (Fig. 41.1). They are usually hypointense to brain tissue on T1-weighted images and hypointense to isointense on T2-weighted images.[2] The tumor location and patient age, not the imaging appearance, are the most important factors in differentiating medulloblastomas from other lesions radiographically. Nonetheless, biopsy is required for definitive diagnosis (maxim 35).

The sharpness of the cerebellar folia and fissures on the midline sagittal MRI is of importance in evaluating a medulloblastoma. Effaced fissures and blurred folia indicate meningeal spread. As with other PNETs, contrast-enhanced MRI demonstrates CSF seeding (Fig. 41.2).[3] During the primary evaluation, both the brain and the entire spine should be evaluated with contrast enhancement.[4]

The most common locations for CSF dissemination are the vermian cistern, subfrontal regions, and lateral fissures. Direct metastasis to the spinal subarachnoid space (thoracic and lumbosacral) and the cauda equina

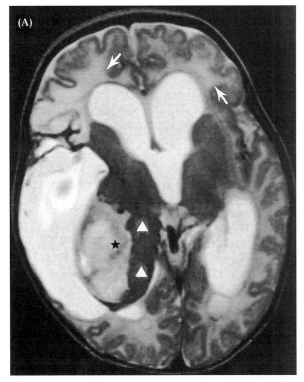

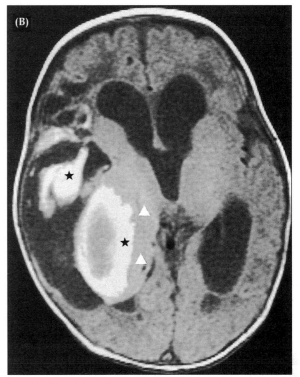

A: T2-weighted image of a 2-month-old girl with lethargy and increased head circumference. A large mixed-signal mass lesion is present in the right temporal lobe. The tumor is the low signal region (white triangles). Pathologic examination was necessary to make the diagnosis. Note the hydrocephalus and transependymal CSF leakage (arrows). The tissue marked with the black star is hemorrhage (see B–D).

B: Noncontrasted T1-weighted image reveals subacute blood with methemoglobin (high-signal areas; black stars). The tumor (white triangles) is isointense to gray matter.

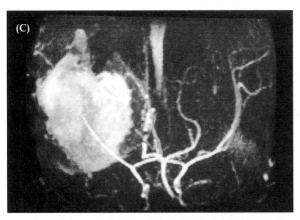

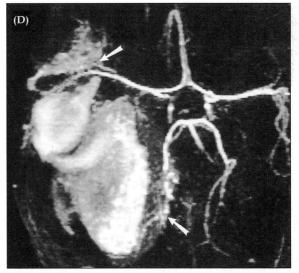

C: A three-dimensional time-of-flight MRA coronal projection image (maxim 51) demonstrates the mass of methemoglobin and distended blood vessels.

D: A three-dimensional time-of-flight MRA axial collapsed image (maxim 51) also demonstrates the mass of methemoglobin and distended blood vessels. Several tortuous vessels from the middle cerebral and posterior cerebral arteries are located on the surface of the tumor (arrows).

Fig. 41.1 PNET.

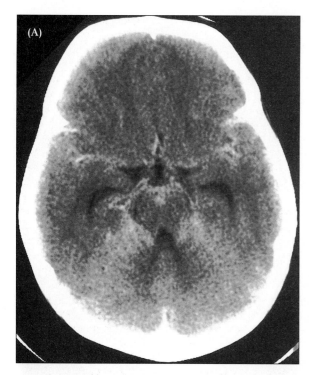

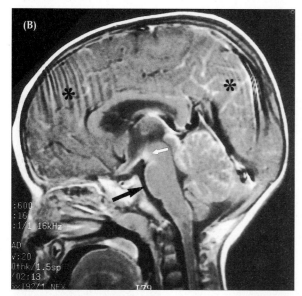

are present in nearly half of affected patients. MRI of the spine is better than conventional myelography and CT myelography for detecting these "drop metastases" (Fig. 41.3). Metastases outside the CNS, in the skeleton, lymph nodes and lung, occur with tumor recurrence.

Radiographic tumor staging should be done before the initial surgery because, during the first few postoperative weeks, the surgery may result in enhancement of the meninges and surgical site, which may be difficult to differentiate from leptomeningeal seeding. If necessary,

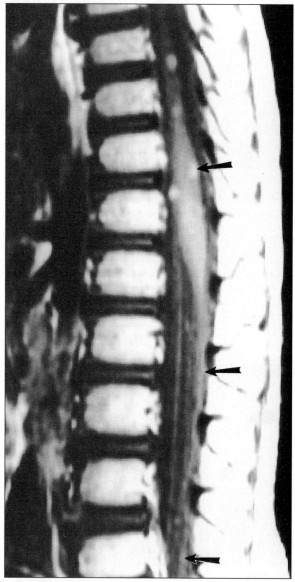

Fig. 41.2 PNET with CSF metastases. **A:** Contrast-enhanced CT of a 20-month-old child with medulloblastoma. The enhancement of the vessels in the circle of Willis is fuzzy, suggesting abnormal meningeal enhancement. Some "fullness" was present in the basal cisterns on the noncontrasted CT (not shown). Otherwise the CT is unremarkable. **B:** MRI 1 day after CT. A T1-weighted contrast-enhanced sagittal image reveals diffuse CSF metastases coating the brain. This appearance, reminiscent of frosting on a cake, gives rise to the term "sugar icing". The coating is most prominent at the ventral surface of the pons (black arrow), interpeduncular cistern (white arrow), optic chiasm, and over the surface of the parietal and occipital lobes and cerebellum. The meningeal enhancement extends into the cervical region. MRI is much more sensitive than CT at detecting meningeal lesions. Note the ghosting artifact in the horizontal (phase-encoded) direction (black asterisks) (maxim 4).

Fig. 41.3 Drop metastases. Sagittal T1-weighted contrast-enhanced image of the lower thoracic and upper lumbar spine in a child with medulloblastoma. Multiple enhancing nodules are present in the intradural space at several levels. The conus medullaris is widened and the cauda equina and nerve roots are nodular (arrows). Drop metastases are most common in medulloblastomas but may also occur in ependymomas, astrocytomas, germinomas, pineal gland tumors, and choroid plexus tumors.

serial postoperative MRIs may be used to distinguish between postsurgical enhancement and CSF seeding.

Primary cerebral neuroblastoma, the most common supratentorial PNET, is still a rare tumor accounting for less than 1% of primary CNS neoplasms. However, it is one of the most common congenital brain tumors. These tumors tend to be located in the frontal or parietal lobes, adjacent to the lateral ventricle. There is no gender predilection, and 80% develop during the first decade of life. The prognosis, as for other PNETs, is generally poor. Local recurrence following surgery is the rule, and CSF dissemination is common.

Neuroblastomas are large hemispheric masses with necrosis, cyst formation, and hemorrhage. Usually, the involved area has prominent vessels. Contrast enhancement on CT is variable and heterogeneous. MRI demonstrates a mixed-signal lesion on T1 and T2 weighted images. Mild-to-moderate contrast enhancement is commonly present on MRI.

References

1. Barkovich AJ. *Pediatric neuroimaging*. New York: Raven Press, 1994: 324–31.
2. Long SD, Kuhn MJ. Primitive neurodermal tumor: CT, MRI, and angiographic findings. *Comput Med Imaging Graph* 1992; **16**: 291–5.
3. Rippe DJ, Boyko OB, Friedman HS. Gd-DTPA enhanced MR imaging of leptomeningeal spread of primary intracranial CNS tumor in children. *AJNR Am J Neuroradiol* 1990; **11**: 329–32.
4. Jennings M, Slatkin N, D'Angelo M, *et al*. Neoplastic meningitis as the sole presentation of an occult CNS primitive neuroectodermal tumor. *J Child Neurol* 1993; **8**: 306–12.

42. *Herniation (subfalcial, uncal, and tonsillar) is well visualized on MRI*

Brain herniation, a secondary effect of many intracranial processes, often has a greater clinical impact than the primary condition. The principal types of brain herniation are: subfalcial (cingulate), lateral (uncal) and central (descending and ascending) transtentorial, and foramen magnum (cerebellum tonsillar). Brain tumors often result in herniation.

The cranial cavity is functionally divided into compartments by bony ridges and dural folds. Brain herniation is the process of mechanical displacement of the brain and vascular structures from one compartment into another. In clinical practice the most common herniations are subfalcial, uncal, and descending transtentorial.[1–3] Although these herniations are best visualized using MRI with three orthogonal image orientations, most acutely ill patients are examined with CT.

Subfalcial herniation is present when the cingulate gyrus is displaced across the midline under the free edge of the falx cerebri (Figs. 36.1B and 87.2B). With large subfalcial herniations the contralateral foramen of Monro may become obstructed, resulting in enlargement of the contralateral lateral ventricle and compression of the ipsilateral ventricle by the primary mass. In pronounced subfalcial herniation, branches of the anterior cerebral artery may be compressed against the falx producing ischemia and infarction in the parasagittal region. Transtentorial herniation may accompany or follow subfalcial herniation.

Lateral transtentorial herniation is caused by an expanding unilateral, often temporal lobe, mass. The initial stage consists of medial and downward displacement of the uncus through the incisura of the tentorium (uncal herniation). The first clinical sign is an ipsilateral dilated ("blown") pupil resulting from compression of the pupillodilatory fibers in the outer part of the third cranial nerve (maxim 63). Ipsilateral suprasellar cistern effacement may be visible at this stage on CT and MRI.

With increasing herniation, the brain stem becomes displaced to such an extent that the contralateral cerebral peduncle is compressed into the free edge of the tentorium, producing Kernohan's notch (a pathologic finding), and resulting in *ipsilateral* hemiparesis (a false localizing sign). As the brain stem is shifted away from the herniating temporal lobe, the ipsilateral cerebellopontine angle cistern may initially enlarge. If the process progresses, the ipsilateral posterior cerebral artery becomes compressed, resulting in occipital lobe and medial temporal lobe infarction. Rarely, aqueduct obstruction, producing hydrocephalus, subsequently occurs. Following this, a sequence similar to the late stages of central descending transtentorial herniation ensues.[4]

Central descending (downward) transtentorial herniation is produced by a large unilateral or bilateral supratentorial mass effect that forces the brain downward into the tentorial opening. Classically, rostral-to-caudal clinical deterioration occurs. This is characterized by progressive impairment of consciousness culminating in deep coma, respiratory deterioration (yawns, Cheyne–Stokes respirations, hyperventilation, shallow and rapid breathing, apneustic breathing, ataxic breathing, and finally apnea), pupillary changes (small, pinpoint, and finally midsize, irregular fixed), and motor dysfunction (purposeful, localizing, flexor posturing, extensor posturing, and finally flaccidity). The degree of horizontal displacement of the pineal, aqueduct, and septum pellucidum on axial CT and MRI corresponds to the level of arousal.[5]

Severe descending transtentorial herniation is characterized by obliteration of the perimesencephalic and quadrigeminal plate cisterns. At this stage, deep coma and extensor posturing are present. Posterior cerebral artery infarction and obstructive hydrocephalus may occur. Central thalamic, midbrain, and pontine hemorrhages (Duret hemorrhages) develop owing to stretching of the paramedian penetrating branches of the basilar artery, which does not descend as far as the brain does because of its attachment at the circle of Willis. Abrupt deterioration, skipping the typical stages, may occur if the intraventricular pressure dramatically increases (e.g. intraventricular extension of hemorrhage) or if the subtentorial pressure drops (e.g. execution of a lumbar puncture, which incidentally may reveal a normal opening pressure because the compartments are isolated by obstruction).

Ascending (upward) transtentorial herniation is the result of mass effect in the posterior fossa. Lowering of the pressure in the supratentorial compartment (with a shunt or surgery) may be the precipitating factor in ascending herniation. This condition can be asymptomatic, although impairment of arousal, and cerebellar and brain stem signs such as ataxia, dysmetria, nystagmus, ophthalmoplegia, and skew deviation, may develop. Occlusion of the superior cerebellar arteries with superior cerebellar infarctions occurs in severe cases. CT reveals distortion to obscuration of the quadrigeminal plate cisterns.[6]

In large posterior fossa mass lesions or with severe descending transtentorial herniation, the cerebellar tonsils may be forced downward (tonsillar herniation). This is best seen on sagittal and coronal MRI studies, although it may also be evident on axial CT. Differentiation from a Chiari I malformation may be difficult, but usually the CSF spaces are obliterated in tonsillar herniation (maxim 88). Deep grooving at the base of the cerebellum is present on pathologic examination. Cerebellar tonsillar herniation is a final common pathway in brain death.

MRI is superior to any other imaging modality, including CT, for detecting and defining herniations. All the different types of herniation and many of the consequences, including changes in ventricular size, CSF flow dynamics, and involvement of intracranial vasculature, can be appreciated on MRI. Quantitative measurements of the relative displacement, upwards or downwards, of the proximal tip of the aqueduct (iter) can be made on midsagittal MRI to estimate the degree of transtentorial herniation.[7] The midsagittal image also may be used to assess cerebellar tonsillar herniation.

References

1. Hahn FJ, Gurney J. CT signs of central descending transtentorial herniation. *AJNR Am J Neuroradiol* 1985; **6**: 844–5.
2. Osborn AG. Diagnosis of descending transtentorial herniation by cranial computed tomography. *Radiology* 1977; **123**: 93–6.
3. Fisher CM. Acute brain herniation: a revised concept. *Semin Neurol* 1984; **4**: 417–21.
4. Plum F, Posner JB. *The diagnosis of stupor and coma*, 3rd edn. Contemporary Neurology Series, 1982. Philadelphia: FA Davis.
5. Ropper AH. A preliminary MRI study of the geometry of brain displacement and level of consciousness with acute intracranial masses. *Neurology* 1989; **39**: 622–7.
6. Osborn AG, Heaston DK, Wing SD. Diagnosis of ascending transtentorial herniation by cranial computed tomography. *AJR Am J Roentgenol* 1978; **130**: 755–60.
7. Reich JB, Sierra J, Camp W, Zanzonico P, Deck MD, Plum F. Magnetic resonance imaging measurements and clinical changes accompanying transtentorial and foramen magnum brain herniation. *Ann Neurol* 1993; **33**: 159–70.

43. *The absence of the fourth ventricle on head CT (or MRI) is a neurologic emergency*

When a patient with a known brain tumor becomes symptomatic, CT or MRI is often performed to determine the cause. Symptoms may be due to tumor recurrence, tumor growth, new metastases, hydrocephalus, or herniation, or a direct complication of chemotherapy or radiation therapy.

Obstruction of the fourth ventricle from a posterior fossa tumor, stroke, or other process is often life-threatening. This sign, visible on CT (and MRI), may occur before brain-stem compression, herniation, and hydrocephalus are evident clinically or radiographically. Emphasizing the CT pattern of obliteration of the fourth ventricle is worthwhile because it is often subtle and thus missed by an inexperienced interpreter. This is especially important because CT is commonly the single imaging study performed in an emergency. A necessary step (with CT or MRI) is identification of the fourth ventricle. Even with substantial beam-hardening artifact, the fourth ventricle should be distinguishable. Symmetric effacement of the fourth ventricle is easiest to overlook. If one cannot identify the fourth ventricle, a crisis is probably in progress (Fig. 43.1).

The CT findings in obliteration of the fourth ventricle, subarachnoid hemorrhage (maxim 58), and epidural hematoma (maxim 86) are among the most important CT patterns because they may require urgent action. Anyone who examines head CT scans (including the emergency physician, internist, family practitioner, and general surgeon) needs to be able to identify even the subtlest of these patterns. Emergency treatment in these three conditions is often life-saving, and delay of treatment can be fatal.

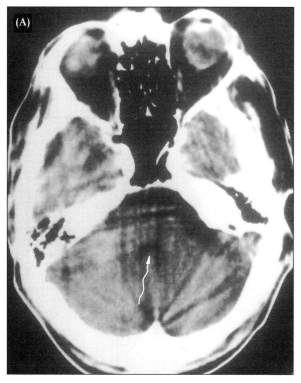

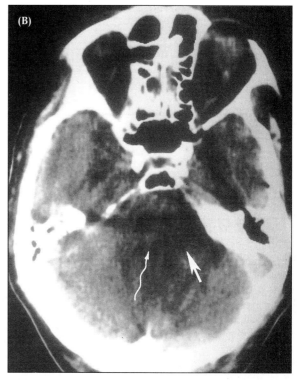

A: Axial CT of a 73-year-old woman on the day she developed the abrupt onset of difficulty walking and left arm ataxia. Note that the fourth ventricle has a normal horseshoe shape (wavy arrow) and there is no evidence for a stroke. The alternating white and dark lines throughout the middle and posterior fossae are beam-hardening artifacts (maxim 4).

B: Axial CT at the same level 2 days later, after the patient's level of consciousness had declined. Note that the fourth ventricle is not visualized. The wavy arrow identifies the expected location of the fourth ventricle. Decreased density is now present in the left anterior cerebellum (straight arrow).

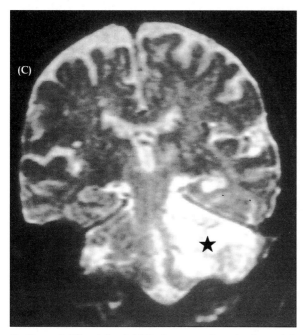

C: Coronal T2-weighted MRI demonstrating the large left cerebellar infarction (black star). Note that MRI produces far better images of the posterior fossa than does CT.

Fig. 43.1 Absence of the fourth ventricle.

44. *Migraine sufferers can develop UBOs and diffuse atrophy*

Migraine is a common chronic disorder with occasional nonspecific imaging abnormalities. CT is usually normal, although diffuse atrophy may occur.[1,2] MRI is also usually normal but small, focal, high-signal lesions on T2-weighted images in the deep or subcortical white matter are present in over one-third of patients with chronic migraine.[3,4] These T2 high-signal lesions are a variety of unidentified bright object (UBO; maxim 16). They are more frequent in older migraine sufferers and uncommon before the age of 40 years. The cause and clinical significance of these focal white-matter lesions is not known. One hypothesis is that they are due to the cumulative effect of transient episodes of focal ischemia. Care must be taken when attributing these white-matter lesions to migraine, because indistinguishable lesions occur in a variety of other conditions, including stroke and multiple sclerosis (maxim 16).

Conventional angiography is not usually performed in acute migraine. However, in a few cases, segmental vasospasm is reported and related to sympathetic activation.[5,6] This phenomenon may play a role in the development of UBOs and atrophy in chronic migraine sufferers. It is probably safe to perform conventional angiography, even during acute attacks, although there is a small risk of precipitating a transient neurologic event, especially in patients with classic migraine.[7]

The mechanism of migraine is not understood. One theory involves so-called "spreading depression". A PET study, coincidentally performed on an individual during the development of a migraine, demonstrated this phenomenon with bilateral hypoperfusion beginning in the occipital lobe and spreading over the cortical surface anteriorly into the temporal and parietal lobes, corroborating the spreading-depression hypothesis.[8]

References

1. du Boulay GH, Ruiz JS, Rose FC, Stevens JM, Zilkha KJ. CT changes associated with migraine. *AJNR Am J Neuroradiol* 1983; **4**: 472–3.
2. Osborn RE, Alder DC, Mitchell CS. MR imaging of the brain in patients with migraine headaches. *AJNR Am J Neuroradiol* 1991; **12**: 521–4.
3. Ferbert A, Busse D, Thron A. Microinfarction in classic migraine? A study with magnetic resonance imaging findings. *Stroke* 1991; **22**: 1010–14.
4. Soges LJ, Cacayorin ED, Petro GR, *et al.* Migraine: evaluation by MRI. *AJNR Am J Neuroradiol* 1988; **9**: 425–9.
5. Schulman EA, Hershey B. An unusual angiographic picture in status migrainosus. *Headache* 1991; **31**: 396–8.
6. Solomon S, Lipton RB, Harris PY. Arterial stenosis in migraine: spasm or arteriopathy? *Headache* 1990; **30**: 52–61.
7. Shuaib A, Hachinski VC. Migraine and the risks from angiography. *Arch Neurol* 1988; **45**: 911–12.
8. Woods RP, Iacoboni M, Mazziotta JC. Brief report: bilateral spreading cerebral hypoperfusion during spontaneous migraine headache. *N Engl J Med* 1994; **331**: 1689–92.

45. *An unexplained change in headache pattern warrants an imaging study*

Be wary of the patient with occasional, relatively stereotyped, migraines who suddenly develops a headache with a different character. Neuroimaging investigation is usually warranted in this situation. A change in headache pattern raises the possibility that an underlying, previously unsuspected, lesion has evolved. (Similarly, a patient with a change in seizure pattern deserves evaluation.) A structural lesion, such as an arteriovenous malformation (that has had a small hemorrhage or produced ischemia via vascular steal), is sometimes present. However, patients with migraine typically have some variability in their headache pattern, and alteration of the side of the head involved with pain is common.[1] In patients with strictly unilateral headaches, an imaging study to rule out a structural lesion should be considered.

Acute findings in migraine are atypical, but transient lesions with characteristics typical of ischemia and cerebral edema are reported.[2] In complicated migraine (migraine with stroke), the typical neuroimaging features of stroke are present. Complicated migraines most frequently involve the posterior circulation, especially the occipital lobes. Because headache may accompany stroke from other causes, the diagnosis of complicated migraine must be made with caution.[3]

Noncontrasted CT is mandatory to rule out subarachnoid hemorrhage in a patient with an abrupt-onset

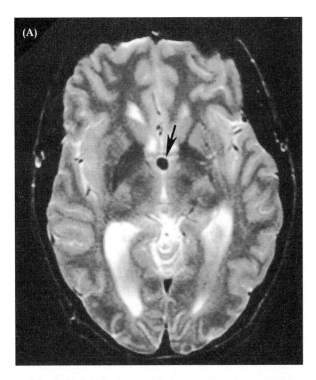

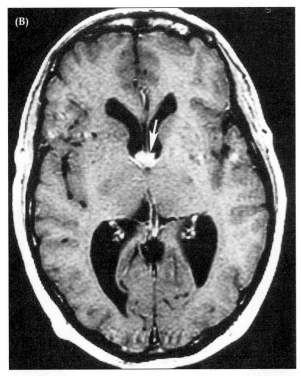

Fig. 45.1 Colloid cyst of the third ventricle. **A:** Axial T2-weighted image of a 33-year-old man with new-onset frequent severe headaches. A low-signal spherical lesion is present in the anterior tip of the third ventricle (arrow). **B:** Axial contrast-enhanced T1-weighted image in the same patient. The lesion exhibited high signal on the noncontrasted T1-weighted image and is not changed with contrast (arrow). The lesion was high density on noncontrasted CT (not shown). These features are typical of a colloid cyst.

severe headache (maxim 58).[4] When headaches occur with an atypical quality, unexpected abnormalities (such as a subdural hematoma, intracerebral hemorrhage, or tumor) are sometimes present.

Neuroimaging is usually indicated in patients with new-onset positional headaches without obvious cause (such as lumbar puncture or acute infection). Occasionally a sequence of symptoms occurs beginning with headache followed by dizziness, weakness, numbness, nausea, vomiting, and decreased level of consciousness. Mass lesions in a variety of locations can be responsible for these symptoms, but this history is classic for a colloid cyst in the third ventricle (Fig. 45.1). Certain postures cause the colloid cyst to block the foramen of Monro, producing transient symptomatic dilation of the obstructed lateral ventricle. Behavioral abnormalities are sometimes present due to involvement of the fornix (part of the limbic system). Colloid cysts are usually hyperdense on noncontrasted CT, high signal on T1-weighted images, and low signal on T2-weighted images. However, the MRI signal characteristics of colloid cysts are variable.

References

1. Dalessio DJ. Diagnosing the severe headache. *Neurology* 1994; **44**(suppl 3): S6–12.
2. Alvarez-Cermeno JC, Gobernado JM, Freije R, Zaragoza E, Gimeno A. Cranial computed tomography in pediatric migraine. *Pediatr Radiol* 1984; **14**: 195–7.
3. Olesen J, Friberg L, Olsen TS, *et al.* Ischaemia-induced (symptomatic) migraine attacks may be more frequent than migraine-induced ischaemic insults. *Brain* 1993; **116**: 187–202.
4. Prager JM, Mikulis DJ. The radiology of headache. *Med Clin North Am* 1991; **75**: 525–44.

46. *Some patients with "trigeminal neuralgia" should be imaged for temporomandibular joint pathology*

Temporomandibular joint (TMJ) disease pain is frequently localized in the face and can mimic trigeminal neuralgia (tic douloureux). The diagnosis of trigeminal neuralgia is straightforward when it presents classically with attacks, precipitated by stimulation of a trigger zone, of paroxysms of unilateral burning facial pain confined to one or more divisions of the fifth cranial nerve. In patients who develop facial pain with chewing, the main differential diagnosis is between TMJ disease, trigeminal neuralgia, and temporal arteritis. The trigger zone for trigeminal neuralgia is stimulated by chewing in some patients. MRI of the TMJ can identify pathology and is indicated in these patients. Temporal arteritis occurs in elderly patients and may present with facial pain precipitated by chewing (jaw claudication) (maxim 53).

TMJ disease is fairly common. The pain is in the face, preauricular region, or ear. Other common complaints include tinnitus, decreased hearing, and diffuse earache. Early treatment of significant TMJ meniscus disease is warranted to avoid development of osteochondritis dissecans and avascular necrosis of the mandibular condyle.[1]

The diagnostic procedure of choice for examination of the TMJ is MRI.[2,3] Arthrogram, an alternative, is especially useful for examining joint dynamics. CT is useful for detecting bone pathology and if MRI is contraindicated. Arthrography and CT are not necessary in most cases.

Proper MRI technique is important for detection of TMJ pathology. Because the mandible has a bilateral articulation, both temporomandibular joints must function in synchrony. Asymmetric movement increases the possibility of TMJ disease. Specially designed dual surface coils with simultaneous image acquisition enables detection of asymmetric motion and examination of the kinematics of joint motion. The optimal signal-to-noise ratio on 1.5 T MRI is achieved with an 8–12 cm field of view. Oblique sagittal images obtained with the mouth both opened and closed and coronal images obtained with the mouth closed only are ideal. The standard imaging protocol should include T2- and PD-weighted images. The fast spin-echo technique results in a better signal-to-noise ratio than does the conventional spin-echo sequence. However, standard spin-echo PD images have better soft-tissue resolution. Slice thickness should be 3 mm or less with 0.5 mm or less interslice gaps.

Internal derangement with meniscus dislocation and degenerative joint disease are the most common pathologic findings in patients with joint pain and dysfunction. Bone marrow edema, joint effusions, and tumors may also be detected. Synovitis and joint effusions are commonly seen accompanying TMJ meniscus dislocation. However, a joint effusion may be present in a healthy, asymptomatic joint.

References

1. Schellhas KP, Wilkes CH, Fritts HM, *et al*. MR of osteochondritis dissecans and avascular necrosis of the temporomandibular condyle. *AJR Am J Roentgenol* 1989; **10**: 3–12.
2. Katzberg RW, Bessette RW, Tallents RH, *et al*. Normal and abnormal temporomandibular joint: MR imaging with surface coil. *Radiology* 1986; **158**: 183–9.
3. Westesson PL, Katzberg RW, Tallents RH, *et al*. CT and MR of the temporomandibular joint: a comparison with autopsy specimens. *AJR Am J Roentgenol* 1987; **148**: 1165–71.

47. *Conventional MRI and CT are usually normal in patients with pseudotumor cerebri*

Pseudotumor cerebri (benign intracranial hypertension) is a syndrome of increased intracranial pressure without a mass lesion or hydrocephalus. It occurs most commonly in obese females and is often idiopathic. Cerebral venous sinus thrombosis, otitis media, mastoiditis, viral infections, and several medications including tetracycline and vitamin A predispose patients to the development of pseudotumor cerebri.

Headache is the primary symptom.[1,2] Transient visual obscurations, blurred vision, diplopia, and dizziness also occur. Examination is typically normal except for dramatic papilledema. Occasionally a sixth nerve palsy may be present. Visual loss, beginning with enlargement of the blind spot followed by constriction of peripheral vision, sometimes leading to blindness, is the major morbidity.

CT and conventional MRI are usually normal. The ventricles are often small (slit-like) (Fig. 47.1). However, small ventricles are a normal variant. Currently the role of structural neuroimaging is to exclude other pathology.

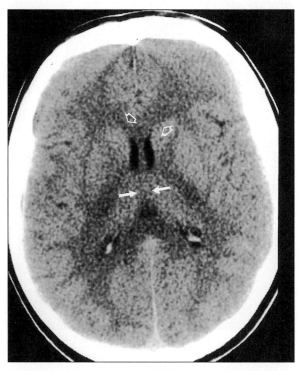

Fig. 47.1 Pseudotumor cerebri. Axial noncontrasted CT of a 30-year-old woman with headaches, blurred vision, and papilledema. The study reveals small lateral ventricles with slit-like anterior horns (open arrows) and an absent (slit-like) third ventricle (straight arrows). The basal cisterns (not shown) were also effaced.

The diagnosis of pseudotumor is confirmed when the neuroimaging is normal and the opening CSF pressure is elevated. The CSF pressure is typically elevated to 250–500 mmCSF, although fluctuations in the CSF pressure may occur and occasionally near-normal values are present. If a causative medication is being used, it should be discontinued. Weight loss can be curative in obese patients. Otherwise, treatment is directed toward decreasing the CSF volume with diuretics (acetazolamide), steroids, or glycerol. In intractable cases, placement of a lumbar peritoneal shunt or, preferably, surgical fenestration of the optic nerve sheath may be necessary.

Several hypotheses have been proposed to explain the pathophysiology of pseudotumor cerebri. In 1937, Dandy proposed increased blood volume as the cause.[3] However, reduced blood flow is present in this condition.[4] A disturbance of CSF circulation, probably due to decreased CSF absorption by the arachnoid villi, is the most likely mechanism.[5–7] Brain biopsy demonstrates increased intracellular and extracellular fluid.[8] Diffusion-weighted MRI (maxim 50) demonstrates increased local water mobility within subcortical white matter,[9–11] and may have a diagnostic role in pseudotumor cerebri in the future.

References

1. Wall M, George D. Idiopathic intracranial hypertension: a prospective study of fifty patients. *Brain* 1990; **114**: 155–80.

2. Giuseffi V, Wall M, Siegel P, Rojas PB. Symptoms and disease associations in idiopathic intracranial hypertension (pseudotumor cerebri): a case–control study. *Neurology* 1991; **41**: 239–44.

3. Dandy WE. Intracranial pressure without brain tumor. *Ann Surg* 1937; **106**: 462–513.

4. Greitz D, Hannerz T, Rahn T, *et al.* MR imaging of cerebrospinal fluid dynamics in health and disease. *Acta Radiol* 1994; **35**: 204–11.

5. Calabrese VP, Selhorst JB, Harbison JW. Cerebrospinal fluid infusion test in pseudotumor cerebri. *Ann Neurol* 1978; **4**: 173 (abstract).

6. Johnston I, Hawke S, Halmagyi M, Teo C. The pseudotumor syndrome: disorders of cerebrospinal fluid circulation causing intracranial hypertension without ventriculomegaly. *Arch Neurol* 1991; **48**: 740–7.

7. Gjerris F, Soelberg Sorensen P, Vorstrup S, Paulson OB. Intracranial pressure, conductance to cerebrospinal fluid outflow, and cerebral blood flow in patients with benign intracranial hypertension (pseudotumor cerebri). *Ann Neurol* 1985; **17**: 158–62.

8. Sahs AL, Joynt RJ. Brain swelling of unknown cause. *Neurology* 1956; **6**: 791–803.

9. Sorensen PS, Thomsen C, Jerris F, *et al.* Increased brain water content in pseudotumor cerebri measured by magnetic resonance imaging of brain water self diffusion. *Neurol Res* 1989; **11**: 160–4.

10. Thomsen C, Henricksen O, Pring P. *In vivo* measurement of water cell diffusion in the human brain by magnetic resonance imaging. *Acta Radiol* 1987; **28**: 353–61.

11. Gideon P, Sorensen PS, Thomsen C, *et al.* Increased brain water cell-diffusion in patients with idiopathic intracranial hypertension. *AJNR Am J Neuroradiol* 1995; **16**: 381–7.

8
Vascular Diseases

48. *The first neuroimaging study in a patient who presents with an apparent stroke should be a noncontrasted head CT*

CT has an essential role in the initial evaluation of the patient who presents with a stroke. It is a rapid and readily available test that is very sensitive for acute hemorrhage, in both the subarachnoid and the parenchymal space, and serves to rule out unexpected lesions. Of patients who present with a stroke-like syndrome, roughly 80% have cerebral infarction, 10% intracerebral hemorrhage, 5% subarachnoid hemorrhage, and 5% another process.[1] Clinical presentations

mimicking a stoke occur in a variety of conditions including tumor, abscess, vascular malformation, subdural or epidural hematoma, trauma, metabolic derangements, and seizures.

The role of neuroimaging in stroke is to assist in the initial diagnosis and to aid in the assessment of brain damage or recovery. 2–5 days after a large ischemic stroke, further deterioration often occurs as local edema becomes maximal. Repeat imaging is commonly performed at this time to exclude hemorrhagic transformation.

On CT, acute blood is hyperdense to the brain (although it is much less dense than bone) (maxim 1). Small petechial hemorrhages indicative of hemorrhagic transformation of a bland infarction are readily apparent on CT and MRI.[2] When petechial hemorrhages are

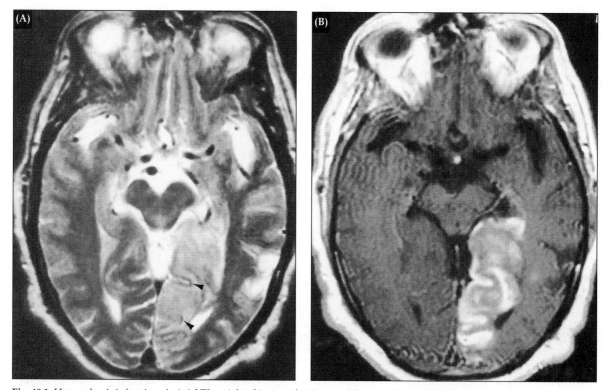

Fig. 48.1 Hemorrhagic infarction. **A:** Axial T2-weighted image of a 50-year-old man with the acute onset of visual symptoms. Edema (high signal) is present in the left occipital lobe in the distribution of the posterior cerebral artery. Decreased signal in several regions of the involved cortex (arrowheads) is indicative of petechial hemorrhage in the subacute stage (intracellular methemoglobin). These hemorrhagic regions were high signal on the PD and T1-weighted images (not shown). **B:** Contrast-enhanced T1-weighted image showing gyral enhancement, a classic pattern for a subacute infarction.

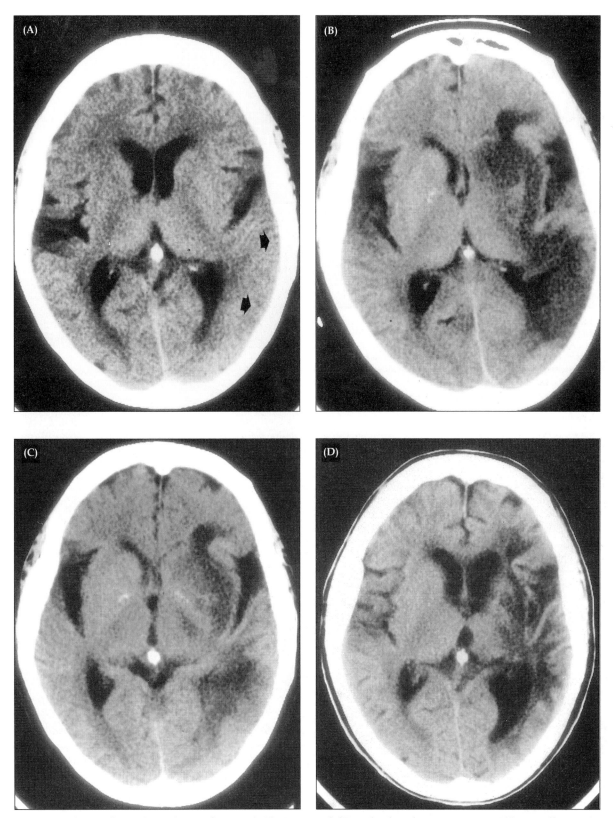

Fig. 48.2 Evolution of an ischemic brain infarction. **A:** Noncontrasted CT on the day of symptom onset. Mild mass effect on the temporoparietal cortex (sulci effaced) is present in the left hemisphere (arrows). **B:** Noncontrasted CT on day 6 after the onset of symptoms. Significant edema is present and mass effect is at its maximum. The lateral ventricle is effaced. **C:** Noncontrasted CT 2 weeks after the onset of symptoms. The edema and mass effect are decreased. **D:** Noncontrasted CT 7 months later showing residual gliosis and encephalomalacia. The left lateral ventricle is enlarged, partially compensating for the tissue loss.

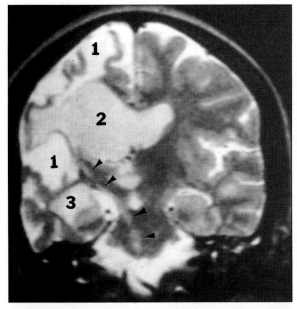

Fig. 48.3 Wallerian degeneration following middle cerebral artery infarction. Coronal T2-weighted image in a patient who suffered a complete right middle cerebral artery territory stroke 4 years earlier. The right hemisphere brain substance is very atrophic with compensatory dilation of the sulci (1) and lateral ventricle (body, 2; temporal horn, 3). The high-signal line traversing the residual brain substance into the brain stem is the gliotic corticospinal tract that has undergone Wallerian degeneration (arrowheads).

present, an embolic cause for the stroke is probable (Fig. 48.1). Careful systemic anticoagulation is generally safe in this situation[3] and indicated, especially if a cardiac embolic source is demonstrable with echocardiography. Transesophageal is more sensitive than transthoracic echocardiography for identifying a cardiac source of emboli. In embolic stroke, intravenous heparin anticoagulation should be initiated gradually (i.e. no bolus should be given) to prevent extension of the hemorrhage.

In ischemic stroke of any cause, the initial CT is often normal, especially if performed within the first 6 hours. Within 24 hours, sulcal effacement, the first structural indication of a cortical infarction, becomes apparent (Fig. 48.2). Over the next few days further evidence of mass effect develops. Depending on the size and location of the stroke, midline shift, subfalcial herniation, transtentorial herniation, and tonsillar herniation are all possibilities (maxim 42). At this time, ischemic strokes become hypodense to brain tissue on CT, due to tissue damage and edema. Lacunar strokes appear as small areas of hypodensity deep to the cortex. Focal cortical atrophy and ventriculomegaly may be present in the chronic phase. In middle cerebral artery (MCA) distribution stokes, ipsilateral Wallerian degeneration of the corticospinal tract and cerebral pyramid atrophy may become visible after several months (Fig. 48.3).

The MCA is occasionally hyperdense on initial noncontrasted CT in a large hemispheric stroke. This finding indicates that an MCA thrombosis is responsible for the stroke (Fig. 48.4).[4,5]

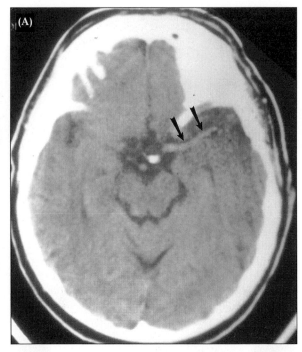

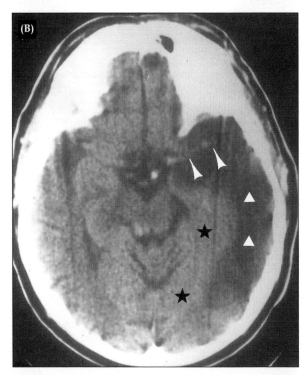

Fig. 48.4 Middle cerebral artery thrombosis. **A:** Noncontrasted axial CT of a 74-year-old man 12 hours after the acute onset of right hemiparesis and aphasia. High density in the left middle cerebral artery is due to thrombosis (black arrows). The left temporal lobe is slightly hypodense and the sulci are effaced. **B:** Noncontrasted axial CT 1 week later reveals low density throughout the left hemisphere middle cerebral artery distribution, indicative of infarction with edema (white triangles). The posterior cerebral artery territory is spared (black stars). The thrombosis in the left middle cerebral artery is less apparent (white arrowheads).

Initially no abnormal contrast enhancement is present within the territory of an ischemic stroke and contrast is generally not indicated.[6,7] Enhancement develops, beginning at the periphery, as early as 3 days, and is maximal at 3–4 weeks after the stroke on CT. This enhancement is due to neovascularity and is termed "luxury perfusion". When the history and previous neuroimaging studies are not available, a stroke at this stage may be misdiagnosed as a tumor or abscess because of the ring-enhancing appearance (maxim 71). On MRI, enhancement may be present within the first day. The initial enhancement is intravascular due to sluggish flow.[8,9] Over the next few days, dural enhancement develops owing to increased collateral blood supply. Later, luxury perfusion with gyral enhancement is apparent on MRI (Fig. 48.1B).

Standard MRI is able to detect intraparenchymal hemorrhage, but hyperacute blood is confusing (maxim 60) and petechial and subarachnoid hemorrhages can be missed. Although new advances in MRI technology may soon change the situation (maxim 50), currently CT should be the first imaging study in patients with suspected brain infarction, especially if anticoagulation is planned. MRI should be performed when the diagnosis is questionable or if invasive treatment is being considered.

References

1. Sacco RL. Classification of strokes. In: Fisher M (ed.), *Clinical atlas of cerebrovascular disorders*. London: Mosby Year Book, 1994.
2. Bryan RN, Levy LM, Whitlow WD, *et al.* Diagnosis of acute cerebral infarction: comparison of CT and MR imaging. *AJNR Am J Neuroradiol* 1991; **123**: 611–20.
3. Pessin MS, Estol CJ, Lafranchise F, Caplan LR. Safety of anticoagulation after hemorrhagic infarction. *Neurology* 1993; **43**: 1298–303.
4. Pressman BD, Tourje EJ, Thomson JR. An early sign of ischemic infarction: increased density in a cerebral artery. *AJNR Am J Neuroradiol* 1987; **8**: 645–8.
5. Launes J, Ketonen L. Dense middle cerebral artery sign: an indicator of poor outcome in middle cerebral artery infarction. *J Neurol Neurosurg Psychiatr* 1987; **50**: 1550–2.
6. Pullicino P, Kendall BE. Contrast enhancement in ischemic lesions: I. Relationship to prognosis. *Neuroradiology* 1980; **19**: 235–9.
7. Kendall BE, Pullicino P. Intravascular contrast injection in ischemic lesions: II. Effect on prognosis. *Neuroradiology* 1980; **19**: 241–3.
8. Crain MR, Yuh WTC, Greene GM, *et al.* Cerebral ischemia: evaluation with contrast-enhanced MR imaging. *AJNR Am J Neuroradiol* 1991; **12**: 631–9.
9. Elster AD, Moody DM. Early cerebral infarction: gadopentate dimeglumine enhancement. *Radiology* 1990; **177**: 627–32.

49. *Ischemic infarctions can be detected early and followed on standard MRI*

Neuroimaging is required to make an accurate early diagnosis of stroke and to follow the evolution, including the effect of therapy. Currently, all patients presenting with an acute neurologic deficit of possible brain origin must undergo initial CT to rule out hemorrhage, tumor, abscess, or another unexpected lesion (maxim 48).

Early changes resulting from ischemia may be visible on MRI within several hours following an acute stroke. T1-weighted images demonstrate subtle, morphologic changes consisting of sulcal effacement and early mass effect secondary to cytotoxic edema (Table 49.1).[1] This is followed by high signal on the T2-weighted images, reflecting water accumulation due to vasogenic edema. Experimentally, increased signal is visible as early as 30 minutes after impaired circulation. In common practice, abnormalities are not discernible for 3–8 hours after the onset of clinical symptoms, paralleling the development of vasogenic edema.[2–4]

Conventional spin-echo images also provide information about flow in the major vessels. With slow flow or thrombosis, the intraluminal signal is hyperintense or isointense within the first few days after infarction. Contrast may result in intravascular enhancement due to sluggish flow within affected arteries.[5]

Meningeal and parenchymal enhancement is apparent earlier on MRI than on CT, and may be present as early as 6 hours following a stroke.[6] The meningeal enhancement is due to collateral flow through the leptomeninges. Parenchymal gyriform enhancement is secondary to abnormalities of blood flow autoregulation, with consequent pooling of contrast within capillaries. Later, breakdown of the blood–brain barrier and neovascularization contribute to enhancement.

Table 49.1 Brain edema on neuroimaging

	Vasogenic	Cytotoxic	Interstitial (hydrocephalic)
Pathogenesis	Increased capillary permeability	Cellular swelling	Block of CSF flow or absorption
Location of edema	Primarily white matter	White and gray matter	Periventricular white matter
Edema fluid composition	Plasma	Increased intracellular water and sodium content	CSF
Cause	Tumor, Abscess, Ischemia/Infarction, Trauma, Hemorrhage, Meningitis	Hypoxia, Hypo-osmolality, Disequilibrium syndromes, Ischemia, Meningitis	Obstructive hydrocephalus

References

1. Fishman RA. Cerebrospinal fluid in diseases of the nervous system, 2nd edn. Philadelphia: WB Saunders, 1992: 116–38.
2. Yuh WT, Crain MR, Loes DJ, Greene GM, Ryals TJ, Sato Y. MR imaging of cerebral ischemia: findings in the first 24 hours. *AJNR Am J Neuroradiol* 1991; **12**: 621–9.
3. Bryan RN, Levy LM, Whitlow WD, *et al.* Diagnosis of acute cerebral infarction: comparison of CT and MR imaging. *AJNR Am J Neuroradiol* 1991; **123**: 611–20.
4. Sipponen JT, Kaste M, Ketonen L, *et al.* Serial nuclear magnetic resonance (NMR) imaging in patients with cerebral infarction. *J Comput Assist Tomogr* 1983; **7**: 585–9.
5. Crain MR, Yuh WTC, Greene GM, *et al.* Cerebral ischemia: evaluation with contrast-enhanced MR imaging. *AJNR Am J Neuroradiol* 1991; **12**: 631–9.
6. Elster AD, Moody DM. Early cerebral infarction: gadopentate dimeglumine enhancement. *Radiology* 1990; **177**: 627–32.

50. *Diffusion-weighted and perfusion MRI may differentiate ischemia from infarction in early stroke*

Diffusion-weighted and perfusion MRI are two recently developed techniques that may help to differentiate between areas of infarcted (dead) brain and ischemic (potentially salvageable) brain within minutes of an acute stroke. These techniques are likely to gain a major role in evaluating the effectiveness of early therapy for stroke. Currently, acute treatment interventions to prevent enlargement of an infarction are largely experimental. However, therapies to re-establish perfusion (e.g. with thrombolytic agents) and protect ischemic, as yet uninfarcted, tissue (e.g. with hyperperfusion and cyto-protective agents) show promise.[1–3]

Diffusion-weighted MRI (DWI) measures the micro-scopic movement (brownian motion) of water protons. A spin-echo sequence with a pair of magnetic field gradient pulses, which effectively mark the location of water protons, is used. Net translational movement of water molecules results in signal loss. The less the net diffusion, the greater the signal.[4] In stroke, initial impair-ment of diffusion is postulated to result from cytotoxic edema causing hyperintense signal corresponding to a decrease in the apparent diffusion coefficient (ADC). The quantitative value of the ADC may help to distinguish infarcted from ischemic tissue.[5] DWI can identify ischemic tissue within minutes following arterial occlu-sion.[6,7] Normalization of the signal can occur within 30 minutes of restoration of blood flow in animal models.[8]

Perfusion MRI uses rapidly acquired images imme-diately after contrast injection to image blood flow. Areas with flow (perfusion) have reduced signal on T2-weight-ed images because of the magnetic susceptibility effect of the contrast.[4] In animal models, reduced perfusion is present in ischemic areas (corresponding to hyperintense regions on DWI), with return of perfusion after removal of the occlusion.[9] Perfusion imaging can also be performed without contrast using special pulses to mark arterial water.

Both diffusion and perfusion MRI are more practical with the rapid image acquisition times made possible by echo-planar imaging (EPI). EPI is an ultrafast MRI technique that uses rapidly switching high-gradient fields to acquire single images in less than 1 second. Specialized hardware and software are required.

SPECT with a perfusion tracer (HMPAO) can demon-strate stroke early in its course but has not achieved wide-spread clinical use. Its limited acceptance may relate to the impracticality of mixing the isotope (not available off-hours in many centers) and its relatively poor resolution. Per-fusion SPECT is a currently available tool that can be used to demonstrate impaired perfusion in acute stroke and rapid reperfusion in spontaneously resolving stroke.[10]

In the near future, with the widespread availability of rapid, high-resolution MRI (facilitated by EPI), a complete acute MRI stroke protocol can be envisioned. This protocol might consist of conventional structural MRI augmented by intracranial and extracranial MRA and diffusion-weighted and perfusion MRI giving a complete picture of the stroke in the acute setting. Data obtained with MR spectroscopy may also be included. This initial information may direct the choice of therapy and be used in comparison with follow-up studies to determine the efficacy of the treatment. Such protocols, which will probably help to revolutionize the clinical approach to stroke, are likely to be routine within the next decade.

References

1. terPenning B. Pathophysiology of stroke. *Neuro Imaging Clin North Am* 1992; **2**: 389–408.
2. Kay R, Wong KS, Yu YL, *et al.* Low-molecular-weight heparin for the treatment of acute ischemic stroke. *N Engl J Med* 1995; **333**: 1588–93.
3. National Institute of Neurologic Disorders and Stroke rt-PA Stroke Study Group. Tissue plasminogen activator for acute ischemic stroke. *N Engl J Med* 1995; **333**: 1581–7.
4. Fisher M, Sotak CH, Minematsu K, Li L. New magnetic resonance techniques for evaluating cerebrovascular disease. *Ann Neurol* 1992; **32**: 115–22.
5. Hasegawa Y, Fisher M, Latour LL, Dardzinski BJ, Sotak CH. MRI diffusion mapping of reversible and irreversible ischemic injury in focal brain ischemia. *Neurology* 1994; **44**: 1484–90.
6. Warach S, Gaa J, Siewert B, Wielopolski P, Edelman RR. Acute human stroke studied by whole brain echo planar diffusion-weighted magnetic resonance imaging. *Ann Neurol* 1995; **37**: 231–41.
7. Reith W, Hasegawa Y, Latour LL, Dardzinski BJ, Sotak CH, Fisher M. Multislice diffusion mapping for 3-D evolution of cerebral ischemia in a rat stroke model. *Neurology* 1995; **45**: 172–7.
8. Mintorovitch J, Moseley ME, Chileuih L, *et al.* Comparison of diffusion and T2-weighted MRI for the early detection of cerebral ischemia and reperfusion in rats. *Magn Reson Med* 1991; **18**: 39–50.
9. Moseley ME, Kucharczyk J, Mintorovitch J, *et al.* Diffusion-weighted MR imaging of acute stroke: correlation of T2-weighted and magnetic susceptibility-enhanced MR imaging in cats. *AJNR Am J Neuroradiol* 1990; **11**: 423–9.
10. Baird AE, Donnan GA, Austin MC, McKay WJ. Early reperfusion in the "spectacular shrinking deficit" demon-strated by single-photon emission computed tomography. *Neurology* 1995; **45**: 1335–9.

51. *MRA enables noninvasive evaluation of the intracranial and neck vessels*

Catheter intra-arterial contrast digital subtraction angiography (DSA), though invasive, is currently the gold standard angiographic technique. Magnetic resonance angiography (MRA) is a noninvasive alternative that is gaining wide acceptance in many situations. MRA is capable of detecting intracranial aneurysms 3 mm or larger in diameter, and is useful as a screening study of the carotid arteries because it tends to exaggerate the degree of stenosis. Therefore the finding of no or mild stenosis on MRA indicates the absence of significant carotid pathology. Computed tomographic angiography (CTA), acquired with helical CT, is another newly developed technique that is being used for the screening evaluation of intracranial vessels. CTA requires intravenous contrast, making it moderately invasive. Ultrasound and Doppler studies are also useful in certain circumstances.

MRA encompasses a group of MRI techniques that detect blood flow. It is noninvasive because intravenous contrast injection is not necessary (although it may be used). Carotid, vertebral, and intracranial arteries and veins can be evaluated with MRA. MRA sequences are designed to emphasize the differences in signals produced by moving *versus* stationary tissue. The two MRA techniques that are in routine clinical use are time-of-

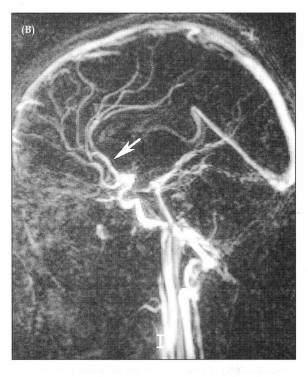

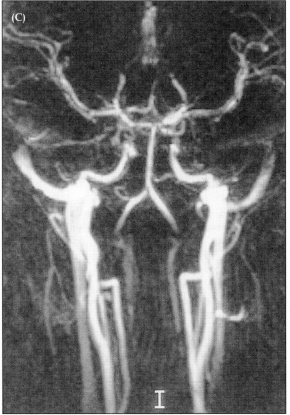

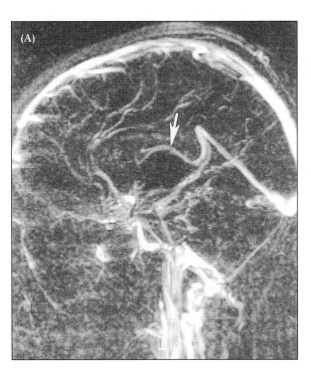

Fig. 51.1 Phase-contrast (PC) MRA. **A:** Midline sagittal two-dimensional PC angiogram (slab = 5 cm) at velocity encoding (VENC) = 30 cm/s. The venous structures are apparent (arrow = internal cerebral vein). **B:** At VENC = 60 cm/s the arteries are visualized (arrow = anterior cerebral arteries). The middle cerebral arteries are not visualized because only a 5 cm wide slab is examined. **C:** Coronal PC angiogram with VENC = 80 cm/s. Normal carotid and vertebral artery flow is demonstrated. This acquisition technique may be used for localization before more time-consuming three-dimensional studies.

flight (TOF) and phase-contrast (PC) angiography. Acquisition in either two dimensions (2D) or three dimensions (3D) is possible with both methods. MRAs are acquired as source data that are reconstructed using a maximum intensity projection (MIP) algorithm. The resulting images allow examination of the vessels from a variety of angles (pseudo-3D). The quality of MRA can be decreased by a variety of factors including patient motion, swallowing, long echo time, and inappropriate voxel size. The MIP post-processing algorithm also may introduce artifact.

TOF MRA uses gradient-echo sequences and relies primarily on the phenomenon of flow-related enhancement to distinguish moving from stationary material. That is, blood flowing into the imaging plane produces high signal because it has previously not been saturated by multiple radiofrequency pulses as has the stationary tissue. The TOF techniques are most sensitive to flow perpendicular to the imaging plane. A modification of the 3D TOF technique uses multiple overlapping thin-slab acquisitions (MOTSA) to improve the flowing blood signal. Substances with short T1 values (e.g. methemoglobin, gadolinium, and fat) also produce high signal on TOF images (Figs. 41.1C and D) and can be mistaken for flowing blood. In the TOF technique, background suppression may be augmented by using magnetization transfer (MT)*. MT further reduces the signal produced by the stationary tissue, permitting visualization of small peripheral vessels.

PC MRA uses velocity-induced phase shifts to distinguish flowing blood from stationary tissue (Fig. 51.1). Usually this technique utilizes a GRE sequence. PC MRA is sensitive to the direction and velocity of the flow. Machine settings may be adjusted to emphasize either venous or arterial flow rates (velocity encoding or VENC). The ability of this technique to demonstrate slow flow makes it useful for evaluating deep intracerebral venous structures, dural sinuses, and cortical veins. This is referred to as MR venography (MRV). MRV is especially valuable in searching for venous sinus thrombosis (maxim 54). The average time to perform a 2D PC study is about 3–4 minutes. The 3D study takes significantly longer. PC MRA images are substantially degraded by patient motion.

Although DSA is the gold standard, the combination of MRI and MRA often yields more information, because no details about the brain parenchyma are present on DSA. Additionally, MRA may reveal a thrombus that

*Magnetization transfer (MT) is a technique in which tissue contrast is altered by transferring energy between macromolecules and water. The technique consists of applying off-resonance radiofrequency saturation pulses selectively to saturate the macromolecules, which in turn saturate the MR signal of the nearby water molecules. The free protons are not affected. MT has its maximum benefit in tissues containing large numbers of macromolecules and water such as in the brain. MT is commonly used in MRA studies because it results in considerable background suppression, improving the clarity of small vessels. MT may be used in conjunction with gadolinium to obtain an augmented contrast effect.

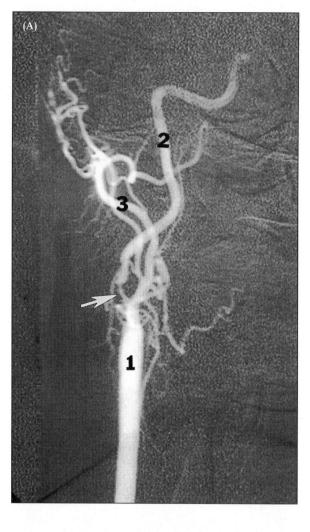

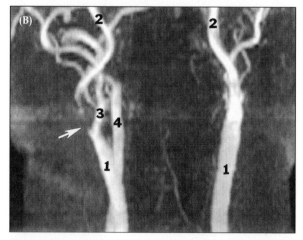

Fig. 51.2 Extracranial atherosclerotic carotid artery stenosis: comparison of DSA (A) and 2D TOF MRA (B) of the carotid bifurcation in a patient with tight stenosis of the proximal internal carotid artery (arrows). Both methods effectively demonstrate the abnormality. The MRA slightly overestimates the degree of stenosis. 1, common carotid artery; 2, internal carotid artery; 3, external carotid artery; 4, jugular vein. Note that both carotid arteries and the venous structures are present in this MRA.

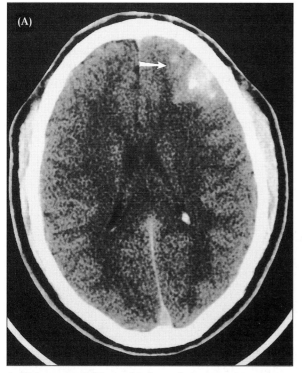

A: Noncontrasted CT reveals a hyperdense area of serpiginous calcifications (arrow).

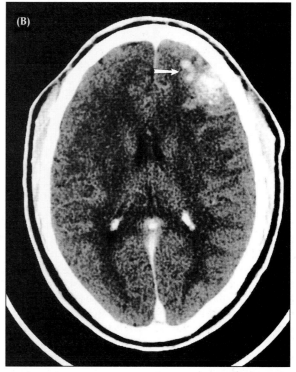

B: Contrast-enhanced CT of the same area demonstrates enhancement of the serpentine lesions consistent with an AVM (arrow). It measures 3 cm.

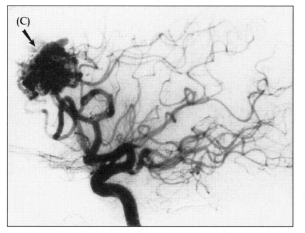

C: Selective left internal carotid artery injection DSA delineates the AVM in the left frontal lobe (arrow).

D: Selective left internal carotid artery injection DSA following a two-stage presurgical embolization. There is complete embolization of a posterior ascending frontal branch of the middle cerebral artery and partial embolization of an internal frontal branch of the left anterior cerebral artery. Essentially no flow is present in the AVM. (Courtesy of Dr Wayne Crow, University of Texas Medical Branch)

Fig. 51.3 Arteriovenous malformation (AVM).

completely obstructs a false lumen. The subtraction in DSA refers to subtraction of a plain scout skull radiograph to eliminate the bone, improving image clarity. DSA is invasive and carries a small risk of complication from the arterial puncture, catheterization, and injection of contrast material. A review of 15 studies involving a total of 8300 patients with mild cerebrovascular disease who underwent conventional catheter arteriography between 1968 and 1987 concluded that neurologic complications occurred in 4% with permanent deficits in 1% and death in less than 0.1%.[1] A more recent study of 1095 patients who had transfemoral DSA reported transient neurologic complications within the first 24 hours in 0.45%, permanent deficits in less than 0.1% and no deaths.[2]

Extracranial vessels

2D TOF MRA is a fast, reliable screening study in evaluation for carotid stenosis (Fig. 51.2). MIP images generated from multiple angles should be reviewed for the highest yield. Severe stenosis is characterized by a signal void within the vessel. Overestimation of the degree of stenosis occurs because turbulent (nonlaminar) flow in the stenotic segment results in intravoxel dephasing, decreasing the signal. The signal recovers where the laminar flow is re-established in the post-stenotic region. The length of the signal void is roughly proportional to the degree of stenosis. A normal MRA excludes significant carotid artery pathology with a high degree of confidence. MRA is as sensitive (96–100%) as conventional angiography but somewhat less specific (64–93%) in evaluating for carotid artery stenosis.[3–5] In the future, when MRI is standard in the evaluation of stroke (maxims 49 and 50), MRA examination of the carotids will be routine.

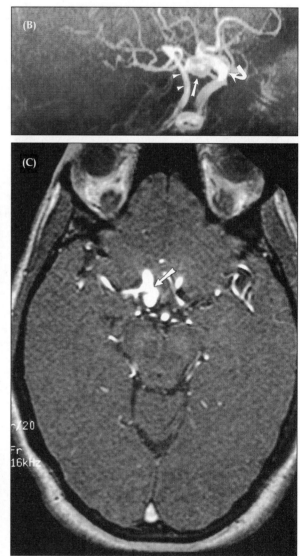

Fig. 51.4 MRA of an aneurysm. The same MRA information may be displayed in a variety of ways for optimal analysis. In this figure an aneurysm is identified on the source image and with two separate post-processing display techniques: compressed and projection. Surface rendering is another display technique. This data set was acquired using the 3D time-of-flight (TOF) method without contrast in a 43-year-old woman who had palsies of the right third, fourth, and fifth cranial nerves.
A: Axial compressed image. A 12 mm aneurysm adjacent to the distal internal carotid artery is present (arrow).
B: Left oblique projection image. The aneurysm (straight arrow), carotid siphon (curved arrow), and basilar artery (arrowheads) are well visualized.
C: The axial source image best demonstrates the neck of the aneurysm and its connection to the parent vessel (arrow).

MRA currently has several pitfalls in carotid imaging. Near-occlusive stenosis of the proximal internal carotid artery, which gives a string sign on DSA, can appear as a complete occlusion on MRA (Fig. 51.2). On rare occasions, a vascular loop or retrograde flow may simulate a severe stenosis. Technological advances will probably eliminate these problems in the future.

When a severe short-segment stenosis is present in the typical location at the carotid bifurcation in conjunction with concordant neurologic symptoms, MRA is the only vascular study needed in the preoperative evaluation for carotid endarterectomy.[6] The usual examination time for the neck vessels is less than 10 minutes. Carotid MRA is typically performed without intravenous contrast.

Intracranial vessels

DSA is the gold standard in the evaluation of intracranial aneurysms and arteriovenous malformations (AVMs) (Fig. 51.3). The patent lumen of the aneurysm or AVM, as well as the adjacent vascular structures and the direction of flow, can be ascertained with DSA. MRI and MRA enable direct evaluation of the intraluminal thrombus and information regarding the adjacent vascular and parenchymal structures.

With the MOTSA technique, aneurysms as small as 2.8 mm can be detected.[7] 3D TOF MRA can accurately depict the arteries of the circle of Willis and the proximal portions of the anterior, middle, and posterior cerebral arteries. Medium-sized aneurysms (3–5 mm) are usually visible on MRA (Fig. 51.4). Small aneurysms may be detectable only on the source images owing to artifacts associated with the MIP post-processing algorithm. Therefore, both projection images and source images must be reviewed to maximize sensitivity. However, even a high-quality MRA is insufficient for the detection of small aneurysms, and conventional catheter angiography is always required in patients with subarachnoid hemorrhage (SAH). MRI and MRA may be used to screen patients at high risk for aneurysms (e.g. polycystic kidney disease), as resection of asymptomatic aneurysms is usually reserved for those greater than 3 mm. Conventional catheter angiography and MRI/MRA are often complementary in the evaluation of moderate and large aneurysms.

In contrast to the successful application of MRI/MRA in identifying and characterizing aneurysms, it is usually not possible to visualize acute SAH on MRI. CT is the appropriate study in the initial evaluation of patients suspected of having SAH. CTA can be performed, when the CT demonstrates SAH, to help to guide the approach during the DSA. However, CTA increases the contrast exposure. MRI/MRA and transcranial Doppler are useful in the subacute state to evaluate for vasospasm (maxim 59).

90% of parenchymal AVMs are located in the supratentorial compartment. Examination of intracranial AVMs should be performed with DSA that includes injections into the internal and external carotid arteries and the vertebrobasilar system. The dural component of AVMs is often supplied by the external carotid artery. Additionally, it is not uncommon for a patient to have both an AVM and one or more aneurysms. DSA effectively demonstrates the blood flow into and away from the AVM nidus. However, the arterial supply to the adjacent parenchyma is usually difficult to determine because of the vascular steal phenomenon. MRI is currently the best noninvasive modality available for evaluating AVMs. CT is better than MRI at detecting calcification and acute hemorrhage. MRI is superior to both CT and DSA at defining the anatomic relationships of the nidus, feeding arteries, draining veins, and associated parenchymal abnormalities.[8] To obtain the most information, especially during preoperative evaluation, all three studies (MRI/MRA, CT, and DSA) are indicated.

MRA is suited for screening for vascular patency in several other conditions such as moya-moya disease (Fig. 51.5). Visualization of the vasculature is especially reliable in children who lack atherosclerotic disease and have a good cardiac output. Low cardiac output with slower flow, a condition often present in the elderly, results in poorer definition of the vessels on MRA.

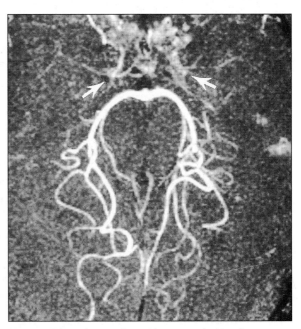

Fig. 51.5 Bilateral internal carotid artery occlusions demonstrated with MRA. Axial 3D TOF (compressed image) of the circle of Willis demonstrating the absent flow in both internal carotid arteries (arrows). There is collateral circulation through the ophthalmic arteries bilaterally. DSA confirmed the findings. This is a precursor lesion to moya-moya disease. MRA is a noninvasive, efficient, and rapid method to screen the intracranial and extracranial circulation for abnormalities.

References

1. Hankey GJ, Warlow CP, Sellar RJ. Cerebral angiographic risk in mild cerebrovascular disease. *Stroke* 1990; **21**: 209–22.
2. Gryska U, Freitag J, Zeumer H. Selective cerebral intraarterial DSA: complication rate and control of risk factors. *Neuroradiology* 1990; **32**: 296–9.
3. Heiserman JE, Dryer BP, Fram EK, *et al*. Carotid artery stenosis: clinical efficacy of 2 dimensional time of flight angiography. *Radiology* 1992; **182**: 761–8.
4. Masaryk AM, Ross JS, DiCello MC, *et al*. 3D FT MR angiography of the carotid bifurcation: potential and limitations as a screening examination. *Radiology* 1991; **179**: 797–804.
5. Polak JF, Bajakian RL, O'Leary DH, *et al*. Detection of internal carotid arterial stenosis: comparison of MR angiography, color doppler sonography and arteriography. *Radiology* 1992; **182**: 35–40.
6. Anson JA, Heiserman JE, Dryer BP, *et al*. Surgical decisions on the basis of magnetic angiography of the carotid arteries. *Neurosurgery* 1993; **32**: 335–47.
7. Blatter DD, Parker DL, Ahn SS, *et al*. Cerebral MR angiography with multiple overlapping thin slab acquisition. Part II. Early clinical experience. *Radiology* 1992; **183**: 379–89.
8. Augustyn GT, Scott JA, Olson E, *et al*. Cerebral venous angiomas: MR imaging. *Radiology* 1985; **156**: 391–3.

52. *Carotid and vertebral artery dissection in the neck can be reliably diagnosed with MRI and MRA*

Dissection of a carotid or vertebral artery can be spontaneous or associated with neck trauma, cervical manipulation (chiropractic), fibromuscular dysplasia, hypertension, vigorous physical activity, sympathomimetic drugs, and pharyngeal infections. Most patients complain of neck discomfort, and usually have a headache and neurologic deficits secondary to brain ischemia or infarction. Occasionally lower cranial nerve (IX, X, XI, XII) dysfunction causing dysphagia, dysarthria, and hoarseness or a Horner syndrome due to direct compression of the nerves by an exophytic pseudoaneurysm occurs in carotid dissection or as a result of vascular compromise in vertebral dissection. An acute onset of headache or neckache associated with cranial nerve palsies should prompt evaluation for dissection of a cervical vessel. Generally, patients with a dissection are treated with anticoagulation to prevent thrombosis within the residual vessel lumen. Sometimes surgical repair of the intimal flap is indicated. Large randomized trials confirming the appropriateness of these treatments are lacking.

The dissection consists of blood penetrating into the arterial wall, splitting the media, and forming a false lumen of variable length. Conventional catheter angiography and MRA usually demonstrate narrowing of the carotid or vertebral artery lumen extending as far as the skull base. The classic angiographic findings of an intimal flap with a false lumen are relatively uncommon. Carotid dissections typically spare the carotid bulb. Vertebral dissections are most commonly located between C2 and the skull base.

Conventional MRI usually reveals the hematoma. A fat-suppression technique should be used so that fat surrounding a vessel is not mistaken for hemorrhage. Axial spin-echo T1-weighted images (with fat suppression) are preferred to demonstrate both the hematoma and the narrowed vessel lumen. MRA is also usually informative; however, the high signal produced by the subacute clot (methemoglobin) may simulate flow in the vessel in the time-of-flight (TOF) technique. Examination of source images or a phase-contrast (PC) MRA can help to avoid this confusion. Occasionally a dissection involves the adventitia instead of the media. In this case the angiogram may be normal but the MRI still demonstrates the hematoma.

Conventional MRI appears to be the most sensitive test for detection of dissections, and is the procedure of choice in suspected dissection.[1-3] Despite this, false-negative MRI is possible, and catheter angiography should be performed when the MRI is negative and dissection is suspected. MRI has several additional advantages compared with catheter angiography. Examination of the brain parenchyma, enabling detection of brain infarction, can (and should) be performed at the same time as examination of the neck. During the course of recovery, follow-up MRI and MRA may be performed to assess vessel recanalization, which may affect the decision of when to discontinue anticoagulation.

References

1. Pacini R, Simon JH, Rubinstein D, Ketonen L, Kido D, Kieburtz K. Direct imaging of spontaneous dissection of the internal carotid artery by chemical shift MRI. *AJNR Am J Neuroradiol* 1991; **12**: 360–2.
2. Quint DJ, Spickler M. Demonstration of artery dissection: report of 2 cases. *J Neurosurg* 1990; **72**: 964–7.
3. Sue DE, Brant-Zawadski MN, Chana J. Dissection of cranial arteries in the neck: correlation of MRI and arteriography. *Neuroradiology* 1992; **34**: 273–8.

53. *MRI usually reveals small strokes in CNS vasculitis*

A wide variety of diseases produce inflammation in the walls of blood vessels. A disease is termed a vasculitis when the vasculature is the target of the inflammation and parenchymal disease results from disordered blood flow.[1] Several conditions with vasculitis affect the intracranial blood vessels, causing neurologic and ophthalmologic disease. Primary vasculitides that affect the CNS include polyarteritis nodosa (PAN), Churg–Strauss angiitis, Wegener's granulomatosis, lymphomatoid granulomatosis, temporal arteritis, Behcet disease, and isolated CNS angiitis. Certain infections (including lymphocytic choriomeningitis, cytomegalovirus, hepatitis B, neisseria, rickettsia, syphilis, tuberculosis, herpes zoster, and mycoplasma), neoplasms, and toxins (cocaine and amphetamines) may produce secondary CNS vasculitis.

Proposed mechanisms for vasculitis include immune complexes, autoantibodies, T-cell immune responses, and direct toxin-mediated damage to endothelium, each of which can elicit vascular inflammation.[1]

The diagnosis of CNS vasculitis may be made when typical findings are present on cerebral angiography and a biopsy specimen from an affected tissue demonstrates vasculitis. In the case of isolated CNS angiitis, the biopsy specimen must be from the brain and leptomeninges. A consistent clinical presentation with headaches and multifocal strokes or encephalopathy must be present. Cranial neuropathies, myelopathy, seizures, pituitary abnormalities, and focal motor and sensory abnormalities may also occur.

High-quality MRI almost always reveals small strokes when the brain is involved in a vasculitic process.[2] These are often located in the white matter of the cerebral hemispheres or deep nuclei on T2- and PD-weighted images (Fig. 53.1A). Larger strokes may also be present. Occasionally the lesions enhance, making differentiation from neoplasm and infection difficult.[3] However, the MRI findings are not specific for vasculitis. Given a compatible clinical presentation, MRI demonstrating multiple small strokes should prompt angiography unless there is another explanation. Usually, the MRI abnormalities are less extensive than those detected with angiography. Nevertheless, when the angiogram is negative, a brain biopsy should be considered in suspicious cases. Negative high-quality head MRI is fairly strong evidence against brain vasculitis.

Four-vessel catheter angiography is the standard radiographic study in suspected CNS vasculitis but is falsely negative in 30% or more of cases because the affected vessels are too small to be resolved.[2] Angiographic features consistent with vasculitis include segmental narrowing (sausage lesions), cuffing, abrupt or tapered areas of obstruction, early venous filling, and regionally prolonged circulation time (Fig. 53.1B). Multiple sites of narrowing along a single vessel are occasionally present and referred to as beading. Unfortunately, there is no angiographic finding that is specific for vasculitis. Indistinguishable lesions are present in atherosclerosis and vasospasm. Additionally, it is not possible with angiography to determine the specific cause or whether the underlying process is primary or secondary.[3]

MR angiography (MRA) is not as useful at present because the resolution of small vessels is not high enough to detect the subtle abnormalities indicative of vasculitis with confidence. In addition, beading artifact is common on MRA and may be overinterpreted. Despite these limitations, evidence of vasculitis (segmental narrowing) can sometimes be detected on MRA (Fig. 53.2).

Brain and leptomeningeal biopsies enable the diagnosis of the primary and secondary causes of vasculitis and the nonvasculitic entities. Because the treatment of CNS angiitis has significant side effects, brain biopsy should be performed in all cases of suspected isolated CNS angiitis before initiating treatment. Brain biopsy

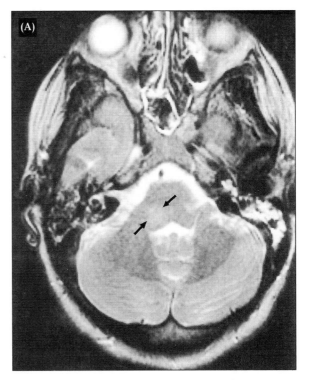

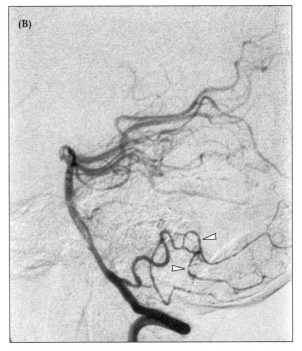

Fig. 53.1 Vasculitis. **A:** Axial T2-weighted image of a 22-year-old woman with the new onset of dysarthria, and ipsilateral facial and contralateral body numbness. A high-signal region is present at the right pontomedullary junction just anterior to the fourth ventricle (arrows). **B:** DSA of selective right vertebral injection demonstrates segmental narrowing in the right posterior inferior cerebellar artery (PICA) and its peripheral branches consistent with vasculitis (arrowheads).

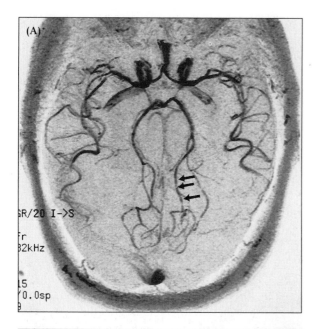

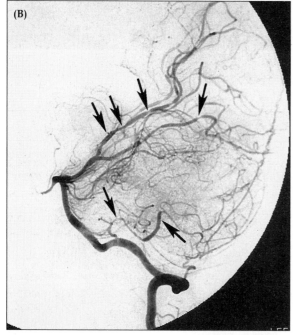

Fig. 53.2 CNS vasculitis. **A:** Compressed image MRA in a 44-year-old woman who presented with a cerebellar infarction. Vasculitic changes of irregularity and beading are present in the left posterior cerebral artery (arrows). **B:** Conventional catheter contrast subtraction angiography in the same patient confirmed the lesions and revealed other involved vessels (arrows).

should also be performed in other forms of suspected vasculitis when the angiogram is positive but vasculitis cannot be confirmed in another tissue. However, brain biopsy is inconclusive in over one-quarter of the cases because patchy involvement results in sampling error. Brain biopsy has a 2% risk of serious morbidity.

PAN serves as a representative vasculitis. It is a systemic necrotizing vasculitis that involves the CNS in approximately one-quarter of cases. Fever, weight loss, renal dysfunction, arthritis/arthralgias, hypertension, and peripheral nerve involvement are the most common clinical manifestations. Angiography and pathologic examination demonstrate small and medium-sized vessel involvement. Microaneurysms are common in the systemic vasculature, but are rare in the CNS. Some of the delayed CNS complications of PAN are due to chronic hypertension and vaso-occlusive changes, and are not directly related to CNS vasculitis. Patients may have a stroke from cardiac valve disease, coagulopathy, or antiphospholipid antibodies. Abdominal angiography with examination of the mesenteric vessels is usually diagnostic. The diagnosis can be confirmed with sural nerve or muscle biopsy. The prognosis is poor. Treatment with corticosteroids and immunosuppressive agents may have a temporary benefit.

Vasculitis occurs as a secondary effect in a number of infections. Bacterial meningitis may produce vessel inflammation at the base of the brain leading to cranial nerve or parenchymal infarction. Tuberculous meningitis with vasculitis classically involves the vertebrobasilar artery and supraclinoid internal carotid arteries. Fungal (mycotic) arteritis also affects the region near the skull base. Actinomyces tends to invade the vessel walls directly, causing multiple hemorrhagic infarctions.

References

1. Moore PM. Neurologic manifestations of vasculitis: update on immunopathogenic mechanisms and clinical features. *Ann Neurol* 1995; **37**(S1): S131–41.
2. Harris KG, Tran DD, Sickels WJ, Cornell SH, Yuh WT. Diagnosing intracranial vasculitis: the roles of MRI and angiography. *AJNR Am J Neuroradiol* 1994; **15**: 317–30.
3. Alhalabi M, Moore PM. Serial angiography in isolated angiitis of the central nervous system. *Neurology* 1994; **44**: 1221–6.

54. *Venous sinus thrombosis occurs in the peripartum period and is readily detected with MRI and MR venography*

Superior sagittal sinus occlusion is a serious and potentially lethal disease that occurs in both late pregnancy and the postpartum period. Other hypercoagulable conditions also predispose to superior sagittal sinus thrombosis. Occlusion of the lateral sinus is usually the result of a bacterial or fungal infection of the ear or mastoid leading to thrombophlebitis. Cavernous sinus thrombosis is characteristically a complication of an infection in the nasal sinuses, orbits, or face. Conditions that predispose to thrombosis within the cerebral venous sinuses are as follows:[1]

- Pregnancy (during the puerperium)
- Behcet disease
- Systemic lupus erythematosus
- Use of oral contraceptives

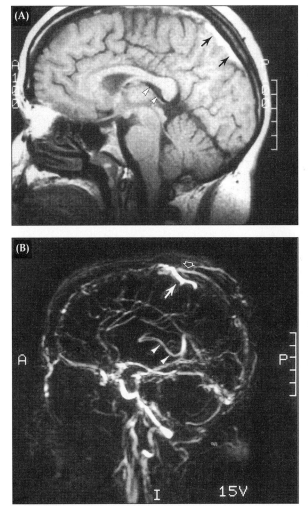

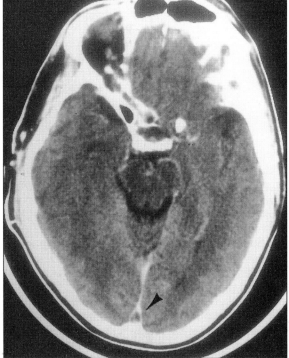

Fig. 54.2 Empty delta sign in superior sagittal sinus thrombosis. Axial contrast-enhanced CT demonstrates a filling defect in the center of the superior sagittal sinus as it approaches the confluence of sinuses (arrowhead). The filling defect is a thrombus. The contrast surrounding the thrombus takes the shape of the Greek capital letter delta.

Fig. 54.1 Superior sagittal sinus thrombosis. **A:** Midline sagittal T1-weighted image of a 22-year-old obese woman on contraceptive hormones. Superior sagittal sinus thrombosis is present. The subacute clot is hyperintense (black arrows). Normally the signal in the sinus should be low, similar to the flow void signal of the flowing blood in the internal cerebral vein (white arrowheads). **B:** 2D PC MRA reveals minimal blood flow in the superior sagittal sinus. A cortical vein with flow is present (straight arrow) with minimal drainage into the superior sagittal sinus (open arrow). Compare with the normal flow in the internal cerebral vein (white arrowheads) and with Fig. 51.1A.

- Polycythemia vera
- Antiphospholipid antibody syndrome
- Resistance to activated protein C
- Antithrombin III deficiency
- Protein C deficiency
- Protein S deficiency
- Sickle cell anemia
- Trauma
- Compressive mass lesions (e.g. tumors)
- Infections
- Severe dehydration

The clinical presentation of acute cerebral venous sinus thrombosis is often confusing. Headache may be the only symptom. Fever may occur, even when the cause is not infectious. When the superior sagittal sinus is affected, focal neurologic signs and seizures indicate involvement of the cerebral veins, usually with focal hemorrhage. Elevated intracranial pressure may develop (due to impaired resorption of CSF through the arachnoid villi), resulting in papilledema, headache, nausea, vomiting, and decreased mental status. Pseudotumor cerebri is a potential acute and long-term complication.

Proptosis, chemosis, retinal hemorrhages, ophthalmoplegia, orbital pain, and constitutional signs of infection occur in septic cavernous sinus thrombosis. Infectious lateral sinus thrombosis most commonly occurs in children with otitis media and presents nonspecifically with signs of infection (fever, chills, and sweats) and increased intracranial pressure.

Cerebral venous sinus thrombosis in neonates is under-recognized.[2–7] It presents with seizures or lethargy, without a predisposing cause in most cases. The examination is usually nonspecific, revealing only hypotonia or hyperreflexia. In most cases the thrombosis resolves spontaneously without sequelae. However, long-term follow-up studies have not been completed.

MRI with MR venography is the best means of making the diagnosis of cerebral sinus thrombosis (Fig. 54.1). If MRI is not available or is contraindicated, CT with contrast enhancement is an alternative. On CT, the classic finding is the "empty delta sign", which consists of a triangular rim of contrast surrounding a clot within the superior sagittal sinus (giving the appearance of the Greek letter delta) (Fig. 54.2). CT may also reveal thrombosed cortical veins ("cord sign"), multiple focal bilateral parasagittal hemorrhages secondary to venous infarctions, intense tentorial enhancement, or gyral enhancement indistinguishable from an infarction.[8]

The MRI findings in cerebral venous sinus thrombosis are:[9,10]

- Brain swelling, manifested by sulcal effacement, without parenchymal changes. The brain swelling is not as severe as in arterial occlusion. Increased T2 signal due to interstitial edema is present in more severe cases. These signs revert to normal when the occlusion resolves or adequate collaterals form.
- Abnormal enhancement in the cortical or deep venous structures due to underlying venous stasis. Because the delivery of arterial blood is not significantly affected, abnormal parenchymal and arterial enhancement is not present.
- In severe cases, intracerebral hemorrhage occurs as a result of the intraluminal pressure exceeding the structural limit of the venous wall.
- Ventriculomegaly due to decreased CSF flow, a consequence of increased resistance to CSF resorption at the arachnoid granulations. The hydrocephalus resolves rapidly when the thrombus liquefies.

MR venography definitively demonstrates the occluded sinuses and reveals collaterals as they develop. The phase-contrast technique, with settings designed to image venous flow, is the best method (Fig. 54.1B). Cerebral catheter angiography may be used to demonstrate venous occlusion but is not necessary if MRI is performed unless thrombolytic therapy is planned.

The choice of treatment partially depends on the underlying cause and the severity of disease. In patients without infection, anticoagulation is indicated, even if small cerebral hemorrhages are present. Patients with septic thrombosis should be treated with antibiotics. Thrombolytic agents, infused directly into the dural venous sinuses, appear to promote recanalization with low morbidity and may become the treatment of choice in noninfectious cases.[11,12]

References

1. Daif A, Awada A, al-Rajeh S, *et al*. Cerebral venous thrombosis in adults: a study of 40 cases from Saudi Arabia. *Stroke* 1995; **26**: 1193–5.
2. Keeney SE, Adcock EW, McArdle CB. Prospective observations of 100 high-risk neonates by high-field (1.5 tesla) magnetic resonance imaging of the central nervous system: I. Intraventricular and extracerebral lesions. *Pediatrics* 1991; **87**: 421–30.
3. Grossman R, Novak G, Patel M, Maytal J, Ferreira J, Eviatar L. MRI in neonatal dural sinus thrombosis. *Pediatr Neurol* 1993; **9**: 235–8.
4. Wong VK, LeMesurier J, Franceschini R, Heikali M, Hanson R. Cerebral venous thrombosis as a cause of neonatal seizures. *Pediatr Neurol* 1987; **3**: 235–7.
5. Shevell MI, Silver K, O'Gorman AM, Watters GV, Montes JL. Neonatal dural sinus thrombosis. *Pediatr Neurol* 1989; **5**: 161–5.
6. Rivkin MJ, Anderson ML, Kaye EM. Neonatal idiopathic cerebral venous thrombosis: an unrecognized cause of transient seizures or lethargy. *Ann Neurol* 1992; **32**: 51–6.
7. Barron TF, Gusnard DA, Zimmerman RA, Clancy RR. Cerebral venous thrombosis in neonates and children. *Pediatr Neurol* 1992; **8**: 112–16.
8. Buonanno FS, Moody DN, Ball MR, Laster DW. Computed cranial tomographic findings in cerebral sinovenous occlusion. *J Comput Assist Tomogr* 1978; **2**: 281–90.
9. Sze G, Simmons B, Krol G, Walker R, Zimmerman RD, Deck MD. Dural sinus thrombosis: verification with spin-echo techniques. *AJNR Am J Neuroradiol* 1988; **9**: 679–86.
10. Yuh WT, Simonson TM, Wang AM, *et al*. Venous sinus occlusive disease: MR findings. *AJNR Am J Neuroradiol* 1994; **15**: 309–16.
11. Horowitz M, Purdy P, Unwin H, *et al*. Treatment of dural sinus thrombosis using selective catheterization and urokinase. *Ann Neurol* 1995; **38**: 58–67.
12. Smith TP, Higashida RT, Barnwell SL, *et al*. Treatment of dural sinus thrombosis by urokinase infusion. *AJNR Am J Neuroradiol* 1994; **15**: 801–7.

55. *When infarctions involve only gray matter or are not confined to vascular territories, suspect MELAS*

Mitochondrial disorders are a clinically heterogeneous group of diseases caused by defects in oxidative metabolism. They are caused by abnormalities in the nuclear DNA (nDNA), which encodes the majority of the mitochondrial proteins, or in the mitochondrial DNA (mtDNA). The circular mtDNA, which is inherited entirely from the mother, is sequenced in its entirety. It encodes 22 transfer RNAs, two ribosomal RNAs, and 13 structural proteins, all subunits of respiratory chain complexes. The mtDNA disorders exhibit heteroplasmy – that is, both normal and mutant mtDNA coexist within the same cell or tissue. The clinical symptomatology is related to the proportion of mutant *versus* normal mtDNA in the different tissues (threshold effect).[1] The specific mechanism by which the genetic abnormality leads to clinical dysfunction is not completely understood. Only 10–15% of normal mtDNA may be all that is needed to maintain adequate function of the respiratory chain.[2] A limited capacity for energy production explains why symptoms tend to develop at times of physical stress, for example during a febrile illness. The high energy requirement of muscle and brain make them particularly vulnerable to deficient energy production.

Clinical features that should raise the suspicion of a mitochondrial disorder include progressive external ophthalmoplegia (PEO), heart block, sensorineural hearing loss, pigmentary retinopathy (e.g. retinitis pigmentosa), multiple lipomas, ataxia, and refractory myoclonic seizures. Lactic acidosis is common to most mitochondrial disorders. Disorders caused by abnormal mtDNA generally have maternal inheritance, although sporadic mutations occur. In many cases ragged red fibers (RRFs) due to mitochondrial proliferation, an indication of "mitochondrial distress", are present on muscle biopsy.[3]

Several mitochondrial disorders have associated neuroimaging abnormalities. High T2 and PD signal in white matter ("leukodystrophy") is present in Kearns–Sayre syndrome, MNGIE syndrome (mitochondrial neuropathy, gastrointestinal disorder, and encephalopathy), and in an autosomal dominant form of PEO. Calcification may occur in the basal ganglia. Kearns–Sayre syndrome is defined as a triad of onset before age 20 years, PEO, and pigmentary retinopathy with either heart block, cerebellar dysfunction, or CSF protein elevated above 100 mg/dl. Almost all have mtDNA deletions. Prognosis is poor despite the use of pacemakers (which thwart future MRI). MNGIE is characterized by PEO, limb weakness, peripheral neuropathy, gastroenteropathy with chronic diarrhea and intestinal pseudo-obstruction, lactic acidosis, and RRFs.

Leigh syndrome (subacute necrotizing encephalomyelopathy) is characterized by impaired vision and hearing, ataxia, weakness, seizures, dementia, PEO, dystonia, and lactic acidosis. It is caused by abnormalities of nDNA (autosomal recessive, X-linked recessive, or sporadic) or mtDNA (maternal inheritance). Relatively symmetric high signal is present on PD- and T2-weighted images in many different brain regions, with the commonest affected areas being the brain stem tegmentum, spinal cord, basal ganglia, and optic pathways. Bilateral involvement of the putamen is typical. Corresponding, nonenhancing, hypodense lesions are present on CT. Paramagnetic material, probably iron, may deposit in the thalamus and globus pallidus bilaterally, resulting in low T2 signal in these areas. The lesions consist of necrosis, demyelination, and vascular proliferation. Leigh syndrome usually presents in infancy or early childhood and has a poor prognosis.

MELAS (mitochondrial encephalomyopathy, lactic acidosis, and stroke-like episodes) has distinctive neuroimaging features. It is a multisystem disease associated with specific point mutations of mtDNA. The most frequent mutation is an A for G substitution at nucleotide 3243 in the transfer RNA$^{Leu(UUR)}$ gene.[4] Several other point mutations in this gene also result in MELAS.

MELAS is characterized by periodic stroke-like episodes beginning before age 40 years with seizures and progressive encephalopathy leading to dementia.[5,6] Exercise intolerance and episodic lactic acidosis are common. Endocrine dysfunction resulting in poor growth, infertility, and diabetes mellitus may occur.

During the stroke-like episodes, CT and MRI reveal multifocal, predominantly cortical, infarction-like lesions. In contrast to typical ischemic strokes, these lesions are not confined to vascular territories (Fig. 55.1).[7-10] Multiple emboli from a cardiac source and CNS vasculitis may also be multifocal and should be considered in the differential diagnosis. Angiography demonstrates a capillary blush and early venous filling in these regions.[11] Calcifications may be present within the basal ganglia.

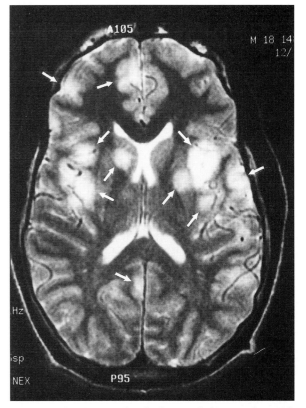

Fig. 55.1 MELAS (mitochondrial encephalomyopathy with lactic acidosis and stroke-like episodes). Axial T2-weighted image of an 18-year-old man who presented with mental status changes. Multiple high-signal foci confined to the basal ganglia and cortical gray matter are present bilaterally (arrows). The lesions are not confined to a specific vascular territory.

The pathophysiology of the stroke-like episodes is not understood. A mitochondrial microangiopathy is described.[12] Deficient energy production stimulating proliferation of mitochondria within the vascular smooth muscle and endothelial cells is a proposed mechanism for the microangiopathy. The abnormality in the vascular endothelium is less pronounced in larger vessels, explaining why the distribution of the infarction-like lesions does not correspond to the major vascular territories. In severe MELAS, the large arteries may occlude.[13] However, metabolic dysfunction as opposed to ischemic stroke as a mechanism for the brain dysfunction is suggested by the results of cerebral blood flow studies with xenon-enhanced CT and metabolic studies with PET.[11,14]

Mitochondrial encephalomyopathies should be included in the differential diagnosis of children and young adults who present with apparent small or large vessel cerebral infarction. Because myocardial involvement with cardiomyopathy occurs in some mitochondrial disorders, emboli must be considered in the differential diagnosis in patients with mitochondrial disorders and stroke.

References

1. Suomalainen A, Majander A, Pihko H, Peltonen L, Syvänen A-C. Quantification of tRNA Leu 3243 point mutation of mitochondrial DNA in MELAS patients and its effects on mitochondrial transcription. *Hum Mol Genet* 1993; **2**: 525–34.

2. Moraes CT, Ricci E, Bonilla E, DiMauro S, Schon EA. The mitochondrial tRNA[Leu(UUR)] mutation in mitochondrial encephalomyopathy, lactic acidosis and stroke-like episodes (MELAS): genetic, biochemical, and morphological correlations in skeletal muscle. *Am J Hum Genet* 1992; **50**: 934–49.

3. DiMauro S, Moraes CT. Mitochondrial encephalomyopathies. *Arch Neurol* 1993; **50**: 1197–208.

4. Goto Y-I, Nonaka I, Horai S. A mutation in the tRNA[Leu(UUR)] gene associated with the MELAS subgroup of mitochondrial encephalomyopathies. *Nature* 1990; **348**: 651–3.

5. Pavlakis S, Phillips P, DiMauro S, *et al*. Mitochondrial myopathy, encephalopathy, lactic acidosis and stroke-like episodes: a distinctive clinical syndrome. *Ann Neurol* 1984; **16**: 481–8.

6. Hirano M, Pavlakis S. Mitochondrial myopathy, encephalopathy, lactic acidosis and stroke-like episodes (MELAS): current concepts. *J Child Neurol* 1994; **9**: 4–13.

7. Barkovich A, Good W, Koch T, Berg B. Mitochondrial disorders: analysis of their clinical and imaging characteristics. *AJNR Am J Neuroradiol* 1993; **14**: 1119–37.

8. Matthews P, Tampieri D, Berkovich S, *et al*. Magnetic resonance imaging shows specific abnormalities in the MELAS syndrome. *Neurology* 1991; **41**: 1043–6.

9. Yamamoto T, Beppu H, Tsubaki T. Mitochondrial encephalomyopathy: fluctuating symptoms and CT. *Neurology* 1984; **34**: 1456–60.

10. Suzuki T, Koizumi J, Shiraishi H, *et al*. Mitochondrial encephalomyopathy (MELAS) with mental disorder: CT, MRI and SPECT findings. *Neuroradiology* 1990; **32**: 74–6.

11. Ooiwa Y, Uematsu Y, Terada T, *et al*. Cerebral blood flow in mitochondrial myopathy, encephalopathy, lactic acidosis, and stroke-like episodes. *Stroke* 1993; **24**: 304–9.

12. Ohama E, Ohara S, Ikuta F, Tanaka K, Nishizawa M, Miyatake T. Mitochondrial angiopathy in cerebral blood vessels of mitochondrial encephalomyopathy. *Acta Neuropathol (Berl)* 1987; **74**: 226–33.

13. Kotagal S, Peterson P, Martens M, *et al*. Impaired NADH-CoQ reductase activity in a child with moyamoya syndrome. *Pediatr Neurol* 1988; **4**: 241–4.

14. Yokoi F, Hara T, Iio M, Nonaka I, Satoyoshi E. 1-[^{11}C]Pyruvate turnover in brain and muscle of patients with mitochondrial encephalomyopathy: a study with positron emission tomography (PET). *J Neurol Sci* 1990; **99**: 339–48.

9

Intracerebral and Subarachnoid Hemorrhage and Cerebral Vascular Malformations

56. The common locations of intracerebral hemorrhages in hypertensive patients are different from those in normotensive elderly patients

Intracerebral hemorrhage (ICH) in the elderly is usually associated with chronic hypertension, coagulopathy, metastasis, primary brain tumor, or cerebral amyloid angiopathy. The common locations for hypertensive ICHs are different from those for the other causes.

The leading cause of spontaneous ICH is chronic hypertension. ICH may be the initial presenting symptom of long-standing untreated hypertension. The most common sites for a hypertensive ICH are the putamen and thalami (Figs. 57.1 and 60.1). The converse is also true: more than 90% of patients with ICH in these nuclei are hypertensive. Hypertensive ICHs also occur in the pons, cerebellum, and head of the caudate. Many lobar hemorrhages, previously ascribed to hypertension, are actually associated with cerebral amyloid angiopathy or angiographically occult vascular malformations. However, a portion of genuine hypertensive ICHs are probably located within the white matter or cortex of the cerebral hemispheres.

Cerebral amyloid angiopathy (CAA) is a common cause of nontraumatic ICH in normotensive patients over 60 years of age. The neuropathologic changes present in CAA are also present in Alzheimer disease. ICHs due to CAA are located superficially within the cerebral cortex and subcortical white matter (Fig. 56.1), and are often multiple. Patients with acute ICH due to CAA may have areas of encephalomalacia in other locations from prior hemorrhages or ischemic infarctions secondary to vessel occlusion. Cerebellar hemorrhage associated with CAA is reported, but is rare.[1] CAA is seldom, if ever, responsible for hemorrhages within the brain stem.

The hematomas caused by CAA are usually midsized or large and multilobulated with irregular borders. Multiple small cortical hemorrhages also occur. CAA presenting as a frontal mass lesion suggestive of brain tumor without enhancement or hemorrhage is reported.[2,3] Although mass effect and narrowing of vessels due to the hemorrhage may be present, no intrinsic vascular abnormalities are visualized with angiography.

CAA may present with concurrent intraventricular or subarachnoid hemorrhage (SAH). Rarely, the only manifestation of CAA is SAH. In one autopsy-confirmed case of CAA and SAH, amyloid deposits were discovered in the leptomeningeal and parenchymal vasculature. Angiograms were repeatedly normal in this case. Thus, CAA should be considered in the differential diagnosis of subarachnoid hemorrhage with negative angiogram in elderly patients (maxim 58).[4]

Spontaneous hemorrhages associated with anticoagulation or a bleeding diathesis are often similar to those that occur in CAA. The clinical history is usually helpful in differentiating between the two. Patients with underlying CAA are at increased risk for ICH when anticoagulants or thrombolytic agents are administered.

Large feeding vessels and draining veins are usually present on both CT and MRI in patients with ICH resulting from arteriovenous malformations (AVMs). Occasionally angiographically occult vascular malformations produce acute ICH, which can be located anywhere within the CNS (maxim 62). MRI may demonstrate multiple cerebral cavernous angiomas, confirming the diagnosis.

Primary and metastatic brain tumors may present with hemorrhage. An underlying lesion is usually discernible, especially when contrast is used. Acute ICH does not enhance with contrast when caused by hypertension, CAA, or coagulopathy. For this reason, contrast should be administered at least once when investigating the cause of an ICH.

CT is the method of choice for demonstrating an acute hemorrhage within the cranium. It should be performed as the initial study in patients presenting with an acute neurologic deficit attributable to brain dysfunction (maxim 48). CT is capable of detecting hemorrhages as small as 2 mm in diameter when they are not adjacent to the skull (maxim 61), and provides information regarding the exact location of the bleed, associated mass effect, and ventricular dilatation. Intraventricular blood, which commonly occurs with medial thalamic hemorrhages, is easily identified on CT. When large, ICH may result in brain herniation, which can be identified on CT but is better delineated with MRI (maxim 42). Care should be taken not to misdiagnose a hemorrhagic infarction as an ICH.

Cerebral angiography may reveal an AVM, aneurysm, neoplasm, or vasculitis, and is indicated when the diagnosis is in doubt, especially if surgical resection is

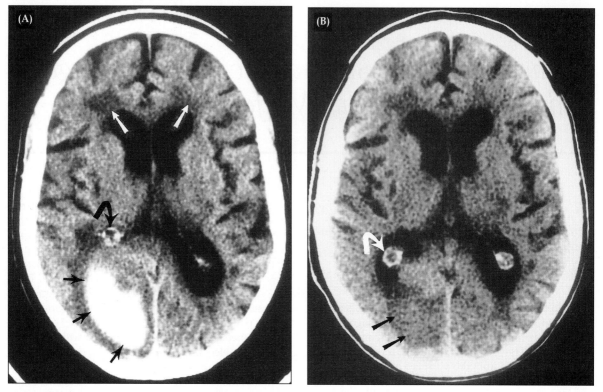

Fig. 56.1 Intracerebral hematoma: cerebral amyloid angiopathy. **A:** Axial noncontrasted CT in a 72-year-old man presenting with an acute headache. A sharply outlined hyperdense clot with surrounding low-density edema is present in the right occipital lobe (straight black arrows). Mass effect on the right trigone with displacement of the calcified choroid plexus is apparent (curved black arrow). This lesion is consistent with cerebral amyloid angiopathy. Note the diffuse atrophy and "leukoaraiosis" (low-density periventricular regions; straight white arrows; maxim 64). **B:** Axial noncontrasted CT at the same level obtained 3 weeks later (subacute phase). The large clot is now isodense to hypodense to the cortical gray matter (straight black arrows). The mass is decreased in size and the choroid plexus calcification is returned to its normal location (curved white arrow). Mass effect on the sulci is still present. In intracerebral hemorrhage, the hyperdensity resolves faster than does the mass effect.

contemplated. Any of these diagnoses may drastically alter the treatment. On occasion, the bleeding vessel may be demonstrated during the angiogram.

References

1. Vinters HB. Cerebral amyloid angiopathy, in: *Stroke: patho-physiology, diagnosis and management*, 2nd edn (ed. Barnett HJM). New York: Churchill Livingstone, 1991, 821–58.
2. Hendricks HT, Franke CL, Theunissen PH. Cerebral amyloid angiopathy: diagnosis by MRI and brain biopsy. *Neurology* 1990; **40**: 1308–10.
3. Briceno CE, Resh CL, Bernstein M. Cerebral amyloid angiopathy presenting as a mass lesion. *Stroke* 1987; **18**: 234–7.
4. Ohshima T, Endo T, Nukui H, *et al.* Cerebral amyloid angiopathy as a cause of subarachnoid hemorrhage. *Stroke* 1990; **21**: 480–3.

57. *Intracranial hemorrhage can complicate drug abuse*

During recent decades, the frequency of recreational drug abuse and its neurologic complications has markedly increased. Drugs of abuse (e.g. cocaine and amphetamines) and several over-the-counter sympath-omimetics (phenylpropanolamine, ephedrine, and pseudoephedrine) may cause intracerebral hemorrhage (ICH), subarachnoid hemorrhage (SAH), and ischemic infarction.[1,2] When any of these acute neurologic conditions occur in young adults, the possibility of drug abuse needs to be considered.[3,4] A variety of unusual vascular events may occur.[5,6] One-half of patients who present with hemorrhage related to drugs have a pre-existing vascular malformation, usually an aneurysm or AVM.[7] Any of these substances may precipitate a seizure or hypertensive crisis and may result in an infarction in a developing fetus.

No feature on neuroimaging is specific for drug-induced stroke. CT and MRI are useful to confirm and localize the sites of hemorrhage or infarction. CT should be the first imaging study (maxim 48), but MRI more consistently reveals underlying vascular pathology. MRI may also demonstrate nasopharyngeal mucosal changes in patients who chronically use intranasal cocaine (maxim 68). Cerebral angiography may be warranted to determine whether an aneurysm, AVM, or vasculitis is present.

The extremely high blood pressures that occur during cocaine use are probably responsible for the associated ICH. These hemorrhages tend to occur in locations similar to those associated with chronic hypertension. The basal ganglia, thalami, pons, cerebellum, and lobar white matter are the most common sites (Fig. 57.1).[8] SAH probably results from the same phenomenon.[9] Conventional angiography is usually indicated to evaluate for an AVM or aneurysm. Segmental stenosis, beading, and vessel occlusions due to vascular instability may be present. Biopsy proven vasculitis due to cocaine is reported, but appears to be rare.[10] In addition to its direct vasoconstrictor and hypertensive effects, cocaine promotes platelet aggregation, further predisposing abusers to arterial thrombosis.

The association of amphetamine abuse with intracranial hemorrhage has been known since 1970, long before the analogous relationship for cocaine was established.[11] Amphetamine abuse results in endothelial damage and is a well-established cause of cerebral vasculitis. Necrosis of the vessel wall media and intima occurs.[12] Because the hypertension induced by amphetamines is short-lived, the finding of a normal blood pressure at the time of presentation does not exclude a drug-related hypertensive bleed.

Phenylpropanolamine, ephedrine, and pseudoephedrine are components of many decongestants, diet pills, and stimulants. These sympathomimetics may also cause acute, transient increases in blood pressure, resulting in vascular complications.

References

1. Jacobs IG, Roszler MH, Kelly JK, Klein MA, Kling GA. Cocaine abuse: neurovascular complications. *Radiology* 1989; **170**: 223–7.
2. Peterson PL, Roszler M, Jacobs I, Wilner HI. Neurovascular complications of cocaine abuse. *J Neuropsychiatr Clin Neurosci* 1991; **3**: 143–9.
3. Tuffol GS, Biller J, Adams HP Jr. Nontraumatic intracerebral hemorrhage in young adults. *Arch Neurol* 1987; **44**: 483–5.
4. Nalls G, Disher A, Daryabagi J, Zant Z, Eisenman J. Subcortical cerebral hemorrhages associated with cocaine abuse: CT and MR findings. *J Comput Assist Tomogr* 1989; **13**: 1–5.
5. Aggarwal S, Byrne BD. Massive ischemic cerebellar infarction due to cocaine use. *Neuroradiology* 1991; **33**: 449–50.
6. Huff JS. Spinal epidural hematoma associated with cocaine abuse. *Am J Emerg Med* 1994; **12**: 350–2.
7. Brown E, Prager J, Lee H-Y, Ramsey RT. CNS complications of cocaine abuse: prevalence, pathophysiology, and neuroradiology. *AJR Am J Roentgenol* 1992; **159**: 137–47.
8. Ramadan NM, Levine SR, Welch KM. Pontine hemorrhage following "crack" cocaine use. *Neurology* 1991; **41**: 946–7.
9. Lishtenfeld PJ, Rubin DB, Feldman RS. Subarachnoid hemorrhage precipitated by cocaine snorting. *Arch Neurol* 1984; **421**: 223–4.
10. Krendel DA, Ditter SM, Frankel MR, Ross WK. Biopsy proven cerebral vasculitis associated with cocaine abuse. *Neurology* 1990; **40**: 1092–4.
11. Goodman SJ, Becker DP. Intracranial hemorrhage associated with amphetamine abuse. *JAMA* 1970; **212**: 480.
12. Citron BP, Halpern M, McCarron M, *et al.* Necrotizing angiitis associated with drug abuse. *N Engl J Med* 1970; **283**: 1003–11.

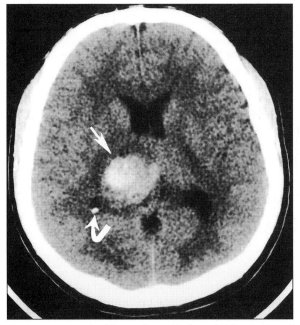

Fig. 57.1 Intracerebral hemorrhage (thalamus). Noncontrasted CT of a 23-year-old drug abuser showing an acute intracranial bleed in the thalamus (straight arrow). The blood is high density, but not as dense as the calcified choroid plexus (curved arrow). Wider window settings can clearly differentiate blood from calcium (maxim 87). Selective internal carotid angiography did not show vascular abnormalities except for mass effect.

58. *CT is the primary test to confirm the presence and location of a subarachnoid hemorrhage*

Subarachnoid hemorrhage (SAH) is a clinical emergency. Rapid diagnosis is important because early aneurysm clipping prevents rebleeding and permits maximal treatment of vasospasm.[1] CT is the primary method to diagnose an acute SAH. In adults, approximately 85% of nontraumatic SAHs are due to a ruptured intracerebral aneurysm; only 5% of SAHs are secondary to AVMs, and 10% are related to intracerebral hemorrhages (ICHs) and other causes.[2] In children, one-quarter of SAHs are due to AVMs.[3]

In general, SAH should be considered in any patient who presents for medical evaluation for a headache because a sentinel bleed may have subtle manifestations. A patient with a suspected SAH (classically presenting with an abrupt-onset "worst headache of his life" – thunderclap headache) should undergo noncontrasted CT to determine whether subarachnoid blood is present. SAH is identified when the subarachnoid spaces (normally occupied by hypodense CSF) are hyperdense to brain (Fig. 58.1). Occasionally intraparenchymal extension is present and may be the only manifestation of a SAH. Approximately 3% of patients with acute-onset headache and normal CT have positive CSF and an aneurysmal SAH.[4] In cases with a reported negative CT, re-examination may reveal subarachnoid blood in the prepontine cistern.[5]

The CT pattern of subarachnoid blood is often symmetric and sometimes missed by inexperienced interpreters. The CT patterns in SAH, epidural hematoma (maxim 86) and obliteration of the fourth ventricle (maxim 43) are among the most important to recognize. Anyone who reads head CTs (including the emergency room physician, internist, family practitioner, and general surgeon) needs to be able to identify even the subtlest of these patterns. Emergency treatment in these conditions is often life-saving, and delay of treatment can be fatal.

CT without contrast enhancement detects the vast majority of acute SAHs. However, the detection rate decreases rapidly over time: 1 week after SAH, 90% of the extravasated blood is gone and only one-half of SAHs are still visible on CT.[6] When a significant amount of subarachnoid blood is present on CT more than 1 week following the initial ictus, rebleeding is suggested.

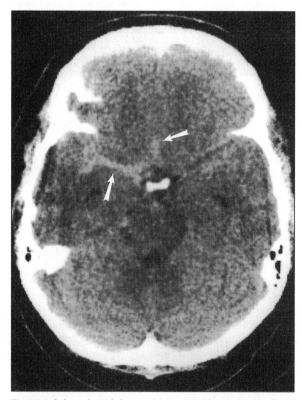

Fig. 58.1 Subarachnoid hemorrhage. Axial noncontrasted CT through the level of the basal cisterns revealing blood in the horizontal fissure and anterior interhemispheric fissure (arrows). DSA confirmed the presence of an aneurysm in the distal internal carotid artery. The subarachnoid blood is sometimes subtle, but hyperdense subarachnoid spaces on noncontrasted CT almost always indicates subarachnoid hemorrhage. Compare this pattern with a normal CT (Fig. 7.5B)

The location of the highest concentration of subarachnoid blood on the initial CT often indicates the site of the ruptured aneurysm. This information can be especially valuable in the approximately 20% of patients with multiple aneurysms (Fig. 58.2).[7] The most specific finding is an isolated hematoma in the lateral fissure, which is characteristic of a ruptured middle cerebral artery aneurysm. Blood in the anterior interhemispheric fissure or frontal lobe suggests an anterior communicating artery aneurysm. Blood concentrated diffusely within the suprasellar cistern implies that the responsible aneurysm is located in the circle of Willis. Interpeduncular cistern blood is usually associated with a basilar tip or posterior communicating artery aneurysm. Blood is usually present within the fourth ventricle in posterior fossa aneurysms. Blood restricted to the ventricular system, concentrating in the fourth ventricle, suggests a posterior inferior cerebellar artery aneurysm.[8] Perimesencephalic blood is a nonspecific finding and not useful for aneurysm localization. In many instances, the blood is located diffusely throughout the subarachnoid space and does not help to identify the location of the aneurysm that bled.

In patients with multiple aneurysms, the largest or most irregular or lobulated aneurysm on conventional catheter digital subtraction angiography (DSA) is most likely to have bled. If present, localized vasospasm is most commonly adjacent to the offending aneurysm. Rarely, active bleeding occurs during the angiography, appearing as focal contrast extravasation.

While awaiting aneurysm clipping surgery, the patient should be kept completely quiet to minimize the risk of rebleeding. In this setting, phenobarbital is often used to reduce agitation and as prophylaxis against seizures. Steroids may be given to relieve meningismus due to the subarachnoid blood. Nimodipine should be started as soon as the diagnosis of SAH is made. Aminocaproic acid, a procoagulant, can reduce the risk of rebleed but its use is controversial because it increases the risk of cerebral vasospasm and systemic thrombotic complications. Adequate quantities of sodium should be given (usually with IV normal saline) to avoid hyponatremia from cerebral salt wasting and to help maintain the cerebral perfusion pressure.

Acute ventricular dilatation is a common complication that often occurs early in the course of SAH. It is characterized by enlargement of the temporal horns, often with relatively normal-sized lateral ventricular bodies. It usually resolves within a few days but may require emergent drainage. In contrast, communicating hydrocephalus (presenting as normal pressure hydrocephalus; maxim 65) is a late complication caused by decreased CSF resorption through the arachnoid granulations. Serial CTs are used to follow the ventricular size to determine the need for CSF shunting in both the acute and the chronic situation.

References

1. Kassel NF, Tornel JC, Haley EC, *et al*. The international cooperative study on the timing of aneurysm surgery. *J Neurosurg* 1990; **73**: 18–47.
2. Inagawa T, Hirano A. Ruptured intracranial aneurysms: an autopsy study of 133 patients. *Surg Neurol* 1990; **33**: 117–23.
3. Hourihan MD, Gates PC, McAllister VL. Subarachnoid hemorrhage in childhood and adolescence. *J Neurosurg* 1984; **60**: 1163–6.
4. van der Wee N, Rinkel GJ, Hasan D, van Gijn J. Detection of subarachnoid haemorrhage on early CT: is lumbar puncture still needed after a negative scan? *J Neurol Neurosurg Psychiatr* 1995; **58**: 357–9.
5. Rinkel GJ, van Gijn J, Wijdicks EF. Subarachnoid hemorrhage without detectable aneurysm: a review of the causes. *Stroke* 1993; **24**: 1403–9.
6. Gonzales CF, Cho YI, Ortega HV, *et al*. Intracranial aneurysms: flow analysis of their origin and progression. *AJNR Am J Neuroradiol* 1992; **13**: 181–8.
7. Rinne J, Hernesniemi J, Niskanen M, Vapalahti M. Management outcome for multiple intracranial aneurysms. *Neurosurgery* 1995; **36**: 31–7.
8. Sadato N, Numaguchi Y, Rigamonti D, Salcman M, Gellad FE. Bleeding patterns in ruptured posterior fossa aneurysms: a CT study. *J Comput Assist Tomogr* 1991; **15**: 612–17.

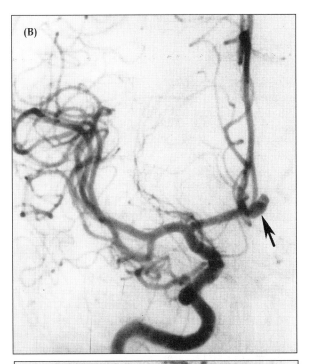

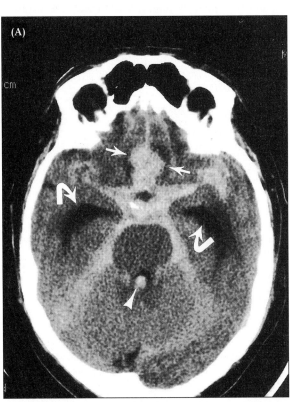

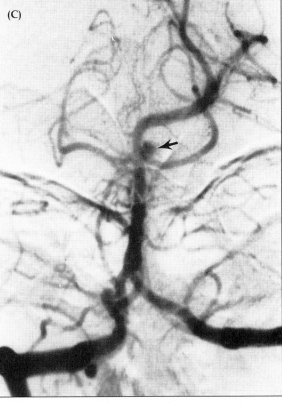

Fig. 58.2 Subarachnoid hemorrhage (multiple aneurysms). **A:** Axial noncontrasted CT of a 39-year-old woman with the acute onset of the "worst headache of her life". An extensive subarachnoid hemorrhage is apparent. The largest accumulation of blood is in the anterior interhemispheric fissure (straight arrows), suggesting that an anterior communicating artery aneurysm is responsible for the hemorrhage. Early hydrocephalus is indicated by dilated temporal horns (curved arrows). Some blood has tracked back into the fourth ventricle (arrowhead). **B:** Selective four-vessel DSA revealed two aneurysms. Injection of the right internal carotid artery demonstrates an anterior communicating artery aneurysm (arrow). This is bilobed and the larger of the two aneurysms. No vasospasm is identified. **C:** Selective left vertebral artery injection demonstrates a basilar artery aneurysm (arrow) arising between the origins of the right superior cerebellar and the posterior cerebral arteries. (Courtesy of Dr Wayne Crow, University of Texas Medical Branch)

59. *Intracranial aneurysm: conventional angiography is the gold standard*

Conventional catheter angiography, to identify aneurysms and characterize the vascular anatomy, should be performed in all patients with SAH (Fig. 58.2). Intravenous contrast is not usually given during the initial CT because, although the aneurysm can sometimes be identified as a small concentration of contrast overlying an artery, it rarely adds important information and increases the contrast exposure. Magnetic resonance angiography (MRA) and computed tomographic angiography (CTA) can outline the intracranial vessels but their role in the acute setting remains to be determined. At present, MRA and CTA do not adequately detect small aneurysms (termed infundibulum when funnel-shaped).

No source for bleeding is found on the first angiogram in up to 20% of acute SAHs.[1,2] These cases are termed "cryptogenic" or "angiogram-negative". In two-thirds of these, the hemorrhage is confined to the cisterns around the midbrain (perimesencephalic pattern). The initial CT is distinctive, with blood centered in the interpeduncular fossa with extension into the ambient cisterns. The perimesencephalic pattern is characterized by an uncomplicated clinical course without rehemorrhage, and these cases do not require repeat angiography.[2,3] Most angiogram-negative perimesencephalic SAHs are probably the result of bleeding from perimesencephalic venous plexus vessels.

In contrast, angiogram-negative SAHs with other patterns of blood on the initial CT are at risk for rehemorrhage. In some of these cases the blood irritates the aneurysm, causing it to temporarily spasm. In this situation, it is common practice to repeat the cerebral angiogram at 2 weeks.[4] If the second angiogram is negative, either the aneurysm was obliterated during the acute hemorrhage or another occult source for hemorrhage was present. The possibility of an arteriovenous malformation within the cervical spinal cord should be considered, especially when prominent nuchal rigidity is present or CT demonstrates the greatest accumulation of blood adjacent to the lower brain stem.

An aneurysm located in the distal middle cerebral artery, or another peripheral site, raises the possibility of infective endocarditis causing a mycotic aneurysm. Mycotic aneurysms usually have a broad base. Although one-third of mycotic aneurysms resolve with appropriate antibiotic therapy, continued follow-up with conventional angiography or MRI/MRA is needed because the remainder eventually require surgical resection.[5,6]

A major complication of acute SAH is vasospasm (maximal between 3 days and 2 weeks after the hemorrhage). The risk of vasospasm is directly related to the quantity of subarachnoid blood. Vasospasm may result in multiple, devastating cerebral infarctions. Daily transcranial Doppler monitoring is effective for detecting and following large-vessel vasospasm. As long as the aneurysm clips are definitely MRI-compatible (maxim 5), MRI can be used to detect both large and small vessel vasospasm by revealing regions of ischemia and infarction. Although invasive, conventional angiography is the most definitive means for detecting vasospasm and may allow direct treatment with intra-arterial antispasm agents. Once an aneurysm is clipped, hypertensive and hypervolemic therapy helps to prevent vasospasm.

The role of diagnostic intra-arterial catheter angiography in the evaluation of patients with intracranial hemorrhage is different in the era of CT, MRI, CTA, and MRA, but it is still the gold standard in searching for aneurysms, and it is required to outline the vascular supply of an arteriovenous malformation. This procedure is likely to have an expanded role in the future as more interventional angiographic techniques are developed.

References

1. Kassel NF, Tornel JC, Haley EC, *et al*. The international cooperative study on the timing of aneurysm surgery. *J Neurosurg* 1990; **73**: 18–47.
2. Rinkel GJ, van Gijn J, Wijdicks EF. Subarachnoid hemorrhage without detectable aneurysm: a review of the causes. *Stroke* 1993; **24**: 1403–9.
3. Rinkel GJ, Wijdicks EF, Vermeulen M, *et al*. Nonaneurysmal perimesencephalic subarachnoid hemorrhage: CT and MRI patterns that differ from aneurysmal rupture. *AJNR Am J Neuroradiol* 1991; **12**: 829–34.
4. Iwagawa H, Wakai S, Ochai C, *et al*. Ruptured cerebral aneurysms missed by initial angiographic study. *Neurosurgery* 1990; **27**: 45–51.
5. Corr P, Wright M, Handler LC. Endocarditis-related cerebral aneurysms: radiologic changes with treatment. *AJNR Am J Neuroradiol* 1995; **16**: 745–8.
6. Ahmadi J, Tung H, Giannotta SL, Destian S. Monitoring of infectious intracranial aneurysms by sequential computed tomographic/magnetic resonance imaging studies. *Neurosurgery* 1993; **32**: 45–9.

Table 60.1 Appearance of blood in an intracranial hematoma[1]

Stage	Time	Compartment	Hemoglobin	T1-weighted image	T2-weighted image
Hyperacute	<24 hours	Intracellular	Oxyhemoglobin	Dark	Bright
Acute	1–3 days	Intracellular	Deoxyhemoglobin	Dark	Black
Subacute					
Early	3+ days	Intracellular	Methemoglobin	Bright	Dark
Late	7+ days	Extracellular	Methemoglobin	Bright	Bright
Chronic					
Center	14+ days	Extracellular	Hemichromes	Bright	Bright
Periphery	14+ days	Intracellular	Hemosiderin	Dark	Black

60. *MRI can diagnose subacute and chronic subarachnoid and intracerebral hemorrhage*

MRI is useful, but not foolproof, in detecting subacute and chronic subarachnoid and intracerebral hemorrhage.[1,2] The changing appearance of an intracranial hematoma with time on MRI is due to the metabolism of hemoglobin. Table 60.1 summarizes the appearance of the various stages of intracranial hemorrhage on MRI.

Subacute blood is hyperintense on T1-weighted images due to the presence of methemoglobin. High signal on a T1-weighted image indicates subacute blood (methemoglobin), contrast, particulate calcium, another paramagnetic material (e.g. iron), or fat. Fat-suppression techniques permit easy detection of fat. Thus, high signal on a noncontrasted, fat-suppressed, T1-weighted image is due to subacute blood, calcium (maxim 13), or heavy metals (maxim 67). Methemoglobin, with T1 high signal, may be present for many months after a bleed. Hemorrhage that occurred more than several weeks

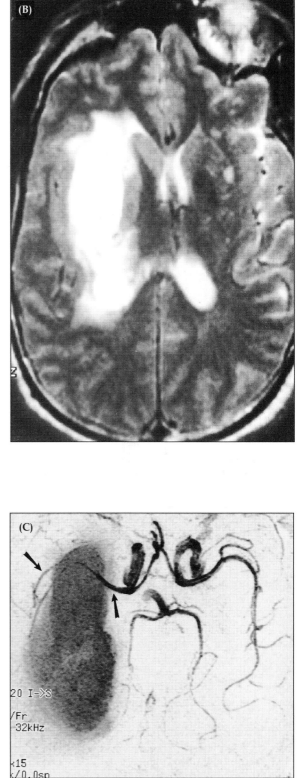

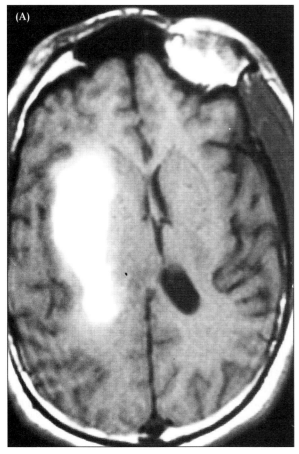

Fig. 60.1 Intracerebral hemorrhage (basal ganglia). **A:** Axial T1-weighted and **B:** T2-weighted image through the basal ganglia in a 36-year-old man with chronic hypertension and acute-onset left hemiparesis and mental status changes. The lesion is high signal on both T1- and T2-weighted images, consistent with a late subacute (at least 7 days) hemorrhage with extracellular methemoglobin. **C:** 3D time-of-flight (TOF) MRA (video reversal of the image) showing decreased flow in the middle cerebral artery branches (arrows). The vessels are stretched around the hematoma. The hematoma is visualized on the MRA image because the 3D TOF is based on T1 images. The phase-contrast MRA technique may be used to view the vessels with no interference from the methemoglobin.

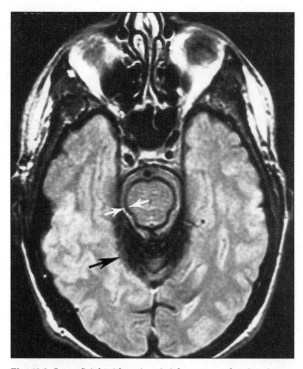

Fig. 60.2 Superficial siderosis. Axial proton density image through the midbrain of a 48-year-old man with hearing loss. The very low signal coating the surface of the pons and vermis (arrows) is diagnostic of superficial siderosis. This condition is caused by repeated chronic subarachnoid or intraventricular hemorrhages depositing hemosiderin and ferritin in the leptomeninges, especially over the cerebellum, brain stem, and vermian cistern, extending to the spinal cord.

before the MRI is characterized by marked hypointensity on T2-weighted spin-echo and especially gradient-echo images due to hemosiderin (maxim 62). The T2 low signal first appears at the periphery of the hematoma. Very low signal on a T2-weighted image usually indicates chronic hemorrhage, flowing blood, acute hemorrhage (deoxy-hemoglobin), air, iron or calcium.

MRI is of limited use in the acute setting in patients with large intracerebral hematomas. Patients are often too sick to tolerate MRI unless placed under anesthesia (which may require intubation, making the study difficult). Although MRI can add additional information to the initial CT, such as detection of underlying small multiple metastases, it is generally reserved for later in the clinical course. If MRI is performed, an accompanying MRA should be included to evaluate for an arteriovenous malformation. With MRA, the hemorrhage can be apparent with the time-of-flight method and may obscure an underlying vascular malformation (Fig. 60.1). This problem can be minimized by using the phase-contrast technique (maxim 51).

Recurrent or chronic SAH may result in superficial siderosis, a rare but probably under-recognized condition.[3,4] It presents with slowly progressive sensorineural hearing loss, cerebellar ataxia, myelopathy, and neuropsychologic dysfunction with xanthochromia and siderophages in the CSF. It may be caused by a vascular malformation or tumor involving the dura mater but it is

often idiopathic.[5] Superficial siderosis is diagnosed when a hypointense "icing" is present on the brain surface on T2-weighted images (Fig. 60.2). The brain surface may be hyperintense on T1-weighted images if subacute blood is present. The cerebellar vermis is usually atrophic.

References

1. Bradley WG. Hemorrhage and hemorrhagic infarctions in the brain. *Neuroimag Clin North Am* 1994; **4**: 707–32.
2. Bradley WG, Schmidt PG. Effect of methemoglobin formation of the MRI appearance of subarachnoid hemorrhage. *Radiology* 1985; **156**: 99–103.
3. Bourgouin PM, Tampieri D, Melancon D, *et al.* Superficial siderosis of the brain following unexplained subarachnoid hemorrhage: MRI diagnosis and clinical significance. *Neuroradiology* 1992; **34**: 407–10.
4. Bracci M, Savoiardo M, Triulzi DF, *et al.* Superficial siderosis of the CNS: MRI diagnosis and clinical findings. *AJNR Am J Neuroradiol* 1993: **14**: 227–36.
5. Stevens I, Petersen D, Grodd W, Poremba M, Dichgans J. Superficial siderosis of the central nervous system: a 37 year follow-up of a case and review of the literature. *Eur Arch Psychiatry Clin Neurosci* 1991; **241**: 57–60.

61. *An isodense subdural or intracerebral hemorrhage can be acute*

CT is the primary neuroimaging method for the diagnosis of acute intracranial hemorrhage. Typically, an acute subdural or intracerebral hematoma appears hyperdense to brain and hypodense to skull (Fig. 61.1). However, because there is a linear relationship between the CT density value and hemoglobin concentration, acute isodense subdural and intraparenchymal hematomas occur in patients with anemia (hemoglobin values below 10 g/dL) (Fig. 61.2).[1–3] Intravenous contrast can be used to highlight the cortical veins and reveal an isodense subdural hematoma (maxim 87) even when acute.[4]

Visualization of an acute hemorrhage on CT depends on the size and location, specific filming parameters, and the hemoglobin value. Small subdural or epidural hematomas that do not produce mass effect may be overlooked if the proper window and level are not used (Fig. 61.1). The optimal setting, referred to as the blood or subdural window, uses a wide window value (window width 150–200; approximate level 60). This setting results in images with different appearances for hematomas, skull, and brain (Fig. 61.1). With the standard brain parenchyma window and level settings, acute blood is white and appears identical to the skull. Similarly, a small amount of subarachnoid blood adjacent to the skull may be missed if images with wide windows are not examined. To examine the skull, bone window settings are required (Fig. 85.1).

Intraparenchymal blood, distant from the skull, is usually readily apparent on CT with the standard brain parenchyma window and level settings (Figs. 56.1 and 57.1). Acute intraparenchymal hemorrhage usually has well-defined borders with surrounding edema. However,

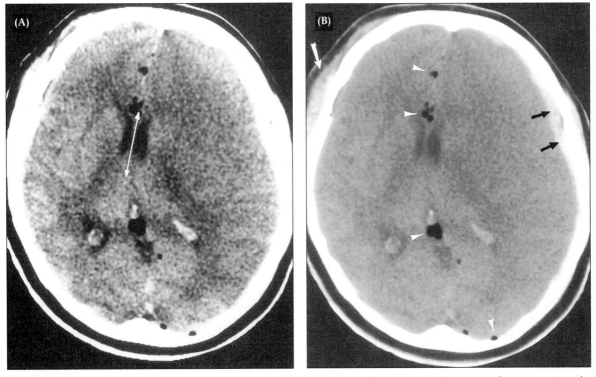

Fig. 61.1 Subdural hematoma and pneumocephalus. **A:** Noncontrasted axial CT filmed with a soft-tissue window in a patient who suffered a major head trauma. Substantial mass effect with midline shift to the right is present (double-headed arrow). It is very difficult to recognize the subdural hematoma because it appears the same density as the skull. **B:** Same study filmed with a wide window (blood window: 150–200 Hounsfield units). The left subdural hematoma is easily identified (black straight arrows). This subdural hematoma is too small to cause any significant mass effect. Note the subarachnoid pneumocephalus (white arrowheads) secondary to a skull fracture, resulting in a communication with a sinus (not shown). Extracranial soft-tissue swelling is also present (white straight arrow). The pneumocephalus and soft-tissue swelling are better visualized with the wider window setting.

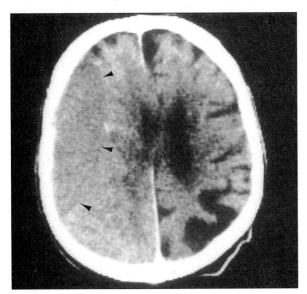

Fig. 61.2 Subdural hematoma. Noncontrasted axial CT reveals a nearly isodense right subdural hematoma (arrowheads). Significant mass effect effacing the cortical sulci with some midline shift is present. The left side demonstrates marked atrophy. In anemic patients, an acute subdural hematoma may be isodense to brain. An isodense subdural hematoma is also shown in Fig. 87.1.

punctate hemorrhages that are smaller than the slice thickness may balance with surrounding edema, creating a falsely isodense region (volume-averaging effect), which may go undetected.

Thin hematomas located on the floor of the skull, parallel to the plane of the CT image, may be difficult to detect. When the hemorrhage fills the full thickness of the slice, it should be easily identifiable. However, if the hemorrhage is thinner than the slice, volume averaging with the skull may obscure the blood.

Sometimes an acute hemorrhage into a confined space is undetectable on CT. That is, when a preexisting CSF collection mixes with blood, the resulting mixture may be isodense to brain tissue.[5] With time the blood settles, forming a CSF-blood level in the dependent portion of the cavity, which makes diagnosis possible. Confined spaces where this phenomenon occurs include the ventricles, cisterns, porencephalic cysts, other brain cysts, and cystic portions of tumors.

CT is less valuable than MRI for diagnosing subacute and chronic hemorrhage. On MRI, the signal intensity difference between an aging hematoma and brain parenchyma increases with time because the metabolic degradation products of hemoglobin have paramagnetic properties (maxim 60).

At 4–21 days after an intraparenchymal hemorrhage, disruption of the blood–brain barrier and neovascularity result in enhancement at the original margins of the hematoma (luxury perfusion, analogous to ischemic stroke; maxim 48). This phenomenon is independent of the cause of the hemorrhage. The ring enhancement may lead to the erroneous conclusion that a neoplasm or abscess is present, especially if the central portion has transformed into an isodense or hypodense region. In this case, MRI more clearly defines the nature of the lesion. Follow-up CT may also clarify the diagnosis.

References

1. Smith WP, Batnidzky S, Rengachary SS. Acute isointense subdural hematomas: a problem in anemic patients. *AJR Am J Roentgenol* 1981; **136**: 543–6.
2. Smith WP, Batnidzky S, Rengachary SS. Acute isointense subdural hematomas: a problem in anemic patients. *AJNR Am J Neuroradiol* 1981; **2**: 37–40.
3. Wilms G, Marchal G, Geusens E. Isodense subdural hematomas on CT: MRI findings. *Neuroradiology* 1992; **34**: 497–9.
4. Boyko OB, Cooper DF, Grossman CB. Contrast-enhanced CT of acute isodense subdural hematoma. *AJNR Am J Neuroradiol* 1991; **12**: 341–3.
5. Kaufman HH, Singer JM, Sadhu VK, Handel SF, Cohen G. Isodense acute subdural hematoma. *J Comput Assist Tomogr* 1980; **4**: 557–9.

62. *Venous malformations, cavernous angiomas, and capillary telangiectasias are best depicted on MRI*

Cerebral cavernous angiomas (CCAs) and capillary telangiectasias are not usually discernible on conventional angiograms. Occasionally a late blush occurs, but in general these lesions are angiographically occult vascular malformations. Completely thrombosed arteriovenous malformations (AVMs) are also angiographically occult. Venous angiomas may be detectable on angiogram, sometimes appearing as dilated normal vessels. All of these lesions have either low or no flow of blood, which probably explains their relatively benign clinical course. These lesions are readily visualized with standard MRI spin-echo and gradient-echo sequences. Calcification, the major finding on CT, is variably present. MRA offers no advantage to conventional angiography for detecting these lesions. When symptomatic, they all have a favorable prognosis with surgical resection as long as eloquent brain is not damaged.

Cavernous angiomas

Cavernous angiomas, one of the most common vascular malformations identified on MRI, are best visualized on gradient-echo and T2- and PD-weighted spin-echo images performed on high-field-strength scanners.[1-3] They are largely spherical (round on planar sections). Classically, they have a "popcorn" appearance, with a mixed signal core on T1- and T2-weighted images (Fig. 62.1). A low signal rim is often present on the T2-weighted images. The MRI signal characteristics are due to the presence of blood at various stages of decomposition. This mixed signal appearance is evidence that these lesions are intermittently active and presumably is the result of recurrent microhemorrhages. At times it is possible to distinguish several smaller spheres within the core corresponding to the "mulberry" appearance on direct examination at surgery or pathology.

Cavernous angiomas exhibit other appearances on MRI that correlate with specific clinical and pathologic patterns.[4] An exclusively high-signal core (methemoglobin) on T1-weighted images with either a high-signal or low-signal core surrounded by a low-signal rim (hemosiderin) on T2-weighted images indicates that the cavernous angioma has recently hemorrhaged (Fig. 62.1). Care must be taken to avoid confusing this appearance with a hemorrhage from another cause or calcium, which is also sometimes high signal on T1-weighted images (maxim 13). (Some cavernous angiomas calcify.) A mixed-signal core indicates that the lesion is a cavernous angioma; the high T1 signal is a subacute hematoma and the surrounding low T2 signal rim is hemosiderin, evidence of an old hemorrhage. Homogeneous low-signal lesions on T2-weighted images that are also low signal on gradient-echo images (isointense or low signal on T1-weighted images) indicate a chronic angioma that has not hemorrhaged for a long time. Small punctate low-signal lesions that are visible only on gradient-echo sequences may represent a primordial stage. These punctate lesions on gradient echo can be numerous in patients with multiple cavernous angiomas (Fig. 62.1C). Most cerebral cavernous angiomas do not enhance with intravenous contrast, although occasionally weak enhancement is present.

Pathologically, cavernous angiomas consist of many dilated sinusoidal channels separated by fibrous strands with no intervening neural tissue. Classic descriptions include vessels lined by a single layer of endothelial cells that are devoid of smooth muscle and elastic tissue. However, in practice, the vessels are often arterialized and may be misdiagnosed as AVMs on pathologic examination.[5] Thrombosis, calcification, and surrounding gliosis and hemosiderin deposition is typical. Microscopic intermittent hemorrhage is probably responsible for these findings and is reflected by the appearance on MRI of blood in various stages of degradation.

Many cerebral cavernous angiomas are asymptomatic and discovered incidentally. Symptomatic CCAs present as focal seizures, focal neurologic deficits (due to hemorrhage), or headaches, with the specific symptoms reflecting the location and size of the lesion. The majority of cavernous angiomas are in the cerebral hemispheres although they also arise in infratentorial structures (especially the brain stem), the spinal cord, optic nerves, and meningeal surfaces. Symptoms most often develop in the second to fifth decades. The natural history of these lesions is not fully established, but a small percentage evolve over time. In patients with multiple lesions the cumulative hemorrhage burden may become significant and new lesions appear to develop.[6]

Approximately one-half of individuals with cerebral cavernous angiomas have an autosomal dominant pattern of inheritance. Multiple cavernous angiomas are much more common in the familial form than in sporadic cases. Retinal and cutaneous angiomas may be present in the familial form. The sporadic cases are more likely to have another type of associated lesion such as a venous angioma or capillary telangiectasia.

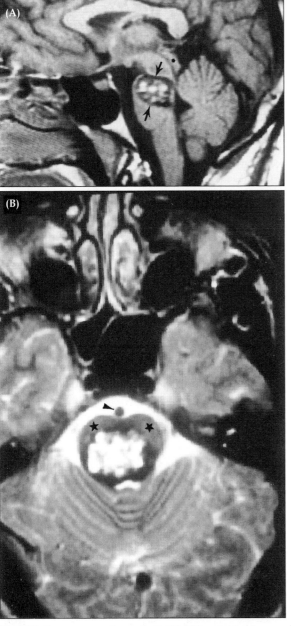

Cavernous angiomas are misdiagnosed as emboli, multiple sclerosis, and tumors (especially those with a tendency to hemorrhage, such as melanoma and renal cell carcinoma; maxim 35). Careful review of all the imaging characteristics should prevent these errors.

High-field-strength MRI with gradient-echo images should be performed on all patients suspected of having cerebral cavernous angiomas.[7] Family members of affected patients, especially when multiple lesions are present, should be screened with high-field-strength MRI. Higher-field-strength images are much more likely to detect small cavernous angiomas because they have a higher sensitivity to iron than low-field-strength imaging.[8,9] This property holds for examination of blood from any source. CT alone is inadequate although it is sometimes useful as an adjunct to detect calcium, which is present in some cavernous angiomas.

Tagged red blood cell-labeled single photon emission CT (SPECT) may prove useful in following cavernous angiomas, especially after surgery.[10] Focal areas of increased activity, corresponding to MRI-detected cavernous angiomas, are present on delayed images, with normal flow on early images (Fig. 62.2). Focal increased activity, present exclusively on the delayed images, implies low but detectable flow, differentiating

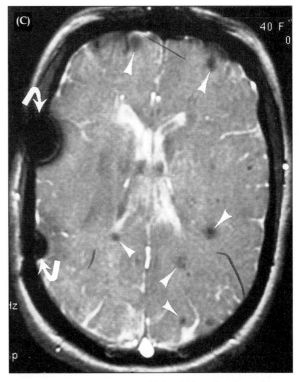

Fig. 62.1 Cavernous angioma. **A:** Sagittal T1-weighted image of a 34-year-old man with only mild headaches. A spherical lesion with a low-intensity rim and a mixed (popcorn)-intensity core occupies nearly the entire pons (arrows). This is the typical appearance of a cavernous angioma. The high signal on the T1-weighted image probably indicates subchronic hemorrhage (methemoglobin). *, quadrigeminal plate dorsal to the aqueduct of Sylvius. **B:** Axial T2-weighted image reveals the cerebral cavernous angioma within the pons, sparing only the periphery (stars). No mass effect is present. Arrowhead, basilar artery. **C:** Axial T2-weighted gradient-echo image in a patient with multiple cerebral cavernous angiomas (CCAs). Note the many punctate, low-signal CCAs scattered throughout the white and gray matter (arrowheads). The two large voids on the right (curved arrows) are caused by surgical wires in the skull after removal of one CCA in the temporal lobe (not shown). The black lines are wax pencil marks left by the interpreting radiologist – the "positive radiological sign"!

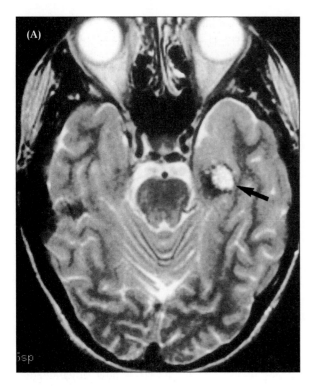

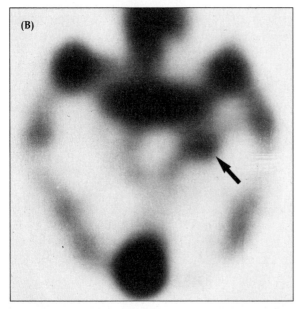

these lesions from old hematomas and AVMs. Often it is not possible to determine whether a residual angioma is present after surgical resection because the postoperative changes on MRI have signal characteristics similar to those of angiomas. This technique, used before and after surgery, may eliminate this problem.

Venous angiomas

Venous angiomas are the most common incidental vascular malformation at autopsy, present in approximately 2% of autopsies. Dilated small veins converge on a larger cerebral or cerebellar vein, producing a caput medusae configuration. These lesions consist only of veins; neither capillaries nor arteries are involved. They are considered to be extreme anatomical variants, consisting of anomalous veins separated by normal brain tissue. Most venous angiomas are clinically silent, but they can present with seizures or headaches. Posterior fossa venous angiomas may cause vertigo, nystagmus, and ataxia. The most common location for a venous angioma is the frontal lobe.

On MRI and contrast-enhanced CT, a stellate collection of small vessels draining to a single dilated vein indicates the presence of a venous angioma (Fig. 62.3). They are sometimes difficult to visualize on CT, and the only evidence for them may be a line of enhancement. Venous angiomas appear as a signal void due to blood flow on spin-echo and high signal on gradient-echo MRI.[11] Both the small peripheral vessels and the larger draining vein are usually apparent on MRI. Approximately one-third of these malformations are discovered only after contrast administration. The typical angiographic appearance is of enlarged veins converging in radial fashion into a larger draining vein. They can be seen on MRA, especially with the contrast-enhanced 3D time-of-flight (TOF) tech-

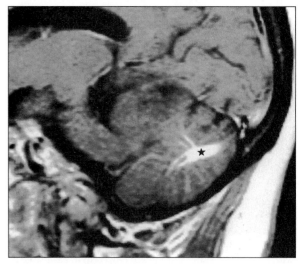

Fig. 62.3 Venous angioma. Sagittal T1-weighted contrast-enhanced image. A radially oriented collection of dilated medullary veins draining into a single vein (star) is present in the cerebellum. This caput medusal pattern is characteristic of a venous angioma. Most venous angiomas remain indolent throughout life; however, venous malformations in the brain stem and cerebellum are more prone to hemorrhage than those in the cerebral hemispheres.

Fig 62.2 Cerebral cavernous angioma (CCA) identified by delayed red blood cell (RBC)-labeled single photon emission CT (SPECT). **A:** Axial T2-weighted MRI revealing a typical CCA in the medial left temporal lobe (arrow). Note the low-signal rim (hemosiderin) surrounding a high-signal core. Numerous small spheres are visible within the core, giving a mulberry appearance. **B:** Axial ⁹⁹ᵐTc-labeled RBC SPECT taken 2 hours after injection, showing increased activity in the medial temporal CCA (arrow). The early SPECT study, performed shortly after the injection, had no significant activity at this location. Thus, time was required for the labeled blood to accumulate in the CCA, consistent with slow flow. The presence of delayed activity indicates that this lesion is not completely thrombosed.

nique. The vessels may appear normal with an adjacent large vein on conventional or MR angiography. The large vein is sometimes termed a varix. Venous angiomas are more common than AVMs but are usually identified only incidentally on neuroimaging because they are seldom symptomatic.

Capillary telangiectasias

Capillary telangiectasias (capillary angiomas) are tangles of small vessels with normal capillary morphology and normal intervening neural tissue. A small focus of a spider-web tangle of small vessels is the typical appearance on MRI. The pons and cerebral cortex are the most common locations. They are often multiple. There may be a continuum between capillary telangiectasias and cavernous angiomas. CT is usually normal or may show a vague blush with contrast. Capillary telangiectasias are not detected on angiography. There are no abnormal feeding arteries. Capillary telangiectasias are rarely symptomatic and are most commonly identified incidentally.

The angiographically occult vascular malformations and venous angiomas can be thought of as low-flow vascular hamartomas that are relatively benign. There may be developmental overlap between these lesions, as different types sometimes occur together. They tend to be encapsulated with surrounding gliosis and they have little tendency to bleed spontaneously or to result in massive intraoperative hemorrhage. Most of these lesions have a benign natural history and do not require therapy, especially when asymptomatic and found incidentally. Surgical intervention should be considered in patients with medically refractory seizures, progressive neurologic deterioration, recurrently symptomatic lesions, or clinically significant hemorrhage. Surgery should also be considered in certain cases for tissue diagnosis.

References

1. Atlas SW, Mark AS, Grossman RJ, Gomori JM. Intracranial hemorrhage: gradient-echo imaging at 1.5 T: comparison with spin-echo imaging and clinical applications. *Radiology* 1988; **168**: 803–7.
2. Atlas S, Mark A, Fram E, Grossman R. Vascular intracranial lesions: applications of gradient-echo imaging. *Radiology* 1988; **169**: 455–61.
3. Gomori J, Grossman R, Goldberg H, Hackney D, Zimmerman R, Bilaniuk L. Occult cerebral vascular malformations: high-field MR imaging. *Radiology* 1986; **158**: 707–13.
4. Zabramski J, Wascher T, Spetzler R, *et al.* The natural history of familial cavernous malformations: results of an ongoing study. *J Neurosurg* 1994; **80**: 422–32.
5. Rapacki T, Brantley M, Furlow T, Geyer C, Toro V, George E. Heterogeneity of cerebral cavernous hemangiomas diagnosed by MR imaging. *J Comput Assist Tomogr* 1990; **14**: 18–25.
6. Sigal R, Krief O, Houtteville J, Halimi P, Doyon D, Pariente D. Occult cerebrovascular malformations: Follow-up with MR imaging. *Radiology* 1990; **176**: 815–19.
7. Valanne L, Ketonen L, Berg MJ. Pseudoprogression of cerebral cavernous angiomas: the importance of proper magnetic resonance imaging technique. *J Neuroimaging* 1996; **6**: 195–6.
8. Norfray J, Couch J, Elble R, Good D, Manyam B, Patrick J. Visualization of brain iron by mid-field MR. *AJNR Am J Neuroradiol* 1988; **9**: 77–82.
9. Gomori J, Grossman R, Yu-Ip C, Asakura T. NMR relaxation times of blood: dependence on field strength, oxidation state and cell integrity. *J Comput Assist Tomogr* 1987; **11**: 684–90.
10. Berg MJ, Cohn FS, Ketonen LM. Detection of low-flow blood in cerebral cavernous angiomas using tagged RBC-SPECT. *Neurology* 1994; **44** (suppl 2): A408 (abstract).
11. Augustyn GT, Scott JA, Olson E, *et al.* Cerebral venous angiomas: MR imaging. *Radiology* 1985; **156**: 391–3.

63. *MRI and conventional angiography are often revealing in unexplained third nerve palsy*

Paralysis of the third cranial nerve (the oculomotor nerve) is a common clinical problem. The diagnostic evaluation includes MRI in many cases and conventional angiography in certain circumstances. The third cranial nerve innervates four of the six extraocular muscles (superior rectus, medial rectus, inferior rectus, and inferior oblique) and the levator palpebrae superioris (one of the two lid elevators – the other is innervated by the sympathetic system). It also carries parasympathetic fibers (which mediate pupillary contraction and accommodation) to the ciliary ganglion. Third nerve dysfunction may be caused by mass lesions in the brain stem, perimesencephalic cisterns, temporal lobe, cavernous sinus, and orbital apex, or by vascular insufficiency.

The motor nucleus of the third nerve resides in the midbrain periaqueductal gray matter, just ventral to the cerebral aqueduct, at the level of the superior colliculus. The Edinger–Westphal nucleus, which supplies the parasympathetic fibers, is located just dorsal to the oculomotor nucleus. The third nerve fibers course ventrally, through the red nuclei and the medial aspect of the cerebral peduncles, to exit near the midline in the interpeduncular cistern at the pontomedullary junction. The third nerves form a V shape as they exit the brain stem (Fig. 63.1). The nerve passes between the posterior cerebral and superior cerebellar arteries and then travels parallel to the posterior communicating artery (where aneurysms of this vessel compress the nerve) and lateral to the dorsum sella. It enters and travels in the superior lateral portion of the cavernous sinus. The nerve exits the cavernous sinus anteriorly and enters the orbital apex through the superior orbital fissure.

Third nerve palsies present with a variable combination of diplopia, ptosis, and mydriasis. The clinical features often suggest the exact location and nature of the lesion:[1]

1. *Brain stem*
   - Infarction
   - Demyelination (multiple sclerosis)
   - Tumor (glioma or metastasis)
   - Shear injury
2. *Brain stem cistern*
   - Aneurysm
   - Infectious and inflammatory lesions (chronic basilar meningitis, e.g. fungal, tuberculous, sarcoid)

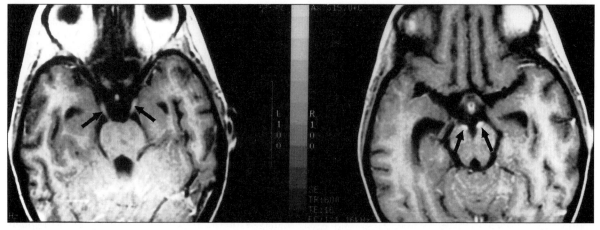

Fig. 63.1 Oculomotor nerves. Contrast-enhanced T1-weighted axial images in a child with leukemia and visual symptoms. The oculomotor nerves enhance bilaterally in the interpeduncular and suprasellar cisterns (arrows). Cranial nerve enhancement is always abnormal, although not always symptomatic.

- Tumor (lymphoma, carcinomatous meningitis, neuroma, craniopharyngioma, pituitary adenoma)

3. *Cavernous sinus*
   - Cavernous carotid artery aneurysm
   - Neoplasm (lymphoma, metastasis, meningioma)
   - Sarcoid
   - Cavernous sinus thrombosis

4. *Orbit*
   - Neoplasm (meningioma, glioma, metastasis, lymphoma, leukemia)
   - Tolosa–Hunt syndrome (orbital pseudotumor).

Third nerve palsies may be either isolated or associated with other deficits.

When MRI is used to investigate third nerve palsies, thin sections through this region, with a small field of view, using a combination of PD, T2, and T1-weighted images in three orthogonal planes with and without contrast enhancement should be performed.[1–4] If the pupil is involved (dilated) and no other cause is identified (such as an Adie tonic pupil or surreptitious use of mydriatics – both distinguishable by topical pharmacologic testing), conventional angiography should be performed to exclude an aneurysm of the posterior communicating artery. Although high-quality MR and CT angiography can detect small aneurysms (Fig. 51.4), further refinement of these techniques is necessary before they will replace conventional angiography (maxim 59). In children with an isolated third nerve palsy and negative imaging evaluation initially, follow-up imaging should be performed to rule out a small tumor.

The primary method for evaluating orbital structures, especially in the superior orbital fissure, is high-resolution MRI. Contrast-enhanced T1-weighted images with fat suppression in the axial and coronal planes are useful in evaluating the orbital apex. Individual cranial nerves are readily distinguished. T2-weighted axial and inversion-recovery techniques are also useful.

CT with orbital cuts is used to discern bone fragments in trauma and enlargement of the extraocular muscles in Graves ophthalmopathy. In Graves ophthalmopathy, the inferior and medial rectus muscles are the first extraocular muscles to enlarge.

The most common causes of an isolated third nerve palsy are ischemia (secondary to diabetes, hypertension, or arteriosclerosis), aneurysm, trauma, and neoplasia (including carcinomatous meningitis; Fig. 63.1). Meningitis (especially chronic basilar), herpes zoster, and sinusitis can also cause a third nerve palsy.[1] Because the parasympathetic fibers travel in the outer superior and medial aspect of the nerve they are susceptible to compression such as may occur with a posterior communicating artery aneurysm, which can present with an isolated ipsilateral enlarged pupil (internal ophthalmoplegia). In contrast, ischemia of the third nerve may spare the pupil (because the outer pupillomotor fibers have a better vascular supply), with microinfarctions damaging the deeper motor fibers, resulting in an external ophthalmoplegia. A pupil-sparing third nerve palsy is the most common presentation for ischemia, especially associated with diabetes. Brain-stem infarctions and intrinsic masses causing third nerve palsy almost invariably also cause other cranial nerve palsies and upper motor neuron dysfunction.

References

1. Trobe JD. Isolated third nerve palsies. *Semin Neurol* 1986; **6**: 135–41.
2. Blake PY, Mark AS, Kattah J, Kolsky M. MR of oculomotor nerve palsy. *AJNR Am J Neuroradiol* 1995; **16**: 1665–72.
3. Laine FJ, Smoker WR. Cranial nerves III, IV and VI. *Neuroimaging Clin North Am* 1993; **3**: 85–103.
4. Mark AS, Blake P, Atlas SW, *et al.* Gd-DTPA enhancement of the cisternal portion of the oculomotor nerve on MR imaging. *AJNR Am J Neuroradiol* 1992; **13**: 1463–70.

10
Neurobehavioral, Neurodegenerative, and Systemic Syndromes

64. White-matter abnormalities help separate multi-infarct dementia from Alzheimer disease

Multi-infarct dementia, better termed cerebrovascular dementia, denotes an acquired cognitive impairment resulting from ischemic or hemorrhagic strokes. It causes approximately one-fourth of all dementias, and trails only Alzheimer disease as the most common cause of dementia. Identification of this diagnosis is important so that appropriate prophylactic treatments can be instituted to decrease the risk of future strokes. The clinical diagnosis of cerebrovascular dementia may be made when a stepwise deterioration in intellectual function (often in the setting of acute strokes) is present in a patient with appropriate risk factors (such as chronic hypertension or diabetes) and consistent neuro-radiologic features.[1]

Large multiple strokes involving the cortex and white matter are easy to detect on CT or MRI. A subset of patients have predominantly subcortical infarctions, resulting in disease confined to the white matter. This condition is best termed subcortical arteriosclerotic encephalopathy (SAE), but is commonly referred to as Binswanger disease.[2,3] There are many causes of SAE, including hypertension, diabetes, amyloid angiopathy, CADASIL (cerebral autosomal dominant arteriopathy with subcortical infarctions and leukoencephalopathy), and the antiphospholipid antibody syndrome.[1]

SAE is not a neuroradiologic diagnosis,[1] but the presence of multiple areas of punctate, partially confluent, or confluent deep-white-matter disease or irregular periventricular disease extending into the deep white matter suggests SAE.[1,4] These lesions appear as hyperintensities on PD and T2-weighted images (Fig. 64.1) and, when severe, are visible on T1-weighted images and CT as dark regions (so-called leukoaraiosis; Fig. 56.1; maxim 16). However, these lesions, especially in isolation, can be caused by other processes (maxims 16–20).

Demyelination, loss of axons, and fibrous thickening of the walls of arteries are present in the white-matter lesions in SAE on pathologic examination.[5,6] Intermittent or chronic insufficient cerebral blood flow may lead to complete or incomplete infarction in the white

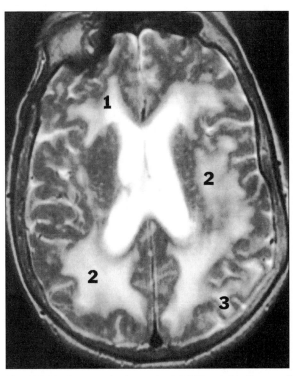

Fig. 64.1 Cerebrovascular dementia. Axial T2-weighted image of a 76-year-old woman with dementia and a history of strokes. Large periventricular (1) and deep (2) confluent white-matter high-signal lesions are evident. The subcortical white matter is involved posteriorly (3). The ventricles are enlarged. Minimal cortical atrophy is present. This pattern is compatible with damage from multiple infarctions and recurrent ischemia, and suggests the diagnosis of subcortical arteriosclerotic encephalopathy (Binswanger disease), a form of multi-infarct dementia.

matter, resulting in these lesions. Although subcortical and periventricular white-matter hyperintensities are common in nondemented elderly people, there is a clear correlation between the presence of these lesions and vascular disease risk factors.

Small white-matter T2 and PD hyperintensities (unidentified bright objects, UBOs) are a frequent MRI finding in the elderly (maxims 17 and 18). The number of hyperintensities increases with age, hypertension, and other vascular risk factors. Some high-signal foci are considered a normal finding in people over 40 years of age. Almost all healthy adults over age 75 years have one or more punctate white-matter hyperintensities.[7,8] The nature, clinical significance, and pathophysiology of these lesions of aging is not well understood. Minor perivascular damage without complete infarction may be responsible for many of these UBOs.[9]

By location, white matter hyperintensities can be divided into three categories: periventricular, subcortical, and deep. Controversy exists about whether periventricular or subcortical lesions correlate with dementia.[10] However, there is a loose correlation between the quantity of white-matter disease detected on MRI and impairment of attention, speed of mental processing, and subjective cognitive dysfunction.[11,12] Nonetheless, small amounts of white-matter disease are often present in normal elderly individuals.

The presence of deep-white-matter lesions is more helpful than periventricular lesions in distinguishing cerebrovascular dementia from Alzheimer disease and normal aging but not diagnostic. Periventricular lesions are often present in normal adults, Alzheimer disease, and vascular dementia (maxim 17). In contrast, deep-white-matter and subcortical lesions are present in essentially all patients with cerebrovascular dementia but in less than one-half of patients with Alzheimer dementia.[10]

No neuroimaging feature reliably differentiates the conditions that cause white-matter disease, and there is a broad differential for MRI T2 white-matter hyperintensities (maxim 16). They can represent normal perivascular spaces, infarctions, cysts, gliosis, postinfectious foci, or demyelination. Small neoplastic lesions and even heterotopic neurons may also be hyperintense on T2-weighted images. For now, the increased sensitivity of MRI has resulted in decreased specificity. The appearance of the lesions on PD-weighted images and the contrast enhancement pattern helps to narrow the differential diagnosis. History and physical examination remain paramount in the evaluation of dementia.

References

1. Caplan LR. Binswanger's disease – revisited. *Neurology* 1995; **45**: 626–33.
2. Fisher CM. Binswanger's encephalopathy: a review. *J Neurol* 1989; **236**: 65–79.
3. Olszewski J. Subcortical arteriosclerotic encephalopathy: review of the literature on the so-called Binswanger's disease and presentation of two cases. *World Neurol* 1965; **3**: 359–73.
4. Fazekas F, Lieinert R, Offenbacher H, *et al.* Pathologic correlates of incidental MRI white matter signal hyperintensities. *Neurology* 1993; **43**: 1683–9.
5. Scheltens P, Barkhof F, Leys D, Wolters E, Ravid R, Kamphorst W. Histopathologic correlates of white matter changes on MRI in Alzheimer's disease and normal aging. *Neurology* 1995; **45**: 883–8.
6. Braffman BH, Zimmerman RA, Trojanovski JQ, *et al.* Brain MR: correlation with gross and histopathology. 2. Hyperintense white-matter foci in the elderly. *AJNR Am J Neuroradiol* 1988; **9**: 629–36.
7. Awad IA, Spetzler RF, Hodak JA, *et al.* Incidental subcortical lesions identified on magnetic resonance imaging in the elderly. I. Correlation with age and cerebrovascular risk factors. *Stroke* 1986; **17**: 1084–9.
8. Kirkpatrick JB, Hayman A. White matter lesions in MR images of clinically healthy brain of elderly subjects: possible pathologic basis. *Radiology* 1987; **162**: 509–11.
9. Fazekas F, Kleinert R, Offenbacher H, *et al.* The morphologic correlate of incidental punctate white matter hyperintensities on MR images. *AJNR Am J Neuroradiol* 1991; **12**: 915–21.
10. Bowen BC, Barker WW, Loewenstein DA, Sheldon J, Duara R. MR signal abnormalities in memory disorder and dementia. *AJNR Am J Neuroradiol* 1990; **11**: 283–90.
11. Ylikoski R, Ylikoski A, Erkinjuntti T, Sulkava R, Raininko R, Tilvis R. White matter changes in healthy elderly persons correlate with attention and speed of mental processing. *Arch Neurol* 1993; **50**: 818–24.
12. Breteler MM, van Swieten JC, Bots ML, *et al.* Cerebral white matter lesions, vascular risk factors, and cognitive function in a population-based study: the Rotterdam study. *Neurology* 1994; **44**: 1246–52.

65. *Large lateral ventricles with normal-sized sulci raise the possibility of normal-pressure hydrocephalus*

Definitive criteria for distinguishing patients with treatable normal-pressure hydrocephalus (NPH) are not yet established. Nevertheless, as some patients with NPH respond to CSF shunting, its recognition is important. NPH is characterized by the clinical triad of gait disturbance, urinary incontinence, and dementia. It is a disorder of adults, usually the elderly. Cases that are most likely to respond to ventricular or lumbar cistern shunting develop within months after a subarachnoid hemorrhage, meningitis, or head trauma, and progress over weeks to months. A positive clinical response to a large-volume lumbar puncture ("tap test") and the presence of prominent ventricular CSF flow (pulsation) on neuroradiologic studies helps to predict the patients most likely to respond to shunting. However, some patients who do not improve after lumbar puncture also respond to shunting.

Although the pathophysiology of NPH is not certain, relatively increased intracranial pressure or pulsations of increased pressure appear to play a role. Most patients with NPH have high normal or slightly elevated CSF pressures at lumbar puncture, but this does not correlate with severity of disease or likelihood of improvement with shunting. The underlying cause of the relatively increased CSF pressure is thought to be CSF outflow obstruction at the arachnoid granulations.

The gait disturbance is often the first presenting sign. Short shuffling steps with moderate loss of postural reflexes are typical, and classified as a frontal lobe gait apraxia. Similar disorders of gait, usually with more chronic development, are present in cerebrovascular dementia (multi-infarct dementia) and some elderly patients without dementia (marche-à-petits-pas). Differentiation from the gait disorder of parkinsonism is usually possible because the upper extremities are not affected in NPH.

The gait abnormality in NPH can improve immediately after removal of a large quantity of CSF (20–30 ml) at lumbar puncture. However, several days of continuous lumbar CSF shunting may be a more reliable test to predict responders to shunting.[1] The best candidates for shunting are patients who respond to lumbar puncture and have only mild cognitive impairment.[2,3] Following shunting, improvement of the gait may be dramatic, but cognitive improvement is generally more modest.[4] Severe dementia suggests irreversible brain damage or coexisting Alzheimer disease or cerebrovascular dementia. When a gait apraxia develops within weeks or months of brain insult and is associated with the typical neuroimaging features of NPH, diagnostic lumbar puncture followed by ventricular shunting may prevent dementia.

The classic neuroradiologic finding in NPH is generalized ventricular dilatation out of proportion to sulcal enlargement. Marked dilatation of the frontal horns and stretching and bowing of the pericallosal arteries and corpus callosum is typical. The temporal horns and third ventricle are rounded, appearing to be ballooned outward under pressure. In contrast, hydrocephalus *ex vacuo*, secondary to brain parenchyma loss in Alzheimer disease and other degenerative dementias, is characterized by ventricular dilatation balanced by sulcal enlargement without deformation of the corpus callosum and pericallosal arteries and with preserved parallel lateral walls of the third ventricle. In some patients with NPH, the lateral fissures become enlarged. This finding suggests obstruction of distal CSF outflow (probably at the level of the arachnoid granulations). After shunting, "paradoxical" collapse of the lateral fissures may occur.

Hyperdynamic CSF flow in the third ventricle and aqueduct, characterized by a flow void on sagittal midline MRI, predicts a good outcome from shunting.[5] Measurements of CSF flow through the aqueduct using a variety of MRI techniques are promising as a noninvasive means of predicting shunt responsiveness and following the course after shunting.[6–8] To assess for shunt responsiveness, some clinicians use intracranial pressure monitoring, with variable results.[4,9]

The presence of periventricular transependymal CSF (high signal on T2- and PD-weighted MRI and low density on CT) with enlarged ventricles is indicative of hydrocephalus and suggests the diagnosis of NPH in the elderly when no lesion compressing the aqueduct or fourth ventricle is present (nonobstructive hydrocephalus). Thicker periventricular halos correlate with better outcome from shunting.[10] The presence of microvascular disease (lacunae and deep-white-matter T2 and PD hyperintensities) is a negative predictor for success of shunting.

There is significant overlap in the neuroradiologic appearance in normal aging, NPH, and other degenerative dementias. Focal temporal lobe atrophy helps to differentiate patients with early Alzheimer dementia from normal age-matched controls.[11–13] Enlargement of the medial temporal lobe CSF spaces is associated with hippocampal, dentate, and amygdalar atrophy. The extent of choroidal and hippocampal fissure dilatation and the size of the temporal horn and hippocampus may also help to differentiate NPH from Alzheimer disease. In patients with NPH, the hippocampal and choroidal fissures are of normal size unless there is a component of Alzheimer disease complicating the process.[14,15] Coronal T1-weighted thin sections are the best images to make this assessment. A reverse (negative)-angle CT may also show the focal atrophy of Alzheimer disease.[16]

Cisternography, using quantitative single photon emission computerized tomography (SPECT) measuring the reflux of a lumbar cistern-injected radioisotope into the lateral ventricles is the traditional study used to detect NPH. It reportedly correlates with improvement from shunting,[17,18] but may not add information to the MRI findings.[19] Decreased global cerebral blood flow, which improves after shunting, is present in NPH on PET[20] and SPECT,[21–23] differentiating NPH from other degenerative dementias. A rise in regional cerebral blood flow after glycerol injection measured with xenon-enhanced CT also correlates with improvement after shunting in patients with NPH.[24] The best test or combination of tests to predict shunt responsiveness remains unclear and to a great degree is dependent on local expertise. The uncertainty about which patients to shunt is of great concern given the significant complication rate of shunts (e.g. meningitis and hemorrhage) in this group of patients, necessitating careful selection of which patients to shunt.[3]

References

1. Chen IH, Huang CI, Liu HC, Chen KK. Effectiveness of shunting in patients with normal pressure hydrocephalus predicted by temporary, controlled-resistance, continuous lumbar drainage: a pilot study. *J Neurol Neurosurg Psychiatr* 1994; **57**: 1430–2.
2. Wikkelso C, Anderson H, Blomstrand C, et al. Normal pressure hydrocephalus: predictive value of the cerebrospinal fluid tap-test. *Acta Neurol Scand* 1986; **73**: 566–73.
3. Vanneste J, Augustijn P, Dirven C, et al. Shunting normal pressure hydrocephalus: do the benefits outweigh the risks: a multicenter study and literature review. *Neurology* 1992; **42**: 54–9.
4. Raftopoulos C, Deleval J, Chaskis C, et al. Cognitive recovery in idiopathic normal pressure hydrocephalus: a prospective study. *Neurosurgery* 1994; **35**: 397–404.
5. Bradley WG, Whitemore AR, Koltman KE, et al. Marked CSF void: indicator of successful shunt in patients with suspected normal pressure hydrocephalus. *Radiology* 1991; **178**: 459–66.
6. Mascalchi M, Arnetoli G, Inzitari D, et al. Cine-MR imaging of aqueductal CSF flow in normal pressure hydrocephalus syndrome before and after CSF shunt. *Acta Radiol* 1993; **34**: 586–92.
7. Gideon P, Stahlberg F, Thomsen C, Gjerris F, Sorensen PS, Henriksen O. Cerebrospinal fluid flow and production in patients with normal pressure hydrocephalus studied by MRI. *Neuroradiology* 1994; **36**: 210–15.
8. Katayama S, Asari S, Ohmoto T. Quantitative measurement of normal and hydrocephalic cerebrospinal fluid flow using phase contrast cine MR imaging. *Acta Med Okayama* 1993; **47**: 157–68.

9. Sahuquillo J, Rubio E, Codina A, *et al*. Reappraisal of the intracranial pressure and cerebrospinal fluid dynamics in patients with the so-called "normal pressure hydrocephalus" syndrome. *Acta Neurochir* 1991; **112**: 50–61.

10. Jack CR Jr, Mokri B, Laws ER Jr, Houser OW, Baker HL Jr, Petersen RC. MR findings in normal-pressure hydrocephalus: significance and comparison with other forms of dementia. *J Comput Assist Tomogr* 1987; **11**: 923–31.

11. Ball MJ, Fishman M, Hachinski V, *et al*. A new definition of Alzheimer's disease, hippocampal dementia. *Lancet* 1985; **i**: 14–16.

12. Convit A, deLeon MJ, Golomb J, *et al*. Hippocampal atrophy in early Alzheimer's disease: anatomic specificity and validation. *Psychiatr Q* 1993; **64**: 371–87.

13. deLeon MJ, George AE, Reisberg B, *et al*. Alzheimer's disease: longitudinal CT studies of ventricular change. *AJNR Am J Neuroradiol* 1989; **10**: 371–6.

14. Golomb J, de Leon MJ, George AE, *et al*. Hippocampal atrophy correlates with severe cognitive impairment in elderly patients with suspected normal pressure hydrocephalus. *J Neurol Neurosurg Psychiatr* 1994; **57**: 590–3.

15. George AE, Holodny A, Golomb J, de Leon MJ. The differential diagnosis of Alzheimer's disease: cerebral atrophy *versus* normal pressure hydrocephalus. *Neuroimaging Clin of North Am* 1995; **5**: 19–31 (review).

16. George AE, deLeon MJ, Stylopoulos LA, *et al*. CT diagnostic features of Alzheimer's disease: importance of the choroidal/hippocampal fissure complex. *AJNR Am J Neuroradiol* 1990; **11**: 101–7.

17. Larsson A, Arlig A, Bergh AC, *et al*. Quantitative SPECT cisternography in normal pressure hydrocephalus. *Acta Neurol Scand* 1994; **90**: 190–6.

18. Larsson A, Moonen M, Bergh AC, Lindberg S, Wikkelso C. Predictive value of quantitative cisternography in normal pressure hydrocephalus. *Acta Neurol Scand* 1990; **81**: 327–32.

19. Vanneste J, Augustijn P, Avies GAG, *et al*. Normal pressure hydrocephalus: is cisternography still useful in selecting patients for a shunt? *Arch Neurol* 1992; **49**: 366–70.

20. Jagust WJ, Friedland RP, Budinger TF. Positron emission tomography with [^{18}F]fluorodeoxyglucose differentiates normal pressure hydrocephalus from Alzheimer-type dementia. *J Neurol Neurosurg Psychiatr* 1985; **48**: 1091–6.

21. Shih WJ, Tasdemiroglu E. Reversible hypoperfusion of the cerebral cortex in normal-pressure hydrocephalus on technetium-99m-HMPAO brain SPECT images after shunt operation. *J Nucl Med* 1995; **36**: 470–3.

22. Granado JM, Diaz F, Alday R. Evaluation of brain SPECT in the diagnosis and prognosis of the normal pressure hydrocephalus syndrome. *Acta Neurochir* 1991; **112**: 88–91.

23. Waldemar G, Schmidt JF, Delecluse F, Andersen AR, Gjerris F, Paulson OB. High resolution SPECT with [^{99m}Tc]-d,l-HMPAO in normal pressure hydrocephalus before and after shunt operation. *J Neurol Neurosurg Psychiatr* 1993; **56**: 655–64.

24. Shimoda M, Oda S, Shibata M, Masuko A, Sato O. Change in regional cerebral blood flow following glycerol administration predicts: clinical result from shunting in normal pressure hydrocephalus. *Acta Neurochir* 1994; **129**: 171–6.

66. *There are characteristic neuroimaging signs in alcohol abuse*

Alcoholics develop generalized cerebral atrophy, cerebellar atrophy (especially involving the medial cerebellar structures), subdural hematomas (often chronic), and other signs of multifocal trauma. The effects of trauma may be more severe in alcoholics, in part due to an alcohol-induced bleeding diathesis.[1] Several disorders with specific neuroimaging findings, including Wernicke–Korsakoff encephalopathy, Marchiafava–Bignami disease, osmotic myelinolysis (central pontine myelinolysis; maxim 21), and liver disease (maxim 67), also occur in alcoholics. Any of these may result in death.[2]

Cerebral atrophy Numerous studies using CT have demonstrated that the cerebral atrophy and accompanying cognitive dysfunction partially reverses over months in alcoholics who stop consuming alcohol.[3–7] Single photon emission computerized tomography (SPECT) studies have revealed decreased cerebral blood flow that is more extensive than the structural change present on CT.[8,9] Structures surrounding the third ventricle and the frontal lobes are preferentially affected, raising the possibility that a mechanism similar to that in Wernicke encephalopathy is responsible.[10,11] Similar cerebral atrophy occurs to a lesser extent in some patients consuming hepatotoxic medications.[12] The cerebral atrophy is not clearly due to dehydration as initially proposed.[13] It may be partially due to loss of dendritic connections and nerve cells, an irreversible process.[14] Nutritional deficiency is a proposed mechanism. However, the actual underlying process has not yet been elucidated.[15]

Cerebellar atrophy Cerebellar atrophy in alcoholics predominantly involves the vermis, deep cerebellar nuclei, and sometimes the anterior lobule. The clinical presentation consists mainly of truncal and gait ataxia. The upper extremities are rarely affected. The clinical syndrome develops over days, weeks, or more chronically. A nutritional deficiency is the proposed mechanism, as cerebellar atrophy may develop in alcoholics who are abstinent for weeks.

Marchiafava–Bignami disease Marchiafava–Bignami disease was originally described in Italian red-wine drinkers, but has subsequently been reported with abuse of other alcoholic beverages and probably with nutritional deficiency. The disease is characterized by demyelination and necrosis (without inflammation) of the middle layers of the corpus callosum. The other commissures and white matter may also be affected.[16–19] The lesions are symmetric and the gray matter is spared. These lesions appear as hypodensities on CT, but are best visualized as high-signal sometimes mixed with low-signal regions on T2-weighted MRI.[20] Edema and enhancement with contrast of the affected areas may occur early during the process.[20–22] Widespread marked decrease of cerebral metabolism, out of proportion to the structural abnormalities on MRI, are detected with positron emission tomography (PET).[23] Clinically only a subtle disconnection syndrome may be present, and the abnormalities on MRI may be the initial evidence for the underlying disorder.[24] Personality changes (especially with apathy), intellectual impairment, severe dementia, and multifocal neurologic deficits may develop. Some cases appear to reverse, but most progress to death over several years.

Wernicke–Korsakoff syndrome. Wernicke encephalopathy and Korsakoff amnesia are clinically distinct entities with similar pathologic findings. Both are the result of nutritional thiamine deficiency and occur mainly but not exclusively in chronic alcoholics.[25] The syndrome develops over days or weeks and may respond promptly to administration of parenteral thiamine (and glucose). However, residual neuroimaging features with or without clinical deficits are common. Untreated, the disease progresses to stupor, coma, and death. Early treatment with thiamine halts the progression.

Wernicke encephalopathy is characterized by the clinical triad of eye movement abnormalities, ataxia, and mental status changes. The eye movement dysfunction ranges from nystagmus to complete ophthalmoplegia. The ataxia is mainly confined to the trunk and lower extremities, impairing gait. The cognitive change usually consists of a global confusional state.

Korsakoff amnesia (psychosis) is characterized by markedly impaired antegrade and some degree of retrograde amnesia. Confabulation may or may not be present. Apathy and impaired insight are typical. It may become evident during recovery from Wernicke encephalopathy.

The pathologic lesions are best visualized on MRI. The involved locations display abnormally high T2 signal and include the periventricular regions, medial thalamic nuclei (including the massa intermedia), floor of the third ventricle, mammillary bodies (often atrophic), periaqueductal gray matter, midbrain reticular formation, and tectal plate.[26] Cerebellar, especially vermian, atrophy may be present. Atrophy of the anterior portions of the diencephalon, mesial temporal lobe structures, and orbitofrontal cortex may be specific for Korsakoff amnesia.[27] The lesions are characteristically symmetric. Asymmetric lesions should raise the possibility of infarctions (especially thalamic) or multiple sclerosis.[28] Resolution of the abnormal T2 signal may follow therapy with thiamine.[29] The findings on CT are less specific; localized atrophy around the third ventricle and midbrain and decreased density in the thalamic region may be present.[30–32] Similar brain lesions may be present in alcoholics without major neurologic deficits, suggesting that Wernicke–Korsakoff syndrome is an extreme of a spectrum.[33,34]

References

1. Ronty H, Ahonen A, Tolonen U, Heikkila J, Niemela O. Cerebral trauma and alcohol abuse. *Eur J Clin Invest* 1993; **23**: 182–7.
2. Skullerud K, Andersen SN, Lundevall J. Cerebral lesions and causes of death in male alcoholics: a forensic autopsy study. *Int J Legal Med* 1991; **104**: 209–13.
3. Marchesi C, De Risio C, Campanini G, *et al*. Cerebral atrophy and plasma cortisol levels in alcoholics after short or a long period of abstinence. *Prog Neuropsychopharmacol Biol Psychiatr* 1994; **18**: 519–35.
4. Mann K, Batra A, Gunthner A, Schroth G. Do women develop alcoholic brain damage more readily than men? *Alcohol Clin Exp Res* 1992; **16**: 1052–6.
5. Cala LA. Is CT scan a valid indicator of brain atrophy in alcoholism? *Acta Med Scand Suppl* 1987; **717**: 27–32.
6. Carlen PL, Wilkinson DA. Reversibility of alcohol-related brain damage: clinical and experimental observations. *Acta Med Scand Suppl* 1987; **717**: 19–26.
7. Gurling HM, Reveley MA, Murray RM. Increased cerebral ventricular volume in monozygotic twins discordant for alcoholism. *Lancet* 1984; **i**: 986–8.
8. Erbas B, Bekdik C, Erbengi G, *et al*. Regional cerebral blood flow changes in chronic alcoholism using Tc-99m HMPAO SPECT: comparison with CT parameters. *Clin Nucl Med* 1992; **17**: 123–7.
9. Melgaard B, Henriksen L, Ahlgren P, Danielsen UT, Sorensen H, Paulson OB. Regional cerebral blood flow in chronic alcoholics measured by single photon emission computerized tomography. *Acta Neurol Scand* 1990; **82**: 87–93.
10. Kato A, Tsuji M, Nakamura M, Nakajima T. Computerized tomographic study on the brain of patients with alcohol dependence. *Jap J Psychiatr Neurol* 1991; **45**: 27–35.
11. Kohlmeyer K, Stober B, Jennen C. Computed tomography in chronic alcoholism. *Acta Radiol* 1986; **369** (suppl): 393–5.
12. Mutzell S. Computed tomography of the brain, hepatotoxic drugs and high alcohol consumption in male alcoholic patients and a random sample from the general male population. *Upsala J Med Sci* 1992; **97**: 183–94.
13. Mann K, Mundle G, Langle G, Petersen D. The reversibility of alcoholic brain damage is not due to rehydration: a CT study. *Addiction* 1993; **88**: 649–53.
14. Harper C, Kril J. If you drink your brain will shrink: neuropathological considerations. *Alcohol Alcoholism Suppl* 1991; **1**: 375–80 (review).
15. Lishman WA, Jacobson RR, Acker C. Brain damage in alcoholism: current concepts. *Acta Med Scand Suppl* 1987; **717**: 5–17.
16. Marjama J, Yoshino MT, Reese C. Marchiafava–Bignami disease: premortem diagnosis of an acute case utilizing magnetic resonance imaging. *J Neuroimaging* 1994; **4**: 106–9.
17. Baron R, Heuser K, Marioth G. Marchiafava–Bignami disease with recovery diagnosed by CT and MRI: demyelination affects several CNS structures. *J Neurol* 1989; **236**: 364–6.
18. Heepe P, Nemeth L, Brune F, Grant JW, Kleihues P. Marchiafava–Bignami disease: a correlative computed tomography and morphological study. *Eur Arch Psychiatr Neurol Sci* 1988; **237**: 74–9.
19. Izquierdo G, Quesada MA, Chacon J, Martel J. Neuroradiologic abnormalities in Marchiafava–Bignami disease of benign evolution. *Eur J Radiol* 1992; **15**: 71–4.
20. Chang KH, Cha SH, Han MH, Park SH, Nah DL, Hong JH. Marchiafava–Bignami disease: serial changes in corpus callosum on MRI. *Neuroradiol* 1992; **34**: 480–2.
21. Caparros-Lefebvre D, Pruvo JP, Josien E, Pertuzon B, Clarisse J, Petit H. Marchiafava–Bignami disease: use of contrast media in CT and MRI. *Neuroradiology* 1994; **36**: 509–11.
22. Ikeda A, Antoku Y, Abe T, Nishimura H, Iwashita H. Marchiafava–Bignami disease: consecutive observation at acute stage by magnetic resonance imaging and computerized tomography. *Jap J Med* 1989; **28**: 740–3.
23. Pappata S, Chabriat H, Levasseur M, Legault-Demare F, Baron JC. Marchiafava–Bignami disease with dementia: severe cerebral metabolic depression revealed by PET. *J Neural Transm Park Dis Dementia Sect* 1994; **8**: 131–7.
24. Berek K, Wagner M, Chemelli AP, Aichner F, Benke T. Hemispheric disconnection in Marchiafava–Bignami disease: clinical, neuropsychological and MRI findings. *J Neurol Sci* 1994; **123**: 2–5.
25. Doraiswamy PM, Massey EW, Enright K, Palese VJ, Lamonica D, Boyko O. Wernicke–Korsakoff syndrome caused by psychogenic food refusal: MR findings. *AJNR Am J Neuroradiol* 1994; **15**: 594–6.
26. Gallucci M, Bozzao A, Splendiani A, *et al*. Wernicke encephalopathy: MR findings in five patients. *AJNR Am J Neuroradiol* 1990; **11**: 887–92.

27. Jernigan TL, Schafer K, Butters N, Cermak LS. Magnetic resonance imaging of alcoholic Korsakoff patients. *Neuropsychopharmacology* 1991; **4**: 175–86.
28. Cole M, Winkelman MD, Morris JC, Simon JE, Boyd TA. Thalamic amnesia: Korsakoff syndrome due to left thalamic infarction. *J Neurol Sci* 1992; **110**: 62–7.
29. Donnal JR, Heinz ER, Burger PC. MR of reversible thalamic lesions in Wernicke syndrome. *AJNR Am J Neuroradiol* 1990; **11**: 893–4.
30. Jacobson RR, Lishman WA. Cortical and diencephalic lesions in Korsakoff's syndrome: a clinical and CT scan study. *Psychol Med* 1990; **20**: 63–75.
31. Shimamura AP, Jernigan TL, Squire LR. Korsakoff's syndrome: radiological (CT) findings and neuropsychological correlates. *J Neurosci* 1988; **8**: 4400–10.
32. McDowell JR, LeBlanc HJ. Computed tomographic findings in Wernicke–Korsakoff syndrome. *Arch Neurol* 1984; **41**: 453–4.
33. Blansjaar BA, Vielvoye GJ, van Dijk JG, Rijnders RJ. Similar brain lesions in alcoholics and Korsakoff patients: MRI, psychometric and clinical findings. *Clin Neurol Neurosurg* 1992; **94**: 197–203.
34. Gebhardt CA, Naeser MA, Butters N. Computerized measures of CT scans of alcoholics: thalamic region related to memory. *Alcohol* 1984; **1**: 133–40.

67. Liver dysfunction and manganism result in high signal on T1-weighted MRI within the globus pallidus

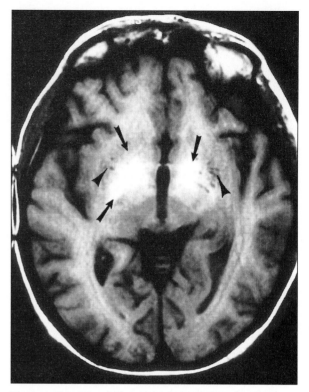

Fig. 67.1 Hepatic encephalopathy. Axial T1-weighted image in a patient with severe liver dysfunction. The globus pallidus is symmetrically hyperintense (straight arrows). No enhancement occurred with contrast (not shown). The abnormal high signal resolved when the liver function returned to normal (not shown). The adjacent punctate hypointensities (arrowheads) are normal Virchow–Robin spaces (maxim 12).

Chronic liver failure with cirrhosis is associated with abnormally high signal in the globus pallidus on T1-weighted MRI.[1–5] This finding is independent of the etiology of the liver disease but correlates with its severity.[6] The amount of portal–systemic shunting probably has a pathophysiologic role. This finding is present in one-half to three-quarters of patients with severe liver failure. It may resolve after liver transplantation.

The globus pallidus abnormality is apparent only on T1-weighted images (Fig. 67.1). It is most prominent when the images are acquired with the inversion recovery technique.[2] The T2-weighted images and CT are normal. The lesions are symmetric and do not enhance with contrast. In some patients, T1 high signal is also present within the putamen, caudate, subthalamic nuclei, anterior pituitary gland, and adjacent to the red nuclei in the mesencephalon.[3]

Deposition of a paramagnetic material, probably manganese, in the globus pallidus and the other affected structures is probably the cause of the increased T1 signal.[7] Long-term parenteral nutrition may also cause T1 hyperintensities, presumably also due to manganese accumulation.[8,9] Manganese toxicity following chronic ingestion causes similar lesions.[10] Manganism presents with parkinsonian features.[11] Chronic exposure to manganese results in degeneration of dopaminergic neurons.[12]

Patients with chronic liver failure often develop cortical atrophy, particularly in the frontal and parietal lobes. Cerebellar volume loss may occur. Laminar cortical necrosis with microcavitary changes at the gray–white matter junction is sometimes visualized on high-quality coronal T2-weighted thin-section MRI.

References

1. Inoue E, Hori S, Narumi Y, *et al.* Portal-systemic encephalopathy: presence of basal ganglia lesions with high signal intensity on MR images. *Radiology* 1991; **179**: 551–5.
2. Pujol A, Graus F, Peri J, Mercader JM, Rimola A. Hyperintensity in the globus pallidus on T1-weighted and inversion-recovery MRI: a possible marker of advanced liver disease. *Neurology* 1991; **41**: 1526–7.
3. Brunberg JA, Kanal E, Hirsch W, Van Thiel DH. Chronic acquired hepatic failure: MR imaging of the brain at 1.5 T. *AJNR Am J Neuroradiol* 1991; **12**: 909–14.
4. Kulisevsky J, Ruscalleda J, Grau JM. MR imaging of acquired hepatocerebral degeneration. *AJNR Am J Neuroradiol* 1991; **12**: 527–8.
5. Kulisevsky J, Pujol J, Deus J, *et al.* Persistence of MRI hyperintensity of the globus pallidus in cirrhotic patients: a 2-year follow-up study. *Neurology* 1995; **45**: 995–7.
6. Pujol A, Pujol J, Graus F, *et al.* Hyperintense globus pallidus on T1-weighted MRI in cirrhotic patients is associated with severity of liver failure. *Neurology* 1993; **43**: 65–9.
7. Barron TF, Devenyi AG, Mamourian AC. Symptomatic manganese neurotoxicity in a patient with chronic liver disease: correlation of clinical symptoms with MRI findings. *Pediatr Neurol* 1994; **10**: 145–8.

8. Mirowitz SA, Westrich TJ, Hirsch JD. Hyperintense basal ganglia on T1-weighted MR images in patients receiving parenteral nutrition. *Radiology* 1991; **181**: 117–20.
9. Reynolds AP, Kiely E, Meadows N. Manganese in long term paediatric parenteral nutrition. *Arch Dis Child* 1994; **71**: 527–8.
10. Nelson K, Golnick J, Korn T, Angle C. Manganese encephalopathy: utility of early magnetic resonance imaging. *Br J Ind Med* 1993; **50**: 510–13.
11. Calne DB, Chu WS, Huang CC, *et al.* Manganism and "idiopathic" parkinsonism; similarities and differences. *Neurology* 1994; **44**: 1583–6.
12. Eriksson H, Tedroff J, Thuomas KA, *et al.* Manganese induced brain lesions in *Macaca fascicularis* as revealed by positron emission tomography and magnetic resonance imaging. *Arch Toxicol* 1992; **66**: 403–7.

68. *Narrowing of the pharynx on head MRI or CT raises the possibility of obstructive sleep apnea, cocaine use, or HIV infection*

Thickening of the mucosa of the oropharynx and nasopharynx may occur following cocaine use or with HIV infection (Fig. 68.1). The changes with cocaine are probably due to direct irritation of the mucosa, in some cases resulting in a burn.[1] In HIV, primary infection of the pharyngeal lymphoid tissue is responsible for this change.[2–4] Because HIV and cocaine use are more common in young adults, thickened pharyngeal mucosa in this age group should raise the possibility of these conditions. A viral upper respiratory infection,

environmental allergens, tumors, and pharyngeal tissue overgrowth can produce similar changes.

Obstructive sleep apnea occurs in patients with small oropharynges. Excessive daytime somnolence is the primary clinical feature. Most patients are obese. Headaches, impaired intellectual performance, and cardiovascular dysfunction may develop. Increases in arterial blood pressure and pulmonary artery pressure, hypoxemia, and respiratory acidosis occur during the episodes of sleep apnea and contribute to the disorder's complications.[5]

In obstructive sleep apnea, the greatest pharyngeal narrowing is usually at the level of the uvula (soft palate) or at the base of the tongue. Imaging with CT, with sagittal or 3D reconstruction,[6] or MRI, with direct sagittal images or ultrafast techniques,[7] enables identification of the location and extent of the narrowing. Multiple areas of obstruction are commonly present, often at different locations in the awake and sleep states.[8] The minimal cross-sectional area of the oropharyngeal lumen correlates with blood oxygen saturation and the number and duration of apneic episodes during sleep.[9] The amount of adipose tissue adjacent to the pharyngeal airway, measurable on T1-weighted images, predicts the presence and degree of obstructive sleep apnea.[10]

Weight loss can dramatically reduce the volume of fat in the pharyngeal tissue and should be the initial treatment in obese patients with obstructive sleep apnea.[11] Other treatments include dental appliances, continuous positive airway pressure (CPAP) masks, and surgery. Nasal CPAP is remarkably effective in most individuals. However, the devices are often uncomfortable and compliance is a problem. Nasal CPAP is not a cure; even continuous use does not result in sustained improvement of the pharyngeal narrowing as measured by imaging.[12] Occasionally clinical improvement occurs when the patient sleeps with the head of the bed elevated (to 45°).

The standard surgical procedure for obstructive sleep apnea is a uvulopalatopharyngoplasty.[13] Usually only severely affected patients, who have failed other therapies, opt for surgery. Careful preoperative imaging of the pharynx can help to guide the resection. Follow-up imaging can be used to assess the success of surgery in relieving the obstruction.[6]

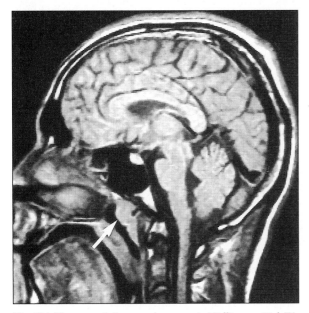

Fig. 68.1 Pharyngeal tissue enlargement. Midline sagittal T1-weighted image in a 43-year-old immunocompromised patient. Prominent adenoid tissue (white arrow) is present, suggesting infection or malignancy.

References

1. Snyderman C, Weissmann J, Tabor E, Curtin H. Crack cocaine burns of the larynx. *Arch Otolaryngol Head Neck Surg* 1991; **117**: 792–5.
2. Pantaleo G, Graziosi C, Demarest JF, *et al.* HIV infection is active and progressive in lymphoid tissue during the clinically latent stage of disease. *Nature* 1993; **362**: 355–8.
3. Burke AP, Benson W, Ribas JL, *et al.* Postmortem localization of HIV-1 RNA by in situ hybridization in lymphoid tissues of intravenous drug addicts who died unexpectedly. *Am J Pathol* 1993; **142**: 1701–13.
4. Barzan L, Carbone A, Tirelli U, *et al.* Nasopharyngeal lymphatic tissue in patients infected with human immunodeficiency virus: a prospective clinicopathologic study. *Arch Otolaryngol Head Neck Surg* 1990; **116**: 928–31.

5. Bradley TD, Brown LG, Grossman RF, *et al.* Pharyngeal size in snorers, known-snorers and patients with obstructive sleep apnea. *N Engl J Med* 1986; **315**: 1327–31.
6. Metes A, Hoffstein V, Direnfeld V, Chapnik JS, Zamel N. Three-dimensional CT reconstruction and volume measurements of the pharyngeal airway before and after maxillofacial surgery in obstructive sleep apnea. *J Otolaryngol* 1993; **22**: 261–4.
7. Shellock FG, Schatz CJ, Julien PM, *et al.* Dynamic study of the upper airway with ultrafast spoiled GRASS MR imaging. *J Magn Res Imaging* 1992; **2**: 103–7.
8. Suto Y, Matsuo T, Kato T, *et al.* Evaluation of the pharyngeal airway in patients with sleep apnea: value of ultrafast MR imaging. *AJR Am J Roentgenol* 1993; **160**: 311–14.
9. Avrahami E, Englender M. Relation between CT axial across the sectional area of the oropharynx and obstructive sleep apnea syndrome in adults. *AJR Am J Roentgenol* 1995; **16**: 135–40.
10. Shelton KE, Woodson H, Gay S, Suratt PM. Pharyngeal fat in obstructive sleep apnea. *Am Rev Resp Dis* 1993; **148**: 462–6.
11. Fleetham JA. Upper airway imaging in relation to obstructive sleep apnea. *Clin Chest Med* 1992; **13**: 399–416.
12. Collop NA, Block AJ, Hellard D. The effect of nightly nasal CPAP treatment on underlying obstructive sleep apnea and pharyngeal size. *Chest* 1991; **99**: 855–60.
13. Conway W, Fujita S, Zorik F, *et al.* Uvulopalatopharyngoplasty in treatment of upper airways sleep apnea. *Mam Rev Respir Dis* 1980; **121**: 121.

69. *Bilateral parieto-occipital edema is characteristic of hypertensive encephalopathy and may occur in eclampsia or cyclosporine toxicity*

Hypertensive crisis is present when a sustained dramatic elevation in blood pressure is associated with encephalopathy, retinal changes, and kidney dysfunction. The clinical presentation usually develops over several days. The first complaint is usually headache followed by drowsiness, confusion, and visual changes. Convulsive seizures and, more rarely, focal neurologic deficits may occur. Fundoscopic examination reveals retinal hemorrhages and disc edema. Proteinuria and hematuria are present on urinalysis.

The clinical features of a hypertensive crisis are non-specific and the diagnosis may be missed, especially when there is concurrent illness[1] or if a careful comparison between the presenting and premorbid blood pressure is not made. Typically the diastolic blood pressure exceeds 130 mmHg, but in patients with low baseline blood pressures (e.g. children and young pregnant women) a hypertensive crisis may occur at a much lower diastolic blood pressure. The differential diagnosis includes stroke, encephalitis, cerebral venous thrombosis, and a wide variety of other disorders. Death may occur if the syndrome is unrecognized or not treated appropriately. Resolution of the encephalopathy, once blood pressure control is achieved, helps to confirm the diagnosis. The neuroimaging findings are characteristic and may be the first indication of the diagnosis of hypertensive crisis.[2]

Bilaterally symmetric vasogenic edema in the parieto-occipital (watershed distribution) white matter is the most prominent feature (Fig. 69.1). CT reveals diffuse symmetric low density, and MRI reveals high T2 signal in these regions. Bilateral, but not necessarily symmetric, cerebellar, brain-stem, subcortical, deep-white-matter, and basal-ganglia T2 hyperintensities are often present. The lesions are isointense to hypointense on T1-weighted images. The parietal and occipital lesions are associated with visual disturbances. The basal-ganglia and deep-white-matter lesions correlate with the encephalopathy.[3] The edema, even when striking, may disappear within a few days of control of the blood pressure. However, lasting damage may remain. Serial MRI and repeated neurologic examination are necessary to distinguish transient edema from permanent zones of infarction.

Multifocal curvilinear high T1 and low T2 signal lesions at the gray–white matter junction represent petechial hemorrhages and are specific for hypertensive encephalopathy. These lesions are not seen on CT.[4] T2-weighted gradient-echo MRI is particularly valuable for visualizing petechial hemorrhages (punctate low-signal lesions). The presence of petechiae suggests an acute or prior episode of hypertensive encephalopathy and differentiates the lesions from nonspecific white-matter disease.[5]

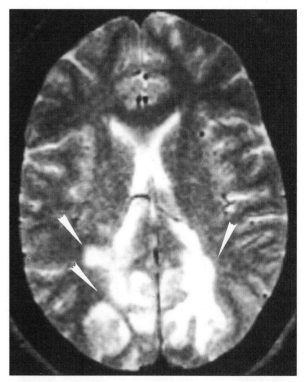

Fig. 69.1 Hypertensive encephalopathy. Axial T2-weighted image of a 23-year-old woman with hypertensive encephalopathy. Increased signal regions are present in the occipital and parietal lobes bilaterally (arrowheads). The abnormal signal completely resolved 5 days later after successful treatment of the hypertension (not shown).

The abnormalities on angiography may be misinterpreted as segmental narrowing with beading and sausage lesions. In fact, focal dilatations are present in the vascular regions where the autoregulation mechanism is overwhelmed by the elevated blood pressure. The narrow segments are the more normal portions of the vessels.[2] A similar, more rapidly developing phenomenon may occur in patients using amphetamines or cocaine and result in vessel rupture. Urine toxic screens should be performed in patients presenting in hypertensive crisis. A less well-developed autoregulation system in the vessels of the posterior circulation makes this territory particularly vulnerable to the effects of a hypertensive crisis.

Eclampsia is present when convulsions or coma develop in association with severe relative hypertension in the third trimester of pregnancy or the early postpartum period. It is usually preceded by preeclampsia consisting of edema, proteinuria, and hypertension.[6] Identical neuroimaging findings are present in eclampsia and hypertensive encephalopathy. The parieto-occipital edema and gray–white matter lesions are not present in preeclampsia although nonspecific foci of increased T2 signal in the deep white matter can occur.[4] Early postpartum patients may not meet the criteria for preeclampsia, but still develop neurologic and neuroimaging changes characteristic of eclampsia.[7]

Cyclosporine toxicity may result in an encephalopathy with clinical and neuroimaging features that are indistinguishable from those present in hypertensive encephalopathy. Many of these patients have relative hypertension. They are probably more susceptible to hypertensive encephalopathy because of partially impaired cerebral autoregulatory function from their underlying disease. Another nonhypertensive mechanism that is yet to be elucidated may also play a role. An increased frequency of seizures in patients taking cyclosporine, independent of blood pressure changes, is evidence that cyclosporine directly affects the brain. The encephalopathy reverses when the cyclosporine levels are decreased.[8–12] Other chemotherapy agents such as vincristine may also produce this condition.[13]

References

1. Healton EB, Brust JC, Feinfeld DA, Thomson GE. Hypertensive encephalopathy and the neurologic manifestations of malignant hypertension. *Neurology* 1982; **32**: 127–32.
2. Schwartz RB, Jones KM, Kalina P, *et al.* Hypertensive encephalopathy: findings on CT, MR imaging, and SPECT imaging in 14 cases. *AJR Am J Roentgenol* 1992; **159**: 379–83.
3. Sanders TG, Clayman DA, Sanchez-Ramos L, Vines FS, Russo L. Brain in eclampsia: MR imaging with clinical correlation. *Radiology* 1991; **180**: 475–8.
4. Digre KB, Varner MW, Osborn AG, Crawford S. Cranial magnetic resonance imaging in severe preeclampsia *vs* eclampsia. *Arch Neurol* 1993; **50**: 399–406.
5. Weingarten K, Barbut D, Filippi C, Zimmerman RD. Acute hypertensive encephalopathy: findings on spin-echo and gradient echo MRI imaging. *AJR Am J Roentgenol* 1994; **162**: 665–70.
6. Mantello MT, Schwartz RB, Jones KM, Ahn SS, Tice HM. Imaging of neurologic complications associated with pregnancy. *AJR Am J Roentgenol* 1993; **160**: 843–7.
7. Raps EC, Galetta SL, Broderick M, Atlas SW. Delayed peripartum vasculopathy: cerebral eclampsia revisited. *Ann Neurol* 1993; **33**: 222–5.
8. Shimizu C, Kimura S, Yoshida Y, *et al.* Acute leukoencephalopathy during cyclosporine A therapy in a patient with nephrotic syndrome. *Pediatr Nephrol* 1994; **8**: 483–5.
9. Mabin D, Fourquet I, Richard P, Esnault S, Islam MS, Bourbigot B. Regressive leukoencephalopathy induced by an overdose of cyclosporine A. *Rev Neurol* 1993; **149**: 576–8.
10. Monteiro L, Almeida-Pinto J, Rocha N, Lopes G, Rocha J. Case report: cyclosporine A-induced neurotoxicity. *Br J Radiol* 1993; **66**: 271–2.
11. McManus RP, O'Hair DP, Schweiger J, Beitzinger J, Siegel R. Cyclosporine-associated central neurotoxicity after heart transplantation. *Ann Thorac Surg* 1992; **53**: 326–7.
12. Reece DE, Frei-Lahr DA, Shepherd JD, *et al.* Neurologic complications in allogeneic bone marrow transplant patients receiving cyclosporine. *Bone Marrow Transplant* 1991; **8**: 393–401.
13. Pihko H, Tyni T, Virkola K, *et al.* Transient ischemic cerebral lesions during induction chemotherapy for acute lymphoblastic leukemia. *J Pediatr* 1993; **123**: 718–24.

11
Infections and Inflammatory Conditions

70. *Edema in the temporal lobe, with or without hemorrhage, in the setting of a progressive encephalopathy, suggests herpes simplex virus encephalitis. CSF analysis using the polymerase chain reaction (PCR) is the gold standard diagnostic test*

Herpes simplex virus type 1 (HSV-1) is a common cause of sporadic encephalitis. Diagnosis of this infection early in its course is critical because there is effective antiviral therapy.[1,2] Headaches, fever, seizures, and behavioral changes or confusion developing over hours to a few days should raise the suspicion of encephalitis. Focal neurologic deficits are characteristic of HSV encephalitis, but occur in only one-half of patients. The level of consciousness deteriorates to coma in days to weeks without treatment. Three-quarters of untreated patients die and most of the remainder are left with severe disability. Early treatment with acyclovir usually results in full recovery. Occasionally a chronic course is present with symptoms developing over months. Reactivation of latent infection in the trigeminal nerve ganglion or penetration of the virus through the olfactory nerve into the temporal lobe are postulated pathophysiologic mechanisms of infection.

Clinical suspicion of HSV encephalitis should result in a CSF examination, EEG, MRI, and empiric treatment with acyclovir. It is crucial to initiate treatment early to prevent hemorrhagic necrosis and resultant permanent neurologic deficits.

In practice, head CT is often the first study obtained. It may be normal during the first week of illness.[3,4] Indeed, normal head CT (or MRI, EEG, or CSF) early in the course does not preclude the diagnosis of HSV encephalitis. Later, CT may reveal low-density unilateral or asymmetric lesions, often with mass effect and hemorrhagic foci, within the temporal lobes (Fig. 70.1).

The MRI may be abnormal on the first day of the illness, or it may take several days to reveal pathologic changes. High signal is present on the T2- and PD-weighted images in the parahippocampal, fusiform,

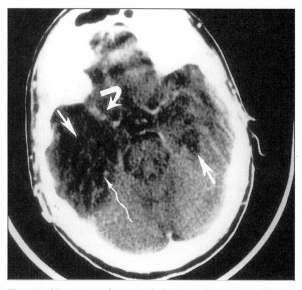

Fig. 70.1 Herpes simplex encephalitis. Axial contrast-enhanced CT of a middle-aged woman 4 days after the onset of a febrile illness and behavioral change. Asymmetric bitemporal low-density areas are present (straight arrows). The right middle cerebral artery is displaced anteriorly by mass effect from the swollen temporal lobe (curved arrow). The involved areas do not respect vascular territories. Minimal strands of enhancement are present (wavy arrow).

inferior, and middle temporal gyri and in the orbital frontal region.[5,6] The hippocampus and amygdala may also be affected. Subacute hemorrhage, commonly present in HSV encephalitis, is better visualized on MRI than on CT.[6]

Enhancement with intravenous iodine or gadolinium contrast occurs early in the course of HSV encephalitis. Patchy enhancement, sometimes following the gyral surface, is more common than ring enhancement. In the chronic stage of the disease, both CT and MRI reveal brain atrophy.[7]

Typically, a predominantly lymphocytic pleocytosis with 10–500 cells is present in the CSF. Red blood cells are frequently but not always present. A modestly elevated protein and opening pressure and normal glucose are usually found. However, the CSF may have normal parameters early in the illness. The EEG initially displays focal slowing over the involved temporal lobe, which develops into an encephalopathic pattern with diffuse slowing and disorganization. Periodic lateralized epileptiform discharges (PLEDs) with a periodicity of 1–4 seconds are a classic, but nonspecific, EEG feature, and may appear as early as the second day or as late as 1 month after onset. Single photon emission computerized tomography (SPECT) may reveal hypermetabolic function in the affected temporal lobe.[8]

Detection of HSV DNA in the CSF using the polymerase chain reaction (PCR) is now the reference standard for diagnosis of herpes simplex encephalitis.[9,10] The PCR is often the first test to become positive sometimes as early as the first day of symptoms.[11] Brain biopsy, formerly the diagnostic procedure of choice, is now only rarely needed. HSV PCR is the most sensitive and specific test and has proved that brain biopsy can be falsely negative.[9] The possibility of contamination of CSF specimens is a potential drawback with this technique, making high-quality laboratories a necessity.

HSV may persist for years within the central nervous system after acute herpes encephalitis, causing either a latent or a low-grade productive infection.[12] The predilection for HSV to cause latent nervous system infection raises the possibility that detection of HSV with PCR of the CSF may not indicate acute infection, especially in atypical cases. Confirmation of an active infection can be made with simultaneous CSF and serum serologic titers, which become positive 3–10 days after the onset of neurologic symptoms.[13,14]

Atypical presentations of HSV encephalitis are probably relatively frequent. In preschool-age children, HSV encephalitis may present with multifocal or diffuse brain involvement.[15] Mesenrhombencephalitis may be caused by HSV, *Listeria monocytogenes*, and probably several other organisms. In this condition, progressive encephalopathy is accompanied by cranial nerve palsies, and motor and sensory dysfunction. MRI reveals T2 hyperintensities and T1 hypointensities in the midbrain, pons, medulla, cerebellum, thalamus, internal capsule, and temporal lobe. These lesions enhance heterogeneously. Focal parenchymal hemorrhage and leptomeningeal enhancement may occur in the temporal lobes. Brain-stem enlargement due to swelling may be present.[16] With widespread use of PCR, numerous other previously unsuspected presentations of HSV encephalitis are likely to be recognized.

References

1. Koskiniemi ML, Vaheri A, Manninen V, *et al*. Herpes simplex virus encephalitis: new diagnostic and clinical features and results of therapy. *Arch Neurol* 1980; **37**: 763–7.
2. Fishman RA. Brain biopsy need not be done in every patient suspected of having herpes simplex encephalitis. *Arch Neurol* 1987; **44**: 1291–2.
3. Cameron PD, Wallace SJ, Munro J. Herpes simplex virus encephalitis: problems in diagnosis. *Dev Med Child Neurol* 1992; **34**: 134–40.
4. Gasecki AP, Steg RE. Correlation of early MRI with CT scan, EEG, and CSF: analyses in a case of biopsy-proven herpes simplex encephalitis. *Eur Neurol* 1991; **31**: 372–5.
5. Schroth G, Gawehn J, Thron A, *et al*. Early diagnosis of herpes simplex encephalitis by MRI. *Neurology* 1987; **37**: 179–83.
6. Demaerel P, Wilms DG, Robberecht W, *et al*. MRI of herpes simplex encephalitis. *Neuroradiology* 1992; **34**: 490–3.
7. Koskiniemi ML, Ketonen L. Herpes simplex virus encephalitis: progression of lesion shown by CT. *J Neurol* 1981; **225**: 9–13.
8. Schmidbauer M, Podreka I, Wimberger D, *et al*. SPECT and MR imaging in herpes simplex encephalitis. *J Comput Assist Tomogr* 1991; **15**: 811–15.
9. Lakeman FD, Whitley RJ. Diagnosis of herpes simplex encephalitis: application of polymerase chain reaction to cerebrospinal fluid from brain-biopsied patients and correlation with disease. National Institute of Allergy and Infectious Diseases Collaborative Antiviral Study Group. *J Infect Dis* 1995; **171**: 857–63.
10. Mertens G, Leven M, Ursi D, Pattyn SR, Martin JJ, Parizel PM. Detection of herpes simplex virus in the cerebrospinal fluid of patients with encephalitis using the polymerase chain reaction. *J Neurol Sci* 1993; **118**: 213–16.
11. Guffond T, Dewilde A, Lobert PE, Caparros-Lefebvre D, Hober D, Wattre P. Significance and clinical relevance of the detection of herpes simplex virus DNA by the polymerase chain reaction in cerebrospinal fluid from patients with presumed encephalitis. *Clin Infect Dis* 1994; **18**: 744–9.
12. Nicoll JA, Maitland NJ, Love S. Autopsy neuropathological findings in "burnt out" herpes simplex encephalitis and use of the polymerase chain reaction to detect viral DNA. *Neuropathol Appl Neurobiol* 1991; **17**: 375–82.
13. Skoldenberg B. Herpes simplex encephalitis. *Scand J Infect Dis Suppl* 1991; **80**: 40–6.
14. Uren EC, Johnson PD, Montanaro J, Gilbert GL. Herpes simplex virus encephalitis in pediatrics: diagnosis by detection of antibodies and DNA in cerebrospinal fluid. *Pediatr Infect Dis J* 1993; **12**: 1001–6.
15. Schlesinger Y, Buller RS, Brunstrom JE, Moran CJ, Storch GA. Expanded spectrum of herpes simplex encephalitis in childhood. *J Pediatr* 1995; **126**: 234–41.
16. Soo MS, Tien RD, Gray L, Andrews PI, Friedman H. Mesenrhombencephalitis: MR findings in nine patients. *AJR Am J Roentgenol* 1993; **160**: 1089–93.

71. *A ring lesion may be due to an abscess, tumor, or resolving stroke*

The principal brain lesions that enhance with contrast to form a ring are abscesses, tumors, and resolving strokes (either ischemic stroke or intracerebral hemorrhage) (Fig. 71.1). Less commonly, other lesions may have a ring-enhancing appearance.[1–3] Although there are certain neuroimaging characteristics that suggest the cause (detailed below), the clinical history or pathologic findings are needed for definitive diagnosis.

Brain abscesses present with fever, headache, and focal neurologic dysfunction. The following conditions are associated with brain abscesses:

1. *Extracerebral infectious focus*
   - Endocarditis
   - Nasal sinusitis
   - Otitis media
   - Mastoiditis
   - Pulmonary infection (lung abscess, bronchiectasis)

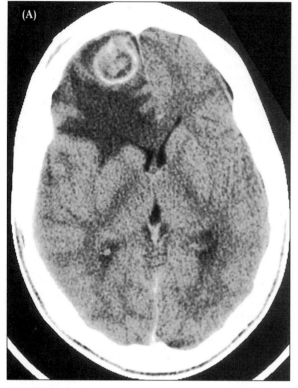

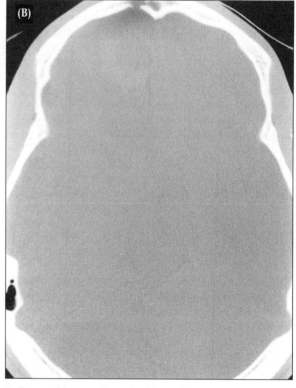

A: Axial contrast-enhanced CT at a soft-tissue window setting. Typical ring-enhancing lesion with surrounding edema and mass effect is present in the right frontal lobe. In this case the lesion is a pyogenic abscess but other processes, such as a tumor, can have an identical appearance (compare with Figs. 35.2 and 38.1).

Fig. 71.1 Ring-enhancing lesion.

B: Same subject at a bone (wide) window setting. Destruction of the posterior wall of the frontal sinus is evident. Frontal sinusitis was present at another level (not shown). Thus, complete examination of the CT suggested the diagnosis and mechanism of illness in this 23-year-old with fever, chills, and severe headache.

- Skull osteomyelitis
- Skull fracture (penetrating trauma)
- Neurosurgical procedure
- Meningitis (rare except in infants)
2. *Arterial-to-venous shunt*
   - Congenital heart disease (right-to-left shunt)
   - Patent foramen ovale
   - Systemic (especially pulmonary) arteriovenous malformation (e.g. commonly present in hereditary hemorrhagic telangiectasia)
3. *Immunosuppression*
   - AIDS
   - Chronic steroid use
   - Chemotherapy
   - Cancer (e.g. chronic lymphogenous leukemia).

The organisms most frequently identified in brain abscesses are:

1. *Bacteria*
   - *Staphylococcus aureus*
   - *Streptococcus* species
   - Anaerobes (e.g. *Bacteroides fragilis*)
   - Gram-negatives (e.g. *Proteus* species and *Escherichia coli*)
   - *Nocardia asteroides*
   - *Actinomyces*

2. *Fungi*
   - *Aspergillus*
   - *Candida*
   - *Mucor*
3. *Parasites*
   - *Entamoeba histolytica*
   - *Toxoplasma gondii.*

In brain abscess, the enhancing ring on CT and MRI tends to have a thin wall (less than 5 mm) with a thicker rim on the side closest to the brain surface. Conventional contrast angiography typically reveals a vague blush within the capsule. A radioisotope uptake study using indium-111 labeled leukocytes or ^{99m}Tc-HMPAO classically demonstrates increased activity within an abscess but not within a tumor or stroke.[4,5] Abscesses in patients on steroids may have normal activity on radioisotope studies and may not demonstrate ring enhancement on structural imaging.[6] Cerebritis precedes abscess formation, although it is infrequently imaged.[7,8] Surgical drainage is required to treat large abscesses, whereas small abscesses may respond to antimicrobial therapy alone.[9]

An abscess involving the brain surface is known as a subdural empyema or an epidural abscess (sometimes termed epidural empyema) depending on the location. These infections are most commonly associated with trauma, although they may result from spread from contiguous infections including sinusitis and meningitis (maxim 74).[10,11] MRI with contrast is the most sensitive study as CT may be falsely negative, especially early in this condition.[12–14] Subdural empyemas are characterized by a thick, enhancing adjacent dura mater, in contrast to an epidural abscess where a portion of the displaced dura mater may not enhance on MRI.[15] Both appear as extra-axial fluid collections. Early surgical drainage and antibiotics are usually required to avoid significant morbidity and death.

Metastatic brain tumors usually occur late in the course of a cancer. Certain tumors have a predilection to metastasize to the brain (maxim 35). The brain metastasis may be clinically silent or present with a focal neurologic deficit or seizure. CT and MRI may reveal either single or multiple ring-enhancing lesions (Fig. 35.2). Metastases tend to be located within the cerebral or cerebellar hemispheres at the gray–white matter junction. However, they can occur at any site within the central nervous system and should always be in the differential diagnosis of a mass lesion. Considerable edema is usually present surrounding a metastasis even when the underlying tumor focus is small. Conventional contrast angiography commonly reveals a tumor blush (stain) or organized neovascularity. A single feeding artery and drainage to a single cortical vein may be present.

Anaplastic astrocytomas and glioblastomas frequently present with progressive neurologic dysfunction developing over weeks or months. CT and MRI usually reveal an irregular enhancing ring. Glioblastomas have a central nonenhancing area (necrosis) that is low density on CT and low signal on T1-weighted MRI (Fig. 36.2). Most glial tumors are located within the white matter and are solitary, although glioblastomas are occasionally multifocal. Conventional contrast angiography typically demonstrates irregular neovascularity, early draining veins, and deep venous drainage.

Resolving intracerebral hematomas are usually preceded by an abrupt onset of headache and neurologic deficits at the time of the initial intracranial bleed. Ischemic strokes also have an abrupt onset. Transient ring enhancement is present between 1 and 4 weeks after the initial event in both of these conditions. During this interval, the ring-enhancing lesion may be mistaken for a brain abscess or tumor, especially when the clinical history is not clear.[16] The core of a resolving hematoma usually has a higher density than normal brain parenchyma on CT. T1-weighted MRI is the best study for differentiating subacute blood; the center of a resolving hematoma has high T1 signal due to methemoglobin (maxim 60). The possibility of an underlying tumor or abscess should always be considered when an intracerebral hemorrhage is present. Follow-up CT or MRI, several weeks to months after the initial event, is useful for determining whether an underlying lesion is present. The evolution of ischemic stroke on neuroimaging is discussed in maxim 48 and of intracranial hemorrhage in maxim 87.

References

1. Salzman C, Tuazon CU. Value of the ring-enhancing sign in differentiating intracerebral hematomas and brain abscesses. *Arch Intern Med* 1987; **147**: 951–2.
2. Dobkin JF, Healton EB, Dickinson PC, Brust JC. Non-specificity of ring enhancement in "medically cured" brain abscess. *Neurology* 1984; **34**: 139–44.
3. Poskitt KJ, Steinbok P, Flodmark O. Methotrexate leukoencephalopathy mimicking cerebral abscess on CT brain scan. *Childs Nerv Syst* 1988; **4**: 119–21.
4. Grimstad IA, Hirschberg H, Rootwelt K. 99mTc-hexamethylpropyleneamine oxime leukocyte scintigraphy and C-reactive protein levels in the differential diagnosis of brain abscesses. *J Neurosurg* 1992; **77**: 732–6.
5. Schmidt KG, Rasmussen JW, Frederiksen PB, Kock-Jensen C, Pedersen NT. Indium-111-granulocyte scintigraphy in brain abscess diagnosis: limitations and pitfalls. *J Nucl Med* 1990; **31**: 1121–7.
6. Black KL, Farhat SM. Cerebral abscess: loss of computed tomographic enhancement with steroids. Case report. *Neurosurgery* 1984; **14**: 215–17.
7. Hatta S, Mochizuki H, Kuru Y, *et al.* Serial neuroradiological studies in focal cerebritis. *Neuroradiology* 1994; **36**: 285–8.
8. Chee CP, Coltheart GJ. Cerebritis preceding cerebral abscess formation: a report of three cases. *Aust N Z J Surg* 1986; **56**: 657–60.
9. Yang SY, Zhao CS. Review of 140 patients with brain abscess. *Surg Neurol* 1993; **39**: 290–6.
10. Dill SR, Cobbs CG, McDonald CK. Subdural empyema: analysis of 32 cases and review. *Clin Infect Dis* 1995; **20**: 372–86.
11. Brennan MR. Subdural empyema. *Am Fam Physician* 1995; **1**: 157–62.
12. Ogilvy CS, Chapman PH, McGrail K. Subdural empyema complicating bacterial meningitis in a child: enhancement of membranes with gadolinium on magnetic resonance imaging in a patient without enhancement on computed tomography. *Surg Neurol* 1992; **37**: 138–41.
13. Hlavin ML, Kaminski HJ, Fenstermaker RA, White RJ. Intracranial suppuration: a modern decade of postoperative subdural empyema and epidural abscess. *Neurosurgery* 1994; **34**: 974–80.
14. McIntyre PB, Lavercombe PS, Kemp RJ, McCormack JG. Subdural and epidural empyema: diagnostic and therapeutic problems. *Med J Aust* 1991; **154**: 653–7.
15. Tsuchiya K, Makita K, Furui S, Kusano S, Inoue Y. Contrast-enhanced magnetic resonance imaging of sub- and epidural empyemas. *Neuroradiology* 1992; **34**: 494–6.
16. Senaati S, Caner H, Oran BM. Resolving cerebral hematoma mimicking cerebral abscess. *AJR Am J Roentgenol* 1992; **159**: 903 (letter).

72. *Thallium-201 SPECT may differentiate cerebral lymphoma from toxoplasmosis encephalitis in patients with AIDS*

Over two million people in the USA and over 16 million people worldwide were infected with HIV as of 1994.[1] Almost all will develop AIDS until an effective treatment is discovered. Two common CNS complications of AIDS are lymphoma and toxoplasma infection. In up to 80% of cases, they are indistinguishable on standard neuroimaging studies. However, as they respond to different treatments, differentiating between the two is essential. Although assessment of the response to empiric treatment for toxoplasmosis is the common initial approach (Fig. 72.1), it is ideal to make the correct diagnosis noninvasively at the outset.

Toxoplasmosis is one of the most common CNS opportunistic infections, responsible for over one-third of neurologically symptomatic illness in AIDS patients. Primary CNS lymphoma accounts for 2–6% and metastatic lymphoma (often involving the meninges and epidural space) is present in 1–5% of these patients. Serological studies are not diagnostic because a positive toxoplasmosis IgG antibody titer occurs in both past and recent infections and does not exclude a second CNS process.

CNS lymphoma and toxoplasmosis both appear as single or multiple focal ring or nodular enhancing mass lesions on standard CT and MRI.[2–4] They are usually located at the cerebral corticomedullary junction or in the region of the basal ganglia. These diseases also occur in the brain stem and cerebellum.

Certain features are more characteristic of lymphomas. On MRI, lymphomas have a rim that is isointense to gray matter on T2-weighted images with a slightly hyperintense core, even when necrosis is present (Fig. 72.2). Lesions that involve the corpus callosum, subependymal region, or periventricular region or have thick, nodular, irregular enhancement are more likely to be lymphomas. Rapid lesion progression is more typical of lymphoma. Lymphomas generally demonstrate significantly greater enhancement than does toxoplasmosis on dynamic contrast enhanced MRI.[5] Toxoplasmosis lesions tend to have more edema. Despite these imaging clues, lymphoma, toxoplasmosis, and other infections (e.g. *Candida* and *Nocardia*) cannot be reliably differentiated by CT or MRI.

Thallium-201 single photon emission computerized tomography (SPECT) has uptake in CNS lymphomas but not in infections (Fig. 72.3).[6–8] An added benefit is that steroids probably do not affect the results of thallium-201 SPECT in lymphoma whereas it may change the appearance on CT (maxim 38).[9] The combination of thallium-201 SPECT, serologic testing, and contrasted MRI eliminates the need for brain biopsy in most cases.

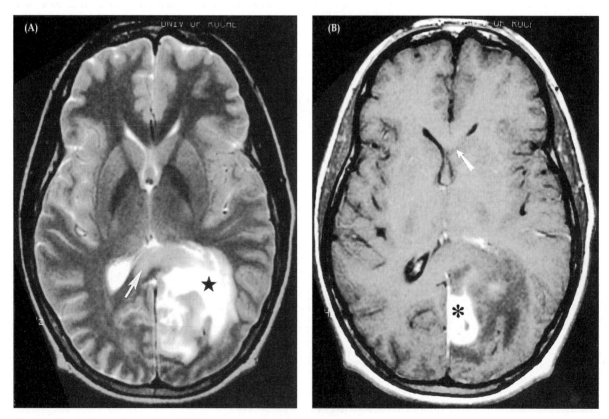

Fig. 72.1 Cerebral toxoplasmosis in AIDS. Remainder of caption is given opposite.

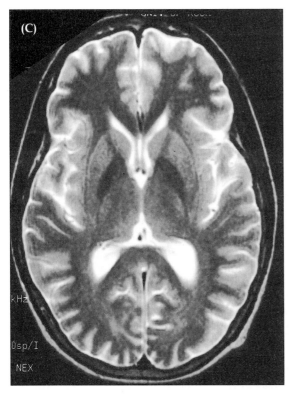

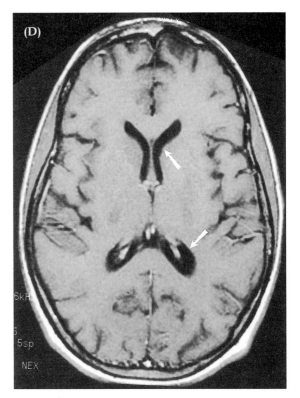

Fig. 72.1 Cerebral toxoplasmosis in AIDS. Cerebral toxoplasmosis can be diagnosed when empiric antitoxoplasmosis treatment results in resolution of the cerebral lesions (Figs **A** and **B** are on the previous page). **A:** Axial T2-weighted image before treatment demonstrates a large zone of edema (black star) and significant mass effect in the left occipital region extending to the splenium of the corpus callosum (straight white arrow). **B:** Axial T1-weighted contrast-enhanced image through the same area. There is striking enhancement in a portion of the lesion. The anterior horn of the lateral ventricle is partially effaced (straight white arrow), a distant effect of the mass. **C:** Axial T2-weighted image after 10 days of antitoxoplasmosis treatment. The lesion has completely resolved. **D:** Axial T1-weighted contrast-enhanced image through the same area. Abnormal enhancement is no longer present. The anterior horn and atrium of the lateral ventricle are now normal (straight white arrows).

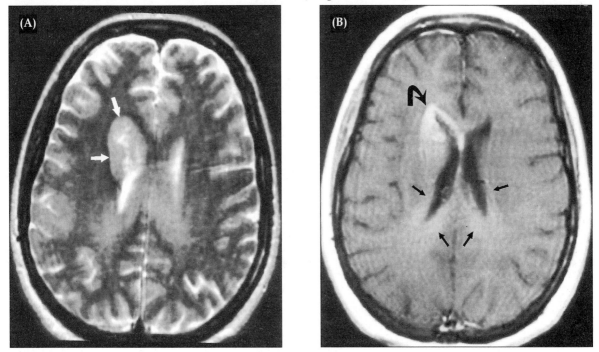

Fig. 72.2 Toxoplasmosis *versus* lymphoma in AIDS. **A:** Axial T2-weighted image in a patient with AIDS. A lesion with signal iso-intense to gray matter is present adjacent to the right anterior horn of the lateral ventricle (straight white arrows). **B:** Axial contrast-enhanced T1-weighted image at the same level. The portion of the lesion cuffing the right anterior horn exhibits striking enhancement (curved black arrow). The posterior periventricular region and splenium of the corpus callosum (straight black arrows) slightly enhance, indicating a more extensive process. These lesions are located in the subependymal area and are typical of lymphoma.

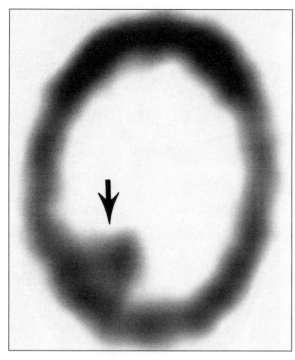

Fig. 72.3 Thallium-201 brain SPECT study in a young man with AIDS and an enhancing mass on CT consistent with toxoplasmosis or lymphoma (not shown). There is focal increased uptake in the right parieto-occipital region (arrow) consistent with lymphoma. (Courtesy of Dr J. Villanueva-Mayer, University of Texas Medical Branch at Galveston)

MRI proton spectroscopy appears to differentiate CNS lymphoma from toxoplasmosis accurately and may become the standard imaging for this purpose in the future.[10–12]

References

1. World Health Organization. WHOSIS world wide web site. http://www.who.ch/whosis/gpa/gpa/htm, 1994.
2. Kupfer MC, Zee CS, Colletti PM, Boswell WD, Rhodes R. MRI evaluation of AIDS-related encephalopathy: toxoplasmosis vs. lymphoma. *Magn Reson Imaging* 1990; **8**: 51–7.
3. Ketonen L, Tuite M. Brain imaging in AIDS infection. *Semin Neurol* 1991; **12**: 57–69.
4. Tuite M, Ketonen L, Kieburtz K. The value of gadolinium in the MRI imaging of the HIV infected patient. *AJNR Am J Neuroradiol* 1992; **14**: 257–63.
5. Laissy JP, Soyer P, Tebboune J, *et al.* Contrast enhanced fast MRI in differentiating brain toxoplasmosis and lymphoma in AIDS patients. *J Comput Assist Tomogr* 1994; **18**: 714–18.
6. Ruiz A, Ganz WI, Post MJ, *et al.* Use of thallium-201 brain SPECT to differentiate cerebral lymphoma from toxoplasma encephalitis in AIDS patients. *AJNR Am J Neuroradiol* 1994; **15**: 1885–94.
7. Ganz WI, Serafini AN. The diagnostic role of nuclear medicine in the acquired immunodeficiency syndrome. *J Nucl Med* 1989; **30**: 1935–45.
8. Vanarthos WJ, Ganz WI, Vanarthos JC, *et al.* Diagnostic uses of nuclear medicine in AIDS. *Radiographics* 1992; **12**: 731–49.
9. Borggreve F, Dierckx RA, Crols R, *et al.* Repeat thallium-201 SPECT in cerebral lymphoma. *Funct Neurol* 1993; **8**: 95–101.
10. Jarvik JG, Lenkinski RE, Grossman RI, *et al.* Proton MR spectroscopy of HIV-infected patients: characterization of abnormalities with imaging and clinical correlation. *Radiology* 1993; **186**: 739–44.
11. Buchthal SD, Chang L, Miller BL, *et al.* Differentiation of brain tumors in AIDS by ^{1}H magnetic resonance spectroscopy. *Proc Soc Magn Reson* 1994; **3**: 1292.
12. Yamagata NT, Miller BL, McBride D, *et al.* In vivo proton spectroscopy of intracranial infections and neoplasms. *J Neuroimaging* 1994; **4**: 23–9.

73. Abnormal high signal on PD-weighted MRI in the fornix–subcallosal area or diffuse atrophy in a young patient suggests AIDS

Pathologic abnormalities are seen in the brains of more than three-quarters of patients who die with AIDS. Neurologic signs and symptoms are the first manifestation of AIDS in 10% of patients.[1] The most common neurologic complication of HIV infection is AIDS dementia complex (ADC, HIV-associated cognitive/motor complex).[2] Affected patients develop progressive intellectual, behavioral, and motor dysfunction and may present only when they can no longer perform activities of daily living. The neuroimaging features of ADC are

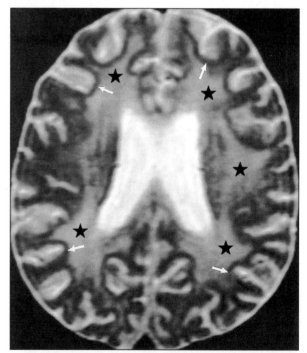

Fig. 73.1 AIDS dementia complex (ADC): atrophy and HIV encephalopathy. Axial T2-weighted image at the level of the bodies of the lateral ventricles. Extensive abnormal high-signal areas are present in the deep white matter (black stars). The hyperintensities spare the subcortical U-fibers (white arrows). Moderate atrophy manifested by enlarged ventricles at this level is also present. This pattern is compatible with HIV-related encephalopathy.

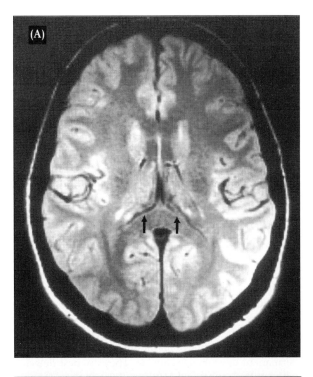

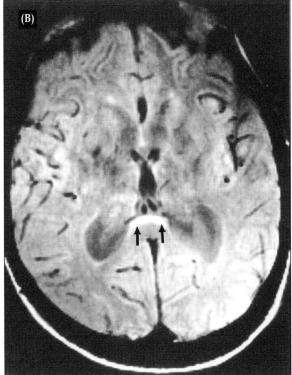

generalized white-matter disease, brain atrophy, and fornix–subcallosal high signal on MRI (Figs 73.1 and 73.2).[3–6] Early presymptomatic HIV changes in the brain are probably not detectable with neuroimaging. The mechanism of this syndrome, which is unique to patients with AIDS, is not fully understood.

Abnormal high signal on PD-weighted images in the fornix–subcallosal region is present in most patients with HIV infection with cognitive impairment (Fig. 73.2).[5,6] Sometimes the abnormal signal involves the splenium of the corpus callosum. Although this finding is sensitive for AIDS, it is not specific as it is also present in patients with vascular disease.

The brain atrophy in patients with ADC is diffuse and involves both deep and superficial brain structures. It is the most common neuroimaging finding in ADC and is associated with the presence of dementia. The CT and MRI findings consist of enlargement of both the subarachnoid spaces and the ventricles. Surprisingly, some patients with severe HIV encephalitis at autopsy do not have atrophy on premorbid neuroimaging studies.[7]

The deep-white-matter abnormalities do not differentiate patients with dementia from those without dementia.[6] The white-matter abnormalities are greater in the deep frontal white matter but often the entire centrum semiovale is involved. Although low-density regions may be present on CT, the T2- and PD-weighted MR images are far superior to CT for detecting the white-matter abnormalities.

References

1. Levy RM, Bredesen DE, Rosenblum ML. Neurological manifestations of the acquired immunodeficiency syndrome (AIDS): experience at UCSF and a review of the literature. *Neurosurgery* 1985; **62**: 475–95.
2. Navia BA, Price RW. The acquired immunodeficiency syndrome dementia complex as the presenting sole manifestation of human immunodeficiency virus infection. *Arch Neurol* 1987; **44**: 65–9.
3. Ekholm S, Simon JH. MRI and the AIDS dementia complex. *Acta Radiol* 1988; **29**: 227–30.
4. Grant I, Atkinson JH, Hesselink JR, *et al*. Evidence for early central nervous system involvement in the acquired immunodeficiency syndrome (AIDS) and other human immunodeficiency virus (HIV) infections: studies with neuropsychologic testing and magnetic resonance imaging. *Ann Intern Med* 1987; **107**: 826–36.
5. Kieburtz KD, Ketonen L, Zettelmaier AE, *et al*. Magnetic resonance imaging findings in HIV cognitive impairment. *Arch Neurol* 1990; **47**: 643–5.
6. Broderick DT, Wippold FJ, Clifford DB, *et al*. White matter lesions and cerebral atrophy on MR images in patients with and without AIDS dementia complex. *AJR Am J Roentgenol* 1993; **161**: 177–81.
7. Navia B, Cho E-S, Petito C, *et al*. The AIDS dementia complex: II. Neuropathology. *Ann Neurol* 1986: **19**; 525–35.

Fig. 73.2 Fornix–subcallosal high signal in AIDS dementia complex. **A:** Axial PD-weighted image through the corpus callosum of a patient with AIDS but normal intellectual function. The fornix–subcallosal region is normal (arrows). **B:** Axial PD-weighted image of a 32-year-old patient with AIDS and progressive intellectual impairment. Abnormal high signal is present in the fornix–subcallosal region (arrows). This finding is strongly correlated with ADC, although it may also occur in patients with vascular disease.

74. *Some of the complications of acute bacterial meningitis can be visualized with neuroimaging*

Bacterial meningitis presents with fever, headache, nuchal rigidity, nausea, vomiting, and sometimes decline in the level of consciousness, seizures, or other focal neurologic deficits. The diagnosis of bacterial meningitis is made by laboratory examination of the CSF. In adults and older children, neuroimaging, usually CT, should be obtained before lumbar puncture to exclude an unexpected process (such as a large abscess) that might preclude a spinal tap. (Some practicing neurologists debate the need for a scan before spinal tap if papilledema is absent, but we firmly believe in neuroimaging before spinal tap in any patient with a potential intracranial mass lesion; Fig. 85.2.) In patients with possible bacterial meningitis, it is prudent to administer a single dose of an antibiotic (commonly a third-generation cephalosporin) while awaiting the scan and CSF analysis. The most common organisms that cause bacterial meningitis in children and adults are *Haemophilus influenzae*, *Neisseria meningitidis*, *Streptococcus pneumoniae*, and *Listeria monocytogenes*.[1] In neonates, *Escherichia coli* and group B streptococci account for most cases.

The initial CT is usually normal in bacterial meningitis. Meningeal enhancement may be present on MRI and occasionally on CT (maxim 15). In some patients, the basal cisterns are obliterated by an accumulation of exudate. Later in the course, complications of bacterial meningitis may occur. These include hydrocephalus, subdural effusion, subdural empyema, venous sinus thrombosis, cerebral infarction, parenchymal abscess, ependymitis, and cerebritis.[2,3] Neuroimaging, after initial head CT, should be reserved for patients who develop progressive neurologic dysfunction despite treatment with antibiotics.[4] MRI with contrast is superior to CT in this situation.

Hydrocephalus associated with bacterial meningitis may be either communicating or obstructive. Ventricular shunting is sometimes indicated. MRI is more sensitive for the presence of transependymal CSF resorption in hydrocephalus, although ventricular size assessment may be made equally well with CT and MRI. Normal-pressure hydrocephalus is a late complication of meningitis in some adults (maxim 65).

Subdural CSF effusions (hygromas) may occur in infants with meningitis due to any organism but are most common with *Haemophilus influenzae*. Bulging fontanelles and persistent neurologic signs should raise the suspicion of this condition. Collections of CSF in the subdural spaces are present on neuroimaging. Subdural percutaneous taps may be needed, but these fluid collections often resolve spontaneously.

Septic cortical venous thrombophlebitis is a rare autopsy finding and probably not a major cause of focal neurologic dysfunction in patients with meningitis.[5] Aseptic cortical venous thrombosis is probably more common. Both result in seizures and other focal neurologic signs. The underlying brain may have edema, hemorrhage, and infarction (maxim 54).

Subdural empyemas are delineated by an inflamed membrane, which almost invariably enhances with contrast (maxim 71). In contrast to effusions, empyemas have an elevated protein content which is reflected in their neuroimaging appearance. When they complicate bacterial meningitis, they are frequently located in the interhemispheric fissure. Empyemas are better visualized on MRI than on CT, especially as the typical enhancement may not be present on CT.[6]

Meningitis can spread to involve the ventricular surface (ependymitis) or brain substance (cerebritis), which may lead to abscess formation. Cerebritis appears as a hypodense area on CT and a high T2 signal region with patchy enhancement on MRI. Cerebritis may cause vasculitis and hemorrhagic infarctions. CT is preferred to determine whether acute hemorrhage is present. Cerebritis, ependymitis, and parenchymal abscess are serious complications of meningitis, which may result in permanent neurologic sequelae. They are best detected with contrast-enhanced MRI.

References

1. Schlech WF, Ward JI, Band JD, *et al.* Bacterial meningitis in the United States, 1978 through 1981: the national bacterial meningitis surveillance study. *JAMA* 1985; **253**: 1749–54.
2. Phillips ME, Ryals TJ, Kambhu SA, *et al.* Neoplastic *vs.* inflammatory meningeal enhancement with Gd-DTPA. *J Comput Assist Tomogr* 1990; **14**: 536–41.
3. Sze G, Zimmerman RD. The magnetic resonance imaging of infections and inflammatory diseases. *Radiol Clin North Am* 1988; **26**: 839–59.
4. Heyderman RS, Robb SA, Kendall BE, Levin M. Does computed tomography have a role in the evaluation of complicated acute bacterial meningitis in childhood? *Dev Med Child Neurol* 1992; **34**: 870–5.
5. DiNubile MJ, Boom WH, Southwick FS. Septic cortical thrombophlebitis. *J Infect Dis* 1990; **161**: 1216–20.
6. Ogilvy CS, Chapman PH, McGrail K. Subdural empyema complicating bacterial meningitis in a child: enhancement of membranes with gadolinium on magnetic resonance imaging in a patient without enhancement on computed tomography. *Surg Neurol* 1992; **37**: 138–41.

75. *Neurosarcoidosis mimics tumor and many other diseases*

Certain diseases, including tuberculosis, syphilis, and most recently HIV, are known as great imitators because they present with protean signs and symptoms and are prone to initial misdiagnosis. Neurosarcoidosis is one of the great imitators in neurologic disease. The differential diagnosis for most intracranial lesions should include sarcoidosis.

Sarcoidosis is a systemic disorder of unknown cause, characterized by noncaseating granulomas usually involving multiple organs. An infectious agent is suspected, but no pathogen has yet been identified. Adult onset is typical, but it may develop in childhood or the elderly. Approximately 5% of patients with sarcoidosis have CNS involvement.[1] It is frequently multifocal. Relapses occur in one-third of patients with neurosarcoidosis.[2] When neurologic dysfunction occurs in a patient with known systemic sarcoidosis, neuroimaging, especially MRI, often reveals abnormalities consistent with the diagnosis of neurosarcoidosis. However, when the presentation involves the CNS without systemic features, the initial differential diagnosis is broad. Isolated CNS sarcoid is rare but does occur.[3,4] Biopsy is usually required for diagnosis.

Neurosarcoidosis usually involves the leptomeninges, either diffusely or focally. Chronic basilar meningitis is typical. Cranial nerve palsies, meningeal signs, and hypothalamic dysfunction with diabetes insipidus are the most frequent clinical CNS features.[5] Brain or spinal cord parenchymal involvement by granulomatous lesions may occur in isolation or with leptomeningeal disease. In addition to the meninges, the regions most frequently affected by sarcoid are the pituitary, hypothalamus, periventricular white matter, optic chiasm, and ependyma. Seizures are associated with more severe and progressive or relapsing forms of neurosarcoidosis and may be an early manifestation.[6] Many uncommon presentations are reported.[7–10]

The MRI findings in neurosarcoidosis include white-matter and periventricular T2 high signal (mimicking multiple sclerosis), leptomeningeal enhancement, enhancing brain parenchymal lesions, lacrimal gland mass, hydrocephalus, periventricular enhancement, extra-axial mass (mimicking meningioma), chiasmal enhancement or swelling, enhancing nerve roots, enlarged pituitary stalk, and enhancing parenchymal spinal cord masses (Fig. 75.1).[11,12] Gadolinium enhancement often demonstrates lesions that are not visible on conventional T1- and T2-weighted images.[13,14] Enhancement is sometimes present on CT, but CT is significantly less sensitive than MRI and may be falsely negative.[15,16] In patients with known or suspected sarcoidosis, contrast-enhanced MRI is required.[17]

Parenchymal lesions, the result of spread along the perivascular spaces, contrast-enhance on MRI. The parenchymal lesions may have associated mass effect and vasogenic edema. However, sometimes large lesions have only minimal mass effect.[18] In many cases sarcoid lesions are indistinguishable from tumors.[19–22]

Diffuse leptomeningeal disease is a nonspecific finding (maxim 15). Tuberculosis, bacterial meningitis, fungal disease, meningeal carcinomatosis, leukemia, and lymphoma may have a similar appearance.

Hydrocephalus is a common complication of sarcoid leptomeningitis because of obstruction of the fourth ventricle outflow. Hydrocephalus may also result from parenchymal lesions compressing the aqueduct.

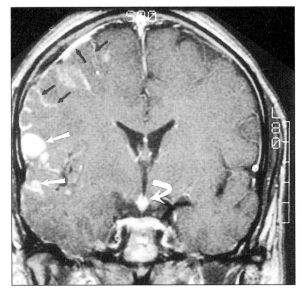

Fig. 75.1 Neurosarcoidosis. Coronal contrast-enhanced T1-weighted image of a patient with systemic sarcoidosis. Abnormal, primarily leptomeningeal, enhancement involving most of the right hemisphere is present (small black arrows). Several parenchymal mass lesions, simulating tumor, also enhance (straight white arrows). Another enhancing mass lesion is apparent in the hypothalamic region (curved white arrow). Multiple other parenchymal mass lesions were present on other images (not shown).

Treatment with steroids and immunosuppressive agents (e.g. cyclophosphamide or cyclosporine) is usually beneficial.[23] The abnormal meningeal, periventricular, and parenchymal enhancement is the neuroimaging feature most likely to regress with treatment.[11]

References

1. Stern BJ, Krumholz A, Johns C, Scott P, Nissim J. Sarcoidosis and its neurological manifestations. *Arch Neurol* 1985; **42**: 909–17.
2. Luke RA, Stern BJ, Krumholz A, Johns CJ. Neurosarcoidosis: the long-term clinical course. *Neurology* 1987; **37**: 461–3.
3. Mayer SA, Yim GK, Onesti ST, Lynch T, Faust PL, Marder K. Biopsy-proven isolated sarcoid meningitis: case report. *J Neurosurg* 1993; **78**: 994–6.
4. Sommer N, Weller M, Petersen D, Wietholter H, Dichgans J. Neurosarcoidosis without systemic sarcoidosis. *Eur Arch Psychiatr Clin Neurosci* 1991; **240**: 334–8.
5. Khaw KT, Manji H, Britton J, Schon F. Neurosarcoidosis: demonstration of meningeal disease by gadolinium enhanced magnetic resonance imaging. *J Neurol Neurosurg Psychiatr* 1991; **54**: 499–502.
6. Krumholz A, Stern BJ, Stern EG. Clinical implications of seizures in neurosarcoidosis. *Arch Neurol* 1991; **48**: 842–4.
7. Berek K, Kiechl S, Willeit J, Birbamer G, Vogl G, Schmutzhard E. Subarachnoid hemorrhage as presenting feature of isolated neurosarcoidosis. *Clin Invest* 1993; **71**: 54–6.
8. Handler MS, Johnson LM, Dick AR, Batnitzky S. Neurosarcoidosis with unusual MRI findings. *Neuroradiology* 1993; **35**: 146–8.
9. Sabaawi M, Gutierrez-Nunez J, Fragala MR. Neurosarcoidosis presenting as schizophreniform disorder. *Int J Psychiatr Med* 1992; **22**: 269–74.

10. Sauter MK, Panitch HS, Kristt DA. Myelopathic neuro-sarcoidosis: diagnostic value of enhanced MRI. *Neurology* 1991; **41**: 150–1.

11. Lexa FJ, Grossman RI. MR of sarcoidosis in the head and spine: spectrum of manifestations and radiographic response to steroid therapy. *AJNR Am J Neuroradiol* 1994; **15**: 973–82.

12. Wilson JD, Castillo M, Tassel P. MRI features of intracranial sarcoidosis mimicking meningiomas. *Clin Imaging* 1994; **18**: 184–7.

13. Zouaoui A, Maillard JC, Dormont D, Chiras J, Marsault C. MRI in neurosarcoidosis. *J Neuroradiol* 1992; **19**: 271–84.

14. Seltzer S, Mark AS, Atlas SW. CNS sarcoidosis: evaluation with contrast-enhanced MR imaging. *AJNR Am J Neuroradiol* 1991; **12**: 1227–33.

15. Kadakia JK, Collette PM, Sharma OP. Role of magnetic resonance imaging in neurosarcoidosis. *Sarcoidosis* 1993; **10**: 98–9.

16. Hayes WS, Sherman JL, Stern BJ, Citrin CM, Pulaski PD. MR and CT evaluation of intracranial sarcoidosis. *AJR Am J Roentgenol* 1987; **149**: 1043–9.

17. Sherman JL, Stern BJ. Sarcoidosis of the CNS: comparison of unenhanced and enhanced MR images. *AJNR Am J Neuroradiol* 1990; **11**: 915–23.

18. Ketonen L, Oksanen V, Kuuliala I. Preliminary experience of magnetic resonance imaging in neurosarcoidosis. *Neuroradiology* 1987; **29**: 127–9.

19. Stubgen JP. Neurosarcoidosis presenting as a retroclival mass. *Surg Neurol* 1995; **43**: 85–7.

20. Ng KL, McDermott N, Romanowski CA, Jackson A. Neurosarcoidosis masquerading as glioma of the optic chiasm in a child. *Postgrad Med J* 1995; **71**: 265–8.

21. Gizzi MS, Lidov M, Rosenbaum D. Neurosarcoidosis presenting as a tumour of the basal ganglia and brain stem: sequential MRI. *Neurol Res* 1993; **15**: 93–6.

22. Ranoux D, Devaux B, Lamy C, Mear JY, Roux FX, Mas JL. Meningeal sarcoidosis, pseudo-meningioma, and pachy-meningitis of the convexity. *J Neurol Neurosurg Psychiatr* 1992; **55**: 300–3.

23. Stern BJ, Schonfeld SA, Sewell C, Krumholz A, Scott P, Belendiuk G. The treatment of neurosarcoidosis with cyclosporine. *Arch Neurol* 1992; **49**: 1065–72.

12
Pediatric Neuroradiology

76. Brain myelination progresses in a predictable manner

Brain myelination is a dynamic process that begins during the fifth fetal month and continues for several decades. Normal myelination occurs in a highly structured order.[1-4] Myelination is best evaluated on T1-weighted MRI during the first 4 months and on T2-weighted MRI thereafter. In children born full-term, the following progression of myelination is normal with 1.5 T MRI:

1. *At birth*
   - Central cerebellar white matter
   - Medulla
   - Dorsal midbrain
   - Ventrolateral thalamus
   - Superior and inferior cerebellar peduncles
2. *By 1 month*
   - Corticospinal tracts
   - Precentral and postcentral gyrus
   - Optic nerves and tracts
3. *By 3 months*
   - Middle cerebellar peduncles
   - Optic radiations
   - Posterior limbs of internal capsule
4. *By 6 months*
   - Splenium of corpus callosum
   - Anterior limbs of internal capsule
   - Centrum semiovale (in part)
5. *By 8 months*
   - Centrum semiovale (complete)
   - Subcortical U-fibers
   - Genu of corpus callosum
6. *By 18 months*
   - Nearly adult appearance except for the peritrigonal region (maxim 9).

MRI has enabled sequential *in vivo* analysis of CNS maturation and especially the characterization of myelination by assessment of the effects of myelin on the T1 and T2 relaxation times. Departures from or delays in the normal progression of myelination are easily identified on MRI. When delayed myelination is suspected, MRI should be performed before 24 months of age because delayed myelination often catches up beyond this age (maxim 77).

In general, myelination progresses caudal to rostral, central to peripheral, and dorsal to ventral. That is, structures in the brain stem myelinate before the cerebellar hemispheres, which myelinate before the cerebral hemispheres. The central regions in the cerebellar and cerebral hemispheres myelinate before the peripheral regions. The posterior structures in the cerebral hemispheres myelinate before the anterior structures (with the exception of the peritrigonal region).

Most myelination is completed during the first 2 years. Excluding the overall brain size (which is unfortunately rarely assessed), the MRI of a normal 2-year-old is essentially indistinguishable from that of a normal adult. The major exception is the parieto-occipital (peritrigonal region) white matter, which myelinates in the second decade (maxim 9). The peritrigonal region usually has a homogeneous (although it can be patchy), modestly high-signal appearance with indistinct borders on T2-weighted images. No other qualitative white-matter differences are normally present between a child and an adult.

The infant brain undergoes rapid chemical changes, especially during the first year, which are reflected in both the T1- and T2-weighted images. A major change involves the water content, which decreases from 88% of brain weight at birth to 82% at 6 months. At the same time the lipid and protein content increase, especially within the white matter (myelin is about 40% water and 30% protein and is relatively hydrophobic). This brain maturation results in different T1 and T2 properties of various tissues. In infants less than 4 months old, myelin is most evident on T1-weighted images. By 9 months the T1-weighted image has an adult appearance. After 4 months, the T2-weighted images are best for evaluating the myelin content. Myelinated regions produce high signal on T1-weighted images and low signal on T2-weighted images. There is little difference between conventional and fast spin-echo acquisitions in the evaluation of myelination.[5]

The medulla, dorsal midbrain, and inferior and superior cerebellar peduncles are myelinated by birth in the normal full-term infant and have relatively high signal on T1-weighted images. At birth, there is minimal differentiation between the gray and white matter, making the cerebral hemispheres nearly featureless. At this age, small amounts of edema may be missed if no mass effect is present. The cortical gray matter is distinguishable from the white matter on standard spin-echo sequences by 4–6 months. The deep cerebellar white matter and middle cerebellar peduncles reach the adult appearance by 3 months. Myelination of the motor tracts is also almost completed by 3 months (Fig. 76.1).

The white and gray matter signal reverses on T2-weighted images during the first year of life. At birth the white matter T2 signal is higher than the gray matter T2 signal. Some patchy T2 low signal is present in the precentral and postcentral gyri by 1 month and in the centrum semiovale by 2 months. By 8–12 months, the white matter produces lower T2 signal than does the gray matter. Peripheral extension of myelination in the cerebellar white matter is completed by 10 months.

The appearance of the posterior and anterior limbs of the internal capsule is very useful for assessing the progress of myelination. The posterior limbs are myelinated by 3 months and the anterior limbs by 6 months. The subcortical white matter within the cerebral hemispheres is one of the last regions of the brain to mature. Myelination progresses from the occipital lobes (9 months) to the frontal lobes (14 months). Myelination is nearly completed by 18 months and the typical white-matter arbor present in adults is visible. At age 4 years, the cortical gray matter reaches its maximal volume and then decreases slightly with advancing age. The cerebral white-matter volume increases steadily until about age 20 years.[6]

References

1. Barkovich AJ, Kjos BO, Jackson DE, *et al.* Normal maturation of the neonatal infant brain: MR imaging at 1.5 T. *Radiology* 1988; **166**: 173–80.
2. Dietrich RB, Bradley WG, Zaragoza EJ, *et al.* MR evaluation of early myelination patterns of normal and developmental delayed infants. *AJNR Am J Neuroradiol* 1988; **9**: 69–76.
3. Holland BA, Haas DK, Norman D, *et al.* MRI of normal brain maturation. *AJNR Am J Neuroradiol* 1986; **7**: 201–8.
4. Osborn AG. *Handbook of neuroradiology.* St Louis: CV Mosby, 1991: 172–3.
5. Prenger EC, Beckett WW, Kollias SS, Ball WS Jr. Comparison of T2-weighted spin-echo and fast spin-echo techniques in the evaluation of myelination. *J Magn Reson Imaging* 1994; **4**: 179–84.
6. Pfefferbaum A, Mathalon DH, Sullivan EV, Rawles JM, Zipursky RB, Lim KO. A quantitative magnetic resonance imaging study of changes in brain morphology from infancy to late adulthood. *Arch Neurol* 1994; **51**: 874–87.

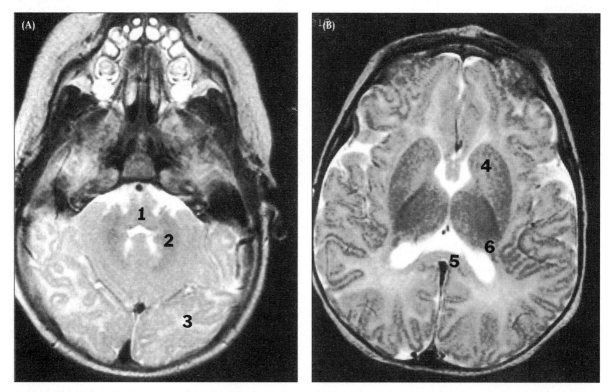

Fig. 76.1 Normal myelination. **A:** Axial T2-weighted image through the posterior fossa of a neurologically normal 4-month-old. The brain stem (1) and deep white matter of the cerebellum (2) are myelinated and thus produce relatively lower signal. The white matter in the occipital lobes (3) is not myelinated and produces higher signal. **B:** Axial T2-weighted image through the lateral ventricles in the same child. Most of the white matter is high signal compared with the cortical gray matter. There is a relatively decreased signal from the basal ganglia with respect to the surrounding brain. The anterior limb of the internal capsule (4) is not myelinated. The splenium of the corpus callosum (5) and the posterior limb of the internal capsule (6) are partially myelinated.

77. Developmental delay is often associated with delayed myelination

Perinatal anoxia, inborn errors of metabolism, genetic abnormalities, and congenital anomalies are frequently associated with delayed or diminished myelination.[1-4] The most common clinical correlates of delayed myelination are developmental delay, West syndrome (infantile spasms), and cerebral palsy.[5,6] No underlying cause is apparent in some children who have delayed myelination.

Delayed myelination is rarely the only abnormality, and other lesions, including cerebral white matter hypoplasia, thinning of the corpus callosum and cortical atrophy, may be present.[7] The degree of delay in myelination and cortical atrophy correlates with the underlying severity of the disorder.[8]

When delayed myelination is suspected, MRI should be performed before the age of 24 months. Eventually the white matter will myelinate and if the imaging is delayed the condition may not be detected (Fig. 77.1). Delayed myelination is identified more readily after 2

months of age.[9] Assessment of myelination in children with head injuries requires caution because trauma may produce abnormalities on MRI that are similar to the earlier stages of myelination.[6]

One of the most common reasons for immature myelination is malnutrition. When the nutritional status is corrected, myelination quickly returns to the appropriate age level and normal myelination progression resumes.

Evaluation of the degree of myelination requires care. In one study, the diagnosis of delayed myelination was inaccurately applied or missed in 15% of patients.[5] In this study, lack of familiarity with the myelination milestones of infancy was the most common reason for a misdiagnosis of delayed myelination, and failure to recognize delayed myelination was due to failure to appreciate the forceps minor (the semicircular white matter "tract" radiating out of the corpus callosum to the occipital pole) as a landmark.

Although myelination can be accurately assessed using routine MRI examinations,[10] other techniques may provide more accurate measures and help to eliminate interpretation errors. Simple measurement of the signal

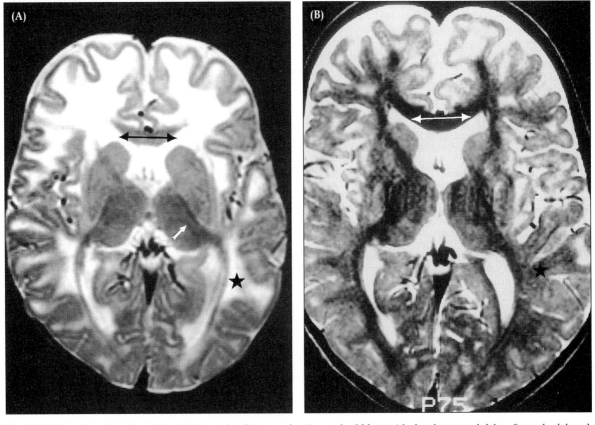

Fig. 77.1 Delayed myelination. **A:** Axial T2-weighted image of a 13-month-old boy with developmental delay. Severely delayed myelination is present. There is abnormal high signal in nearly all of the white matter in the frontal and temporal (black star) lobes. The poorly myelinated region includes the genu of the corpus callosum (double-headed black arrow). Normal myelination is present in the posterior limbs of the internal capsules (white arrow). **B:** Axial T2-weighted image in the same patient at 2 years of age. There is significant progression of the myelination in the temporal lobes (black star) and genu of the corpus callosum (double-headed white arrow). It is important to image patients with developmental delay before the age of 2 years to facilitate the diagnosis of delayed myelination.

intensity ratio of gray to white matter is a practical way to evaluate delayed myelination.[11] Quantitative assessment of T2 values gives a more objective analysis than does visual inspection, and can be used to determine the rate at which abnormal myelination is progressing.[12] Diffusion-weighted imaging detects brain myelination patterns earlier than conventional T1- and T2-weighted images do and is likely to have a future role in assessment of maturation in early infancy.[13]

References

1. Kolodny EH. Dysmyelinating and demyelinating conditions in infancy. *Curr Opin Neurol Neurosurg* 1993; **6**: 379–86.
2. Kendall BE. Inborn errors and demyelination: MRI and the diagnosis of white matter disease. *J Inherit Metab Dis* 1993; **16**: 771–86.
3. Konishi Y, Hayakawa K, Kuriyama M, *et al.* Developmental features of the brain in preterm and fullterm infants on MR imaging. *Early Hum Dev* 1993; **34**: 155–62.
4. Rademakers RP, van der Knaap MS, Verbeeten B Jr, Barth PG, Valk J. Central cortico-subcortical involvement: a distinct pattern of brain damage caused by perinatal and postnatal asphyxia in term infants. *J Comput Assist Tomogr* 1995; **19**: 256–63.
5. Squires LA, Krishnamoorthy KS, Natowicz MR. Delayed myelination in infants and young children: radiographic and clinical correlates. *J Child Neurol* 1995; **10**: 100–4.
6. Staudt M, Schropp C, Staudt F, *et al.* MRI assessment of myelination: an age standardization. *Pediatr Radiol* 1994; **24**: 122–7.
7. Candy EJ, Hoon AH, Capute AJ, Bryan RN. MRI in motor delay: important adjunct to classification of cerebral palsy. *Pediatr Neurol* 1993; **9**: 421–9.
8. Christophe C, Clercx A, Blum D, Hasaerts D, Segebarth C, Perlmutter N. Early MR detection of cortical and subcortical hypoxic–ischemic encephalopathy in full-term infants. *Pediatr Radiol* 1994; **24**: 581–4.
9. Fujii Y, Konishi Y, Kuriyama M, *et al.* MRI assessment of myelination patterns in high-risk infants. *Pediatr Neurol* 1993; **9**: 194–7.
10. Staudt M, Schropp C, Staudt F, Obletter N, Bise K, Breit A. Myelination of the brain in MRI: a staging system. *Pediatr Radiol* 1993; **23**: 169–76.
11. Maezawa M, Seki T, Imura S, Akiyama K, Takikawa I, Yuasa Y. Magnetic resonance signal intensity ratio of gray/white matter in children: quantitative assessment in developing brain. *Brain Dev* 1993; **15**: 198–204.
12. Ono J, Kodaka R, Imai K, *et al.* Evaluation of myelination by means of the T2 value on magnetic resonance imaging. *Brain Dev* 1993; **15**: 433–8.
13. Nomura Y, Sakuma H, Takeda K, Tagami T, Okuda Y, Nakagawa T. Diffusional anisotropy of the human brain assessed with diffusion-weighted MR: relation with normal brain development and aging. *AJNR Am J Neuroradiol* 1994; **15**: 231–8.

78. *Many congenital malformations are demonstrated with MRI*

MRI is the best neuroimaging method for characterizing congenital malformations, and permits detection of some abnormalities that in the past could be identified only at autopsy.[1] Several common and interesting

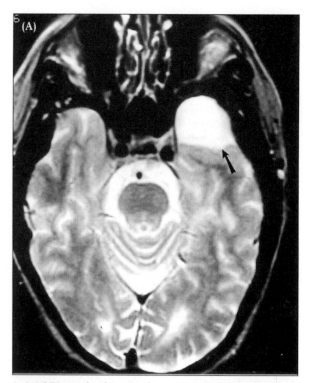

A: Axial T2-weighted image of an adult man with headaches. A left middle cranial fossa arachnoid cyst is present (arrow). The cyst is isointense to CSF with clearly marginated borders. There is some mass effect on the displaced temporal lobe tip.

Fig. 78.1 Arachnoid cysts.

malformations are considered below.[2] See also maxim 25 (neuronal migration abnormalities), maxim 29 (Dyke–Davidoff–Mason syndrome), and maxims 88 and 89 (Chiari malformations).

Arachnoid cysts

Arachnoid cysts are fluid-filled cavities that arise between weeks 6 and 8 of gestation from abnormal splitting of the membrane. Some communicate with the subarachnoid space. One-half are located in the middle cranial fossa, one-third in the posterior fossa, and 10% in the suprasellar region (Fig. 78.1). Occasionally they enlarge, producing mass effect and other symptoms such as seizures. Surgical drainage may be required.

Hydranencephalus

In hydranencephalus the cerebral hemispheres are largely replaced by a CSF-containing sac (Fig. 78.2). The most common mechanism is destruction of the brain from bilateral stroke. Although the infant may appear normal at birth, failure to thrive develops quickly and death usually ensues within months. Neuroimaging is important to exclude a treatable condition such as severe hydrocephalus, large porencephalic cysts, or bilateral subdural effusions.

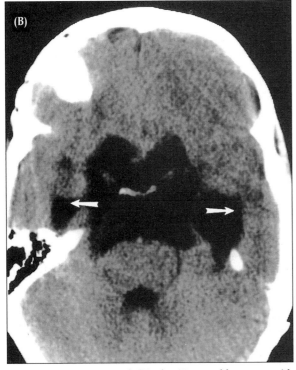

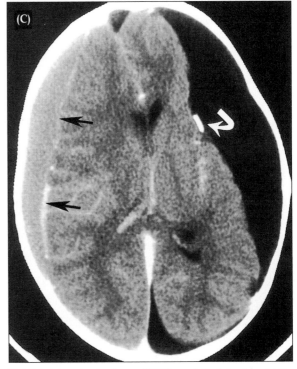

78.1B: Axial noncontrasted CT of a 20-year-old woman with seizures and shunted hydrocephalus. A large suprasellar low-density lesion, which extends laterally into the middle cranial fossa displacing the medial temporal structures (arrows), is present. This suprasellar arachnoid cyst also extends into the posterior fossa, flattening the pons.

78.1C: Axial contrast-enhanced CT through the lateral ventricles of a 3-year-old boy with a large head circumference. The initial CT identified the large left-sided arachnoid cyst with mass effect shifting the brain to the right (not shown). Shunting (curved white arrow) resulted in decreased pressure with return of the brain to the midline, but a subdural hematoma developed on the right (straight black arrows). At the time of CT, the subdural hematoma was nearly isodense. Contrast enhancement reveals the cortical surface (maxim 87).

Agenesis of the corpus callosum

The corpus callosum develops during the first trimester of fetal life. Agenesis of the corpus callosum (Fig 78.3) is an abnormality of the development of the midline telencephalon. It may be the only abnormality but more commonly is found in association with other developmental anomalies, including heterotopic gray matter, Chiari malformations, Dandy–Walker malformation, holoprosencephaly, and midline facial deformities. Patients often have mild to severe mental retardation and seizures. The clinical features are usually due to the associated developmental abnormalities and not directly caused by the corpus callosum agenesis (*see also* Figs 25.1 and 78.4).

Dandy–Walker malformation

The Dandy–Walker malformation consists of a triad of agenesis of the cerebellar vermis (complete or partial), dilation of the fourth ventricle, and an enlarged posterior fossa with upward displacement of the transverse sinuses, tentorium, and torcula (Fig. 78.4). It probably results from failure of neural tube closure at the cerebellar level at 4 weeks gestation. Associated anomalies of the brain and spinal cord are common. Corpus callosum dysgenesis is present in almost one-third of affected patients.

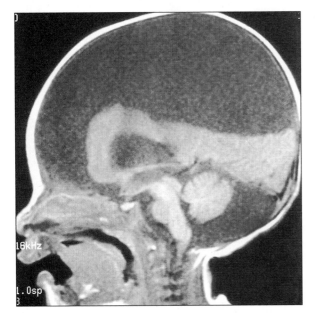

Fig. 78.2 Hydranencephalus. Midline sagittal T1-weighted image of a 5-week-old infant reveals the absence of most of the cerebrum. Portions of the anterior inferior frontal lobe and the occipital lobe are present. The brain-stem structures are remarkably spared. This pattern is diagnostic of hydranencephalus.

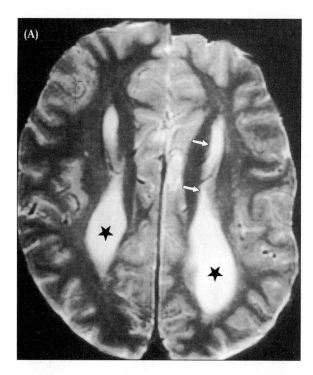

Congenital malformation of the spinal cord

Some patients with congenital malformations have associated spinal cord abnormalities. If spinal cord pathology is suspected, examination of the complete spine is often necessary to appreciate the full extent of a malformation (Fig. 78.5).

Cranial synostosis

Premature closure of the cranial sutures may be primary or secondary to a wide variety of conditions. When one or more sutures is prematurely closed, the term cranial synostosis is used (Fig. 78.6). Three-dimensional reconstructions of a conventional CT are useful in diagnosis and surgical planning.

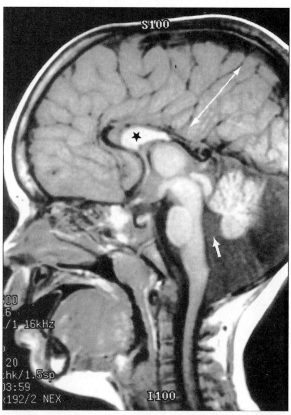

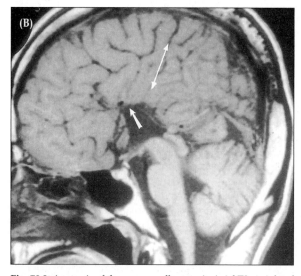

Fig. 78.3 Agenesis of the corpus callosum. **A:** Axial T2-weighted image of a 26-year-old man with seizures and mild retardation. The lateral ventricles are widely separated. The anterior medial borders are concave (opposite of normal; white arrows). The posterior horns are enlarged (colpocephaly; black stars). **B:** Midline sagittal T1-weighted image in the same patient. The corpus callosum is absent (compare with Fig. 7.3). The third ventricle is dilated and enlarged superiorly (white arrow). The sulci have a radial arrangement (double-headed white arrow). The foramen of Monro is elongated (not shown). Angiography would demonstrate loss of curvature of the anterior cerebral and pericallosal arteries with elevated internal cerebral veins and an anteriorly positioned great vein of Galen.

Fig. 78.4 Dandy–Walker malformation. Midline sagittal T1-weighted image. The posterior fossa is enlarged. The cerebellum is moderately hypoplastic. The inferior vermis is absent. The fourth ventricle communicates with a large posterior fossa cyst (arrow). This patient also has partial agenesis of the corpus callosum. Only the genu and anterior body of the corpus callosum developed (black star). The rostrum and splenium did not form. The midline posterior gyri radiate toward the third ventricle (white double-headed arrow).

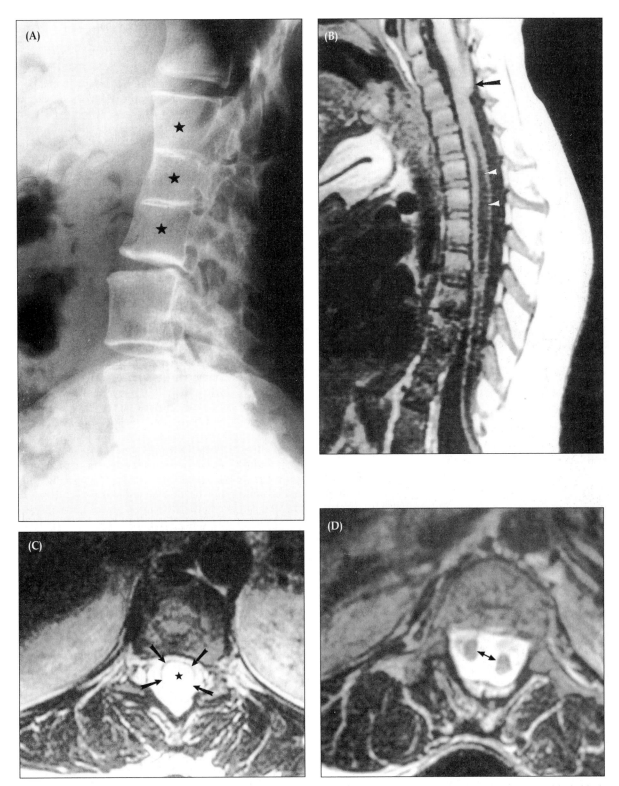

Fig. 78.5 Congenital malformation of the spinal cord. **A:** Plain radiograph demonstrates three vertebral bodies forming a block (black stars). The sagittal diameter of this block is less than that of the normal vertebral bodies. **B:** Sagittal T1-weighted image. The cerebellar tonsils descend to C4 (black arrow), a severe Chiari I malformation. A syrinx (white arrowheads) and tethered cord with distal dilatation of the spinal canal are present. **C:** Axial T2-weighted image in the thoracic region. The syrinx (high-signal fluid; black star) occupies almost the entire area of the spinal cord. Only a rim of spinal cord tissue remains (black arrows). Outside the rim of spinal cord tissue is the normal subarachnoid CSF. **D:** Axial T2-weighted image at the upper lumbar level of the block vertebrae. The spinal canal is dilated. Diastematomyelia, a double spinal cord, is present (black double-headed arrow). Thus, this patient has a Chiari I malformation, syringomyelia, tethered spinal cord, block vertebral bodies, and diastematomyelia.

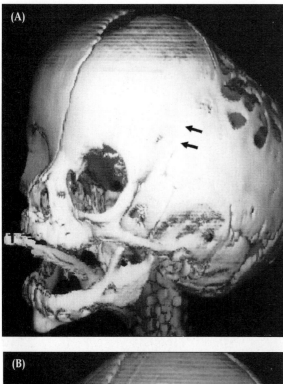

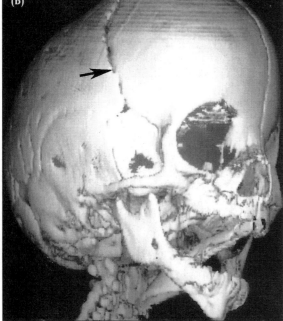

Fig. 78.6 Cranial synostosis. **A:** In this 3-month-old, the left coronal suture is prematurely closed (small black arrows). **B:** Note the open right coronal suture (large black arrow). This child also has other facial and skull deformities.

References

1. Barkovich AJ, Maroldo TV. Magnetic resonance imaging of normal and abnormal brain development. *Top Magn Reson Imaging* 1993; **5**: 96–122.
2. Menkes JH, Till K. Malformations of the central nervous system. In: Menkes JH (ed.), *Textbook of child neurology*, 5th edn. Baltimore: Williams & Wilkins, 1995: 240–324.

79. *Low back pain in preadolescent children often indicates a serious problem*

Low pack pain is uncommon in preadolescent children and, when present, is frequently the result of a significant abnormality. In contrast to adults, in whom back pain without overt pathology is common, a child rarely complains of back pain unless structural disease is present. The diagnostic evaluation should include a complete history, physical examination, standard laboratory testing (including white blood cell count and sedimentation rate), and plain spine radiographs. In many cases a CT, MRI, or a technetium-99 nuclear medicine study of the spine is warranted, especially if physical signs are present. If imaging is deferred, repeated clinical evaluations should be performed because the pathology may develop over time.[1-7]

Intervertebral disc space infections (discitis) occur in preadolescent children. *Staphylococcus aureus* and diphtheroids are the most commonly identified organisms. Back, abdominal, hip, or leg pain, with or without fever, are presenting features. The child may refuse to walk or stand. The diagnosis is frequently delayed because of a nonspecific presentation. The lumbosacral spine is most often affected. Plain radiographs demonstrate narrowing of the disc space with erosion of the end plate and vertebral body. Soft-tissue abnormalities may be present. MRI is the most useful study, although CT may identify the lesions. MRI offers the advantage of demonstrating other findings, such as an epidural abscess or paraspinal soft-tissue mass. Discitis in children is usually self-limited and the need for antibiotics is debated. Immobilization relieves the pain. Residual spinal disease with some restricted movement is a frequent long-term sequela.

Primary osseous neoplasms are uncommon but may occur in childhood.[8] The most common primary vertebral body tumors in children are osteoid osteomas, benign osteoblastomas, and Ewing sarcomas. Aneurysmal bone cysts also occur. Spinal cord tumors, including ependymomas, schwannomas, neurofibromas, and neurofibrosarcomas may present with back pain.

Juvenile osteoporosis may result from metabolic defects or nutritional deficiencies, or may be idiopathic. Presentation includes long-bone fractures, back pain, and gait difficulties. Compression of the vertebrae and metaphyses of the long bones is typically present on radiographs. Metabolic and nutritional causes usually respond to appropriate treatment. Most patients with idiopathic juvenile osteoporosis have spontaneous recovery.[9]

Congenital anomalies, such as spondylolysis or spondylolisthesis, may first become symptomatic during childhood play or sports activity. Spondylolysis is a defect in the pars interarticularis (the weakest part of the vertebra located near the junction of the lamina and pedicle). Unilateral spondylolysis may result in ipsilateral pedicle and facet hypertrophy. When the pars interarticularis defect is bilateral, spondylolisthesis may result (Fig. 79.1). (Spondylolisthesis also results from trauma and degenerative disease, usually in the elderly.) Spondylolisthesis is displacement of a vertebra, usually L4 or L5, anteriorly relative to the vertebrae below. Spondylolisthesis is graded by dividing the vertebral body into quarters: grade I is displacement of the vertebral body up to one-quarter over the adjacent vertebral body, grade II between one-quarter and one-half, grade III from one-half to three-quarters, and grade 4 more than three-quarters. A "pseudo-disc-bulge" may be present on axial images, the result of the anteriorly displaced vertebra uncovering the annulus.

Spondylolysis and spondylolisthesis are more common in males and typically become symptomatic in young athletic boys. Sports commonly implicated in the appearance of these conditions include gymnastics, diving, and football. Plain radiographs of the lumbosacral spine in lateral and both oblique views are usually sufficient to demonstrate a pars interarticularis defect and the degree of spondylolisthesis. If the plain radiographs are inconclusive, CT is usually definitive. Symptomatic pars defects are frequently associated with increased scintigraphic activity best detected and localized by single photon emission computerized tomography (SPECT).[10,11] Nonoperative treatment is successful in a majority of adolescents with mild-to-moderate disease.[12] Surgery is only required in severe spondylolisthesis.[13]

Other congenital spinal abnormalities, including scoliosis, vertebral body fusions, and congenital absence of a lumbar pedicle,[14] may result in back pain. Spinal stenosis is infrequent in children but may cause back pain, myelopathy, or radiculopathies.

Disc herniations are rare in children and young adults. When they occur they are usually traumatic rather than degenerative in origin. Most commonly the lower lumbar region is affected. In children, the disc consists of healthy tissue and a small tear in the annulus fibrosus can result in a large herniation which can mimic an extradural tumor. MRI is the preferred diagnostic study, but, as in adults, CT or conventional myelography can also demonstrate the abnormality.

Scheuermann disease, also termed osteochondrosis of the spine or epiphyseal ischemic necrosis, can be a painful disorder with collapse of the anterior vertebral body, irregular end plates, and narrowed disc spaces. It is characterized by degeneration and regeneration of the growth centers. It is most common in the thoracic and lumbar region. Herniation of the intervertebral disc into the adjacent vertebral body, resulting in Schmorl's nodes, is common. Plain radiographs usually demonstrate this condition. Scheuermann disease may be incidental, and further evaluation for the cause of the back pain is sometimes warranted.

In contrast to young children, adolescents have back pain as frequently as adults. One-third have back pain at some time, with a third of these requiring activity restriction.[15] Most back pain in adolescents is related to spinal malformations, rapid growth, or excessive exertion.

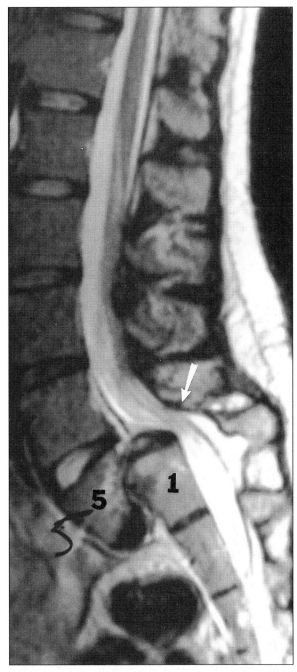

Fig. 79.1 Spondylolisthesis. Sagittal T2-weighted image of a 17-year-old cheerleader with back pain. Severe (grade 4) spondylolisthesis is present. The L5 vertebral body (5) is displaced anteriorly by almost the entire vertebral body diameter over the S1 vertebral body (1). Local spinal stenosis is present with stretching of the nerve roots (arrow).

References

1. King HA. Back pain in children. *Pediatr Clin North Am* 1984; **31**: 1083–95.
2. King HA. Evaluating the child with back pain. *Pediatr Clin North Am* 1986; **33**: 1489–93.
3. Turner PG, Green JH, Galasko CS. Back pain in childhood. *Spine* 1989; **14**: 812–14.

4. Svenson J, Stapczynski JS. Childhood back pain: diagnostic evaluation of an unusual case. *Am J Emerg Med* 1994; **12**: 334–6.
5. Rosenblum BR, Rothman AS. Low back pain in children. *Mount Sinai J Med* 1991; **58**: 115–20.
6. Burger EL, Lindeque BG. Sacral and non-spinal tumors presenting as backache: a retrospective study of 17 patients. *Acta Orthop Scand* 1994; **65**: 344–6.
7. Bunnell WP. Back pain in children. *Orthop Clin North Am* 1982; **13**: 587–604.
8. Delamarter RB, Sachs BL, Thompson GH, Bohlman HH, Makley JT, Carter JR. Primary neoplasms of the thoracic and lumbar spine: analysis of twenty-nine consecutive cases. *Clin Orthop Related Res* 1990; **256**: 87–100.
9. Smith R. Idiopathic juvenile osteoporosis: experience of twenty-one patients. *Br J Rheumatol* 1995; **34**: 68–77.
10. Collier BD, Johnson RP, Carrera GF, *et al.* Painful spondylolysis or spondylolisthesis studied by radiography and single-photon emission computed tomography. *Radiology* 1985; **154**: 207–11.
11. Bellah RD, Summerville DA, Treves ST, Micheli LJ. Low-back pain in adolescent athletes: detection of stress injury to the pars interarticularis with SPECT. *Radiology* 1991; **180**: 509–12.
12. Morita T, Ikata T, Katoh S, Miyake R. Lumbar spondylolysis in children and adolescents. *J Bone Joint Surg [Br]* 1995; **77**: 620–5.
13. Pizzutillo PD, Hummer CD III. Nonoperative treatment for painful adolescent spondylolysis or spondylolisthesis. *J Pediatr Orthop* 1989; **9**: 538–40.
14. Polly DW Jr, Mason DE. Congenital absence of a lumbar pedicle presenting as back pain in children. *J Pediatr Orthop* 1991; **11**: 214–19.
15. Olsen TL, Anderson RL, Dearwater SR, *et al.* The epidemiology of low back pain in an adolescent population. *Am J Public Health* 1992; **82**: 606–8.

80. *Children have more posterior fossa tumors than do adults*

CNS tumors are the second most common pediatric cancer, exceeded in frequency only by leukemia. Tumors located in the posterior fossa predominate from ages 4 to 11 years. Supratentorial tumors are more common under 4 years of age and during adolescence and adulthood. Approximately 10% of tumors involve multiple compartments at presentation.[1]

The four most common posterior fossa tumors in children are medulloblastoma (primitive neuroectodermal tumor, PNET), cerebellar astrocytoma, ependymoma, and brain-stem glioma (maxim 36).[2,3] These four account for over 95% of childhood posterior fossa tumors. Metastases and extra-axial neoplasms such as acoustic and trigeminal neuromas, meningioma, chordoma, dermoid, and epidermoid are rare in children.

In general, MRI with contrast should be performed before surgery in the diagnostic work-up of a posterior fossa lesion.[4] CT is useful to assess for calcification and to follow tumor progression and degree of hydrocephalus. Single photon emission computerized tomography (SPECT) thallium-201 studies are becoming important for diagnosis and especially treatment assessment. This technique is more than 75% sensitive and 90% specific for detection of pediatric brain tumors and is less expensive and more readily available than positron emission tomography.[5]

Medulloblastomas account for almost 40% of posterior fossa neoplasms in children (maxim 41). They are highly malignant tumors arising from poorly differentiated germ cells in the roof of the fourth ventricle. Patients usually present within 1 month of symptom onset. Medulloblastomas may arise anywhere along the path of neuronal migration (from the fourth ventricle to the cerebellar cortex), but they are usually located in the midline. Approximately two-thirds of medulloblastomas in children are located in the inferior vermis. They can grow through the fourth ventricular foramina and extend into the cisterna magna, the upper spinal canal, or the cerebellar pontine angle cistern.[6] Vermian tumors usually cause partial or complete obstruction of CSF flow with resultant hydrocephalus. Meningeal seeding through the CSF is almost always present (maxim 41).

Cerebellar astrocytomas are slightly less common than medulloblastomas. Nearly one-half have a cystic component with a mural nodule within the cyst wall. On CT, the "cyst fluid" has a slightly higher density than CSF, due to a higher protein content. The mural nodule enhances homogeneously on both CT and MRI. Most cerebellar astrocytomas in children are pilocytic, a specific histologic type. Juvenile pilocytic cerebellar astrocytomas are one of the most benign neoplasms of the CNS. Long-term survival rates exceed 90%. Malignant astrocytomas (anaplastic or glioblastoma multiforme) are rare in the posterior fossa in children.

Posterior fossa ependymomas are most common in the first 5 years. On CT they have increased density and patchy enhancement. Occasionally they have strong homogeneous enhancement. Ependymomas are typically calcified.

Brain-stem gliomas are highly morbid tumors, which usually occur before 10 years of age (Fig. 80.1). They originate in the pons or, less commonly, in the midbrain or medulla. They are usually slow-growing fibrillary or pilocytic astrocytomas. Less commonly they are anaplastic. The lesion enlarges diffusely and sometimes eccentrically, eventually compressing the fourth ventricle, aqueduct, and adjacent CSF spaces. Brain-stem gliomas are hyperdense on CT and hyperintense on T2-weighted MRI. One-half enhance. Calcification and hemorrhage are uncommon. Exophytic extension occurs in over one-half.

Treatment of pediatric brain tumors usually involves neurosurgical intervention with biopsy and sometimes resection. Use of chemotherapy or radiation therapy depends on tumor type. With improved long-term survival, the effect of the treatment regimen on the developing brain is an important consideration. High-dose whole-brain irradiation treatment in children results in decreased intellectual ability later in life (maxim 39). The choice of therapy must be weighed against the risk of future long term sequelae.[7–11]

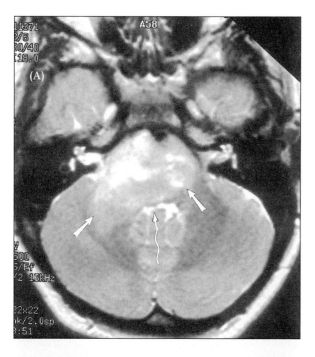

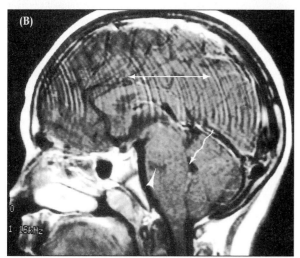

Fig. 80.1 Brain-stem glioma. **A:** Axial T2-weighted image of a 5-year-old boy with ataxia and cranial nerve palsies. Diffuse inhomogeneous increased signal is present in an enlarged pons (straight arrows). The fourth ventricle is mildly deformed (wavy white arrow). **B:** Sagittal contrast-enhanced T1-weighted image in the same patient. The anterior wall of the fourth ventricle is displaced (bowed) posteriorly by the tumor (wavy arrow). A small necrotic area (low signal) is present in the middle of the tumor (arrowhead). No contrast enhancement is present within the tumor. Note the artifact within the cerebral hemispheres (double-headed arrow) in the phase-encoded direction (maxim 4).

It may be possible to treat certain small tumors routinely with stereotactic radiotherapy in the near future. Stereotactic radiosurgery consists of thin external beams of ionizing radiation, which are projected from multiple directions and focused onto a sharply defined target. The intent is to avoid large doses to the surrounding healthy tissue, minimizing unwanted damage. The desired goal, of destroying neoplastic tissue without damaging brain parenchyma, is particularly relevant to children in whom brain development is not complete.[12] Unfortunately, long-term complications may develop. Currently, only a small portion of lesions are detected while they are still small enough for radiosurgery.

References

1. Gilles FH, Leviton A, Hedley-Whyte ET, *et al.* Childhood brain tumors that occupy more than one compartment at presentation: multiple compartment tumors. *J Neurooncol* 1992; **14**: 45–56.
2. Gusnard DA. Cerebellar neoplasms in children. *Semin Roentgenol* 1990; **25**: 264–78.
3. Kramer ED, Vezina LG, Packer RJ, Fitz CR, Zimmerman RA, Cohen MD. Staging and surveillance of children with central nervous system neoplasms: recommendations of the Neurology and Tumor Imaging Committees of the Children's Cancer Group. *Pediatr Neurosurg* 1994; **20**: 254–62.
4. Rippe DJ, Boyko OB, Friedman HS, *et al.* Gd-DTPA-enhanced MR imaging of leptomeningeal spread of primary intracranial CNS tumor in children. *AJNR Am J Neuroradiol* 1990; **11**: 329–32.
5. Nadel HR. Thallium-201 for oncological imaging in children. *Semin Nucl Med* 1993; **23**: 243–54.
6. Mueller DP, Moore SA, Sato Y, Yuh WT. MRI spectrum of medulloblastoma. *Clin Imaging* 1992; **16**: 250–5.
7. Albright AL. Pediatric brain tumors. *CA Cancer J Clin* 1993; **43**: 272–88.
8. Donahue B. Short- and long-term complications of radiation therapy for pediatric brain tumors. *Pediatr Neurosurg* 1992; **18**: 207–17.
9. Duffner PK, Cohen ME. Changes in the approach to central nervous system tumors in childhood. *Pediatr Clin North Am* 1992; **39**: 859–77.
10. Mulhern RK, Hancock J, Fairclough D, Kun L. Neuropsychological status of children treated for brain tumors: a critical review and integrative analysis. *Med Pediatr Oncol* 1992; **20**: 181–91.
11. Glauser TA, Packer RJ. Cognitive deficits in long-term survivors of childhood brain tumors. *Childs Nerv Syst* 1991; **7**: 2–12.
12. Benassi M, Begnozzi L, Carpino S, Valentino V. Magnetic resonance guided radiosurgery in children: tridimensional extrapolation from isodose neuroimaging superimposition. *Childs Nerv Syst* 1994; **10**: 115–21.

81. *Suspect child abuse when intracranial blood is present*

Tragically, child abuse is a major health problem. Every effort should be made to identify children who are victims of abuse to facilitate appropriate treatment and prevent future injuries. Despite its violent nature, the diagnosis of child abuse is often difficult. External signs of trauma may be absent and the history is invariably difficult to obtain. Radiology has played an important role in the diagnosis of child abuse since the early descriptions by Caffey[1] and Silverman.[2] Radiographic patterns characteristic of abuse in a child include multiple fractures in various stages of healing, spiral fractures, subdural hematomas (often chronic), and subarachnoid hemorrhage.[3] When any of these conditions are present, the possibility of child abuse should always be raised by the radiologist.

Head injury is the major cause of morbidity and death in victims of child abuse. Brain damage can result from direct impact, shaking, asphyxia, or vascular occlusion from strangulation. With the advent of CT, the "shaken baby syndrome" was identified.[4] This syndrome is characterized by retinal hemorrhages, subdural hematomas, subarachnoid hemorrhages, intracerebral contusions, and diffuse brain edema (Fig. 81.1).[5] Usually there is minimal or no external evidence of trauma, although blunt trauma may be more common than is recognized.[6] Retinal hemorrhages in a child less than 3 years old are virtually pathognomonic of nonaccidental trauma.[7,8]

Vigorous shaking of a young child tears blood vessels, resulting in subdural hematomas (often interhemispheric), subarachnoid hemorrhages, and retinal hemorrhages, or it may shear neurons causing diffuse axonal injury. Cortical contusions are usually the result of a blow and may occur with or without a skull fracture. Because the infant skull is elastic, it can undergo significant deformation without fracturing. The presence of multiple, bilateral, depressed, or complex fractures without a history of significant trauma is almost always the result of child abuse. Cerebral infarctions induced by trauma typically involve tissue in several vascular territories – in contrast to idiopathic infarctions of childhood, which are usually in a single vascular territory.

The shaken child usually presents with confusion or decreased level of consciousness. The findings may mimic infection, intoxication, or metabolic abnormalities. The diagnosis depends on a high index of suspicion and the physical findings of a bulging fontanelle, head circumference greater than the 90th percentile, and retinal hemorrhages. Bloody CSF is also highly suggestive. Neuroimaging usually helps to confirm the diagnosis. Prognosis in the shaken baby syndrome is poor.[9]

CT is the primary neuroimaging tool for identifying damage due to nonaccidental trauma because it depicts both intracranial hemorrhages and skull fractures. CT is specifically superior to MRI in detecting subarachnoid hemorrhages, which are present in up to three-quarters of victims of shaking. However, CT has limitations as it may miss over 70% of small subdural hematomas in the skull base and high parietal region.[10] Although small subdural hematomas are generally clinically insignificant, their identification may be crucial for the diagnosis of child abuse. Subacute (isodense) subdural hematomas are also easily overlooked (maxim 87) and it is difficult to differentiate a chronic subdural hematoma from a postinfectious effusion on CT. Thus, in cases of potential child abuse, where the head CT is negative or equivocal, MRI should be performed (Fig. 81.2).[11,12] MRI is more sensitive for identifying and estimating the age of a subdural hematoma and detecting coexisting primary parenchymal lesions, such as cortical contusions and shear injuries. Hematomas of various ages should increase

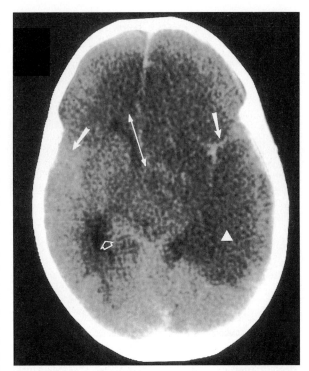

Fig. 81.1 Child abuse. Axial noncontrasted CT of a 3-year-old child. Hyperdense lateral fissures (arrows) are definite evidence of a subarachnoid hemorrhage. Diffuse cerebral edema is manifested by absent sulci and decreased parenchymal density, which is worse in the left temporal lobe (white triangle). The severe edema is resulting in mass effect, shifting the midline structures to the right (double-headed arrow approximates the septum pellucidum) and compressing the foramen of Monro causing dilation of the right temporal horn (open white arrow). The parenchymal abnormalities suggest ischemia. The subarachnoid hemorrhage is indicative of shaken baby syndrome.

the suspicion of child abuse and prompt evaluation of the social situation. Ultrasonography also can reveal brain damage resulting from child abuse in infants.[13]

Upper spinal cord injuries may occur in a shaken baby. Subdural hematomas, epidural hematomas, or contusions near the cervical medullary junction may cause severe morbidity or death.[14] MRI is superior to CT for evaluating the cervical spine and craniocervical junction.

Anoxic or ischemic brain injury may result from strangulation. Diffuse brain edema with loss of the gray–white matter interface occurs with significant damage. The cerebellum, brain stem, and thalami are usually spared.

References

1. Caffey J. Multiple fractures in the long bones of infants suffering from chronic subdural hematoma. *AJR Am J Roentgenol* 1946; **56**: 163–73.
2. Silverman FN. The radiologic manifestations of unrecognized skeletal trauma in infants. *AJR Am J Roentgenol* 1953; **69**: 413–27.
3. Merten DF, Carpenter BL. Radiologic imaging of inflicted injury in the child abuse syndrome. *Pediatr Clin North Am* 1990; **37**: 815–37.

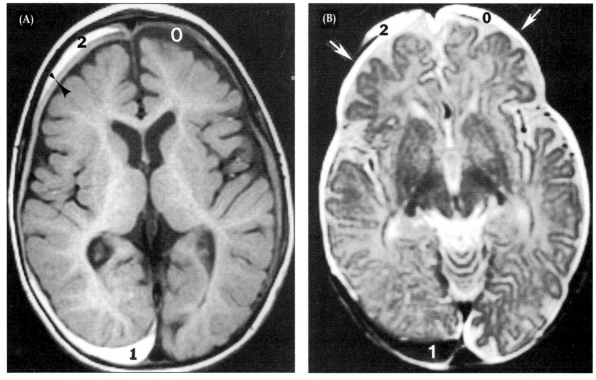

Fig. 81.2 Nonaccidental trauma in a child. **A:** Axial T1-weighted image in an 8-month-old child. A membrane (black arrowheads) separates two subdural hematomas. Their different signal characteristics indicate that they occurred at different times. The deeper subdural hematoma extends from the frontal to the occipital pole. **B:** Axial T2-weighted image reveals bilateral subdural hematomas (white arrows) with different signal characteristics. Regions that have low T1 signal (gray crescent just anterior to O in Fig. A) and high T2 signal (0) consist of intracellular oxyhemoglobin and are less than 1 day old. Regions that have high T1 signal and low T2 signal (1) consist of intracellular methemoglobin and are at least 3 days old. Regions that have high T1 and T2 signals (2) consist of extracellular methemoglobin and are at least 7 days old (maxim 60). Subdural hematomas of differing ages in a child strongly suggest child abuse.

4. Caffey J. The whiplash shaken infant syndrome: manual shaking by the extremities with whiplash-induced intracranial and intraocular bleeding, linked with residual permanent brain damage and mental retardation. *Pediatrics* 1974; **54**: 396–403.
5. Alexander R, Sato Y, Smith W, Bennett T. Incidence of impact trauma with cranial injuries ascribed to shaking. *Am J Dis Child* 1990; **144**: 724–6.
6. Duhaime AC, Gennarelli TA, Thibault LE, Bruce DA, Margulies SS, Wiser R. The shaken baby syndrome: a clinical, pathological, and biomechanical study. *J Neurosurg* 1987; **66**: 409–15.
7. Buys YM, Levin AV, Enzenauer RW, *et al*. Retinal findings after head trauma in infants and young children. *Ophthalmology* 1992; **99**: 1718–23.
8. Munger CE, Peiffer RL, Bouldin TW, Kylstra JA, Thompson RL. Ocular and associated neuropathologic observations in suspected whiplash shaken infant syndrome: a retrospective study of 12 cases. *Am J Forensic Med Pathol* 1993; **14**: 193–200.
9. Ludwig S, Warman M. Shaken baby syndrome: a review of 20 cases. *Ann Emerg Med* 1984; **13**: 104–7.
10. Sato Y, Yuh WT, Smith WL, Alexander RC, Kao SC, Ellerbroek CJ. Head injury in child abuse: evaluation with MR imaging. *Radiology* 1989; **173**: 653–7.
11. Levin AV, Magnusson MR, Rafto SE, Zimmerman RA. Shaken baby syndrome diagnosed by magnetic resonance imaging. *Pediatr Emerg Care* 1989; **5**: 181–6.
12. Bernardi B, Zimmerman RA, Bilaniuk LT. Neuroradiologic evaluation of pediatric craniocerebral trauma. *Top Magn Reson Imaging* 1993; **5**: 161–73.
13. Zepp F, Bruhl K, Zimmer B, Schumacher R. Battered child syndrome: cerebral ultrasound and CT findings after vigorous shaking. *Neuropediatrics* 1992; **23**: 188–91.
14. Hadley MN, Sonntag VK, Rekate HL, Murphy A. The infant whiplash-shake injury syndrome: a clinical and pathological study. *Neurosurgery* 1989; **24**: 536–40.

82. *Transcranial ultrasonography identifies intracranial hemorrhage in the newborn*

Cranial ultrasonography (US) is the primary screening method for evaluating the newborn brain and especially for predicting neurologic outcome in the asphyxiated term and preterm infant. It is very effective at detecting and grading intraventricular hemorrhage, other intracranial hemorrhages, periventricular leukomalacia, tumors, hydrocephalus, and polycystic encephalomalacia. US is performed in the newborn intensive care unit, eliminating the risks of transport, sedation, intravenous contrast, and ionizing radiation (CT). Coronal and sagittal images are obtained with a real-time transducer imaging through the anterior fontanelle. Vascular flow can be quantified with Doppler US, which is useful for evaluating vascular malformations and venous sinus thrombosis (maxim 54). All these factors and its relatively low cost make US ideal for screening. Its major limitations are poor visualization of the brain periphery, brain stem, and fourth ventricle, inadequate differentiation between gray and white matter structures, and relative insensitivity to calcium and edema.

Preterm infants commonly develop periventricular (germinal matrix) hemorrhages. Intraventricular, subarachnoid, and other intracerebral hemorrhages may also occur. In contrast, intracranial hemorrhages are infrequent in full-term infants. Symptomatic intracranial hemorrhages in the full-term infant are usually subdural, cortical, or intraparenchymal, and are associated with birth trauma, arteriovenous malformations, emboli, or hypoxia. Surprisingly, 5–10% of apparently normal full-term infants may have an intracranial hemorrhage.[1]

One-half of germinal matrix hemorrhages occur within the first day of life and 90% develop by day 6. Germinal matrix hemorrhages tend to occur with hypoxia or ischemia, which produces relative cerebral hypertension secondary to hypercapnea-induced increased cerebral blood flow.[2] The increased arterial hypertension leads to rupture of fragile arteries and capillaries within the germinal matrix.

Grading neonatal intraventricular hemorrhages is useful (Table 82.1), although prognostication needs to be done with caution.[3,4] The presence of periventricular leukomalacia, parenchymal echodensities/lucencies, ventricular enlargement, and, to a lesser extent, intraventricular hemorrhage correlates with poor neurodevelopmental outcome in preterm infants.[5,6] Serial US studies are useful to follow the resolution of the hematoma and enable assessment of ventricular size and the need for a shunt.

Usually grade I and II hemorrhages completely resolve within a few days to a few weeks. Although grade II hemorrhages sometimes progress and moderate ventriculomegaly develops, shunting is rarely required. It may be difficult to define a grade II hemorrhage because blood in the normal-sized ventricle may be hard to visualize. Grade III intraventricular hemorrhages are more easily identified because the ventricle is dilated. Grade III intraventricular hemorrhages usually resolve within a few weeks to a few months. They occasionally leave intraventricular septations.

Ventriculomegaly, which develops within 2 weeks of the hemorrhage, spontaneously resolves or arrests in three-quarters of patients. Ventriculomegaly is enlargement of the ventricles with the head circumference increasing less than 2 cm per week and no clinical signs of increased intracranial pressure. In the remaining one-quarter, the ventricular dilatation progresses to obstructive hydrocephalus (more common in grade IV

Table 82.1 Neonatal intraventricular hemorrhage grades

Grade	Criteria
I	Germinal matrix (subependymal) hemorrhage
II	Germinal matrix hemorrhage with intraventricular hemorrhage but normal ventricular size
III	Germinal matrix hemorrhage with intraventricular hemorrhage and ventricular dilatation
IV	Germinal matrix hemorrhage with intraventricular hemorrhage and intraparenchymal hemorrhage

hemorrhage). Obstructive hydrocephalus is defined as progressive enlargement of the ventricles with increasing head size and clinical symptoms of increased intracranial pressure. The differentiation between ventriculomegaly and hydrocephalus is clinical.

US does not accurately differentiate parenchymal hemorrhage from edema, infarction, or necrosis, and underestimates cortical and basal ganglia lesions. US is also relatively insensitive to subarachnoid hemorrhage, which may be significant in the newborn because it can result in communicating hydrocephalus. CT is very useful in the newborn because it is sensitive to acute hemorrhage in any compartment and to other pathologies including extra-axial fluid collections, mass lesions, and congenital malformations. CT will detect choroid plexus hemorrhage, which may be missed on US. CT is warranted in the neonate if shunt placement is anticipated. In asphyxiated babies, 1–2 weeks after birth, CT may demonstrate vascular proliferation and hyperperfusion in the basal ganglia, brain stem, cerebellum, periventricular region, depth of the cortical sulci, and hippocampus. However, abnormalities in the basal ganglia and thalamus may develop in a variety of conditions and may be detectable only on US.[7] US, with higher-energy transducers, appears to be accurate for assessing damage in neonatal hypoxia–ischemia.[8]

Extracorporeal membrane oxygenation (ECMO) treatment increases the risk of intracerebral hemorrhage and infarctions because of heparinization, hypoxia, and impaired autoregulation.[9,10] Small, grade I intraventricular hemorrhages do not usually progress on ECMO, but the presence of higher grades of intraventricular hemorrhage or other evidence of hypoxic–ischemic damage increases the likelihood of a poor outcome.[11,12]

Although MRI provides superior anatomic information and inherent tissue sensitivity, the normally high water content of the immature brain limits resolution during the first week of life. Additionally, the time required to produce high-quality MRI and the need for patient sedation and MRI-compatible life support devices limit the role of MRI in fragile preterm neonates. CT is a good compromise when additional imaging beyond US is needed, especially when the neonate is on life support. MRI is more appropriately used in follow-up, beginning at 1–2 months of age, to delineate brain damage and myelination pattern (maxim 76).[13]

References

1. Heibel M, Heber R, Bechinger D, Kornhuber HH. Early diagnosis of perinatal cerebral lesions in apparently normal full-term newborns by ultrasound of the brain. *Neuroradiology* 1993; **35**: 85–91.
2. Volpe JJ. Current concepts of brain injury in the premature infant. *AJR Am J Roentgenol* 1989; **153**: 243–51.
3. Lin JP, Goh W, Brown JK, Steers AJ. Heterogeneity of neurological syndromes in survivors of grade 3 and 4 periventricular hemorrhage. *Childs Nerv Syst* 1993; **9**: 205–14.
4. Volpe JJ. Current concepts in neonatal medicine: neonatal intraventricular hemorrhage. *N Engl J Med* 1981; **304**: 886–8.
5. Pinto-Martin JA, Riolo S, Cnaan A, Holzman C, Susser MW, Paneth N. Cranial ultrasound prediction of disabling and nondisabling cerebral palsy at age two in a low birth weight population. *Pediatrics* 1995; **95**: 249–54.
6. van de Bor M, den Ouden L, Guit GL. Value of cranial ultrasound and magnetic resonance imaging in predicting neurodevelopmental outcome in preterm infants. *Pediatrics* 1992; **90**: 196–9.
7. Cabanas F, Pellicer A, Morales C, Garcia-Alix A, Stiris TA, Quero J. New pattern of hyperechogenicity in thalamus and basal ganglia studied by color Doppler flow imaging. *Pediatr Neurol* 1994; **10**: 109–16.
8. Eken P, Jansen GH, Groenendaal F, Rademaker KJ, de Vries LS. Intracranial lesions in the fullterm infant with hypoxic ischaemic encephalopathy: ultrasound and autopsy correlation. *Neuropediatrics* 1994; **25**: 301–7.
9. Bulas DI, Taylor GA, Fitz CR, Revenis ME, Glass P, Ingram JD. Posterior fossa intracranial hemorrhage in infants treated with extracorporeal membrane oxygenation: sonographic findings. *AJR Am J Roentgenol* 1991; **156**: 571–5.
10. Lazar EL, Abramson SJ, Weinstein S, Stolar CJ. Neuroimaging of brain injury in neonates treated with extracorporeal membrane oxygenation: lessons learned from serial examinations. *J Pediatr Surg* 1994; **29**: 186–90.
11. von Allmen D, Babcock D, Matsumoto J, *et al*. The predictive value of head ultrasound in the ECMO candidate. *J Pediatr Surg* 1992; **27**: 36–9.
12. Radack DM, Baumgart S, Gross GW. Subependymal (grade 1) intracranial hemorrhage in neonates on extracorporeal membrane oxygenation: frequency and patterns of evolution. *Clin Pediatr* 1994; **33**: 583–7.
13. de Vries LS, Eken P, Groenendaal F, van Haastert IC, Meiners LC. Correlation between the degree of periventricular leukomalacia diagnosed using cranial ultrasound and MRI later in infancy in children with cerebral palsy. *Neuropediatrics* 1993; **24**: 263–8.

83. Infants with rapidly increasing head circumference may have benign enlargement of the subarachnoid spaces

The head circumference of some infants increases much more rapidly then expected with no clinical or radiographic signs of hydrocephalus or macrocephaly (enlarged brain). CT and MRI usually demonstrate enlarged bifrontal CSF spaces with normal ventricular size (Fig. 83.1). This enlargement of the frontal subarachnoid space is relatively benign.[1] It must be distinguished from pathologic subdural collections to avoid unnecessary surgery.[2] CT is frequently not adequate to

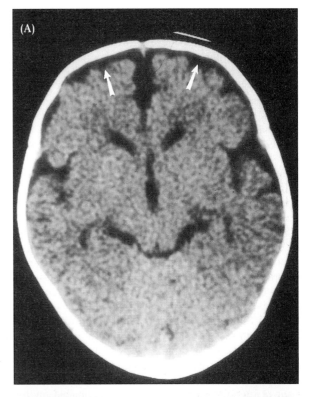

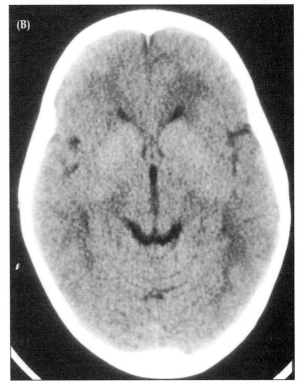

Fig. 83.1 Benign enlargement of the subarachnoid spaces in infants. **A:** Axial CT in a 2-month-old infant with macrocephaly but no signs of increased intracranial pressure or developmental delay. Enlarged frontal subarachnoid spaces (arrows) with normal-sized ventricles are present. No treatment was given. **B:** Axial CT in the same child at age 2.5 years. The scan is normal, as is the child.

determine the location of the extra-axial fluid collections (subarachnoid vs. subdural) or to differentiate CSF from pathologic subdural collections.[3] MRI readily differentiates between these two conditions because the signal characteristics of CSF, blood products, and high-protein-content collections are very different (Fig. 81.2).

The benign enlargement of the subarachnoid space in infants with normal development has several names: external hydrocephalus, benign communicating hydrocephalus, benign subdural effusions, benign subdural collections of infancy, and benign enlargement of the subarachnoid spaces. The best term is the descriptive name "benign enlargement of the subarachnoid spaces",[4] which implies dilatation of the subarachnoid space without dilatation of the ventricles and preserved CSF circulation.

In affected children, the head circumference increases dramatically during the first few months of life. Later in infancy, the head circumference stabilizes. The extra-axial fluid collections typically resolve within 24 months (Fig. 83.1B).[5] This form of "macrocephaly" is familial in over 85% of cases.[6] It is also associated with steroids, chemotherapy, and malnutrition. In contrast, abnormal subdural collections are usually postinfectious or post-traumatic (including birth trauma). A transient CSF resorption disturbance resulting from delayed maturation of the arachnoid villi may be responsible for the accumulation of the subarachnoid fluid and explains the resolution over time.

The presence of symmetrically enlarged bifrontal subarachnoid spaces, with near-normal ventricles, in a patient less than 2 years of age usually indicates a benign condition, and neurosurgical treatment is rarely warranted. However, when a two-compartment lesion consisting of an inner CSF layer and an outer pathologic layer with higher signal on T1 or T2 images is present on MRI, drainage is often warranted.

References

1. Odita JC. The widened frontal subarachnoid space: a CT comparative study between macrocephalic, microcephalic, and normocephalic infants and children. *Childs Nerv Syst* 1992; **8**: 36–9.
2. Andersson H, Elfverson J, Svendsen P. External hydrocephalus in infants. *Childs Brain* 1984; **11**: 398–402.
3. Wilms G, Vanderschuren G, Demaerel PH, *et al.* CT and MRI in infants with pericerebral collections and macrocephaly: benign enlargement of the subarachnoid spaces *versus* subdural collections. *AJNR Am J Neuroradiol* 1993; **14**: 855–60.
4. Barkovich J. *Pediatric neuroimaging.* New York: Raven Press, 1990: 220–1.
5. Maytal J, Alvarez LA, Elkin CM, Shinnar S. External hydrocephalus: radiologic spectrum and differentiation from cerebral atrophy. *AJR Am J Roentgenol* 1987; **148**: 1223–30.
6. Alvarez LA, Maytal J, Shinnar S. Idiopathic external hydrocephalus: natural history and relationship to benign familial macrocephaly. *Pediatrics* 1986; **77**: 901–7.

84. *Multiple sclerosis is in the differential diagnosis of childhood white-matter diseases*

Multiple sclerosis (MS) is a rare, but not unknown, disorder in children and adolescents. The proportion of MS in children ranges from 0.5% to over 5% of all MS cases.[1–3] The female-to-male ratio in adults is 2 : 1, but in children it is higher at 3 : 1 to 5 : 1.[4,5] Because of the rarity of the disease and the transient and often vague symptoms, MS is seldom considered in the differential diagnosis of childhood diseases.[6] In children with MS, cerebellar and brain-stem symptoms are most typical, but the distribution and appearance of MS plaques on MRI are similar in children and in adults.[7,8]

The differential diagnosis for MRI T2 and PD high-signal white-matter foci is given in maxim 16. In children, acute disseminated encephalomyelitis (ADEM), a monophasic episode of inflammatory demyelination, is more common than MS. ADEM often follows vaccination or viral illness and is probably the result of an immune-mediated cross-reaction with CNS myelin. Although the MRI appearance of ADEM may be identical to that of MS, the lesions in ADEM tend to be more peripheral (adjacent to the gray–white matter junction) and have ill-defined borders. ADEM lesions resolve within weeks or months, making serial imaging useful. In children, mitochondrial encephalomyopathy, herpes simplex encephalitis, mucopolysaccharidosis, adrenoleukodystrophy, metachromatic leukodystrophy, Krabbe disease, Canavan disease, Alexander disease, orthochromatic leukodystrophy, progressive multifocal leukoencephalopathy, HIV, CNS lymphoma, and other brain tumors and abscesses are also considerations.[9]

The MRI findings in children, like those in adults, are nonspecific.[10] MS lesions may be confined to the spinal cord (termed Devic syndrome when accompanied by optic neuritis) or may be solitary.[11] Most, but not all, solitary MS lesions have no mass effect, aiding in the differentiation from a tumor or abscess. Intravenous contrast is not very helpful, as acute MS plaques, tumors, and abscesses may all enhance. The finding of additional, even subtle, white-matter lesions in a child suggests MS because vascular lesions and UBOs are extremely uncommon. The presence of multiple lesions may obviate the need for a biopsy. Age is not a helpful feature because MS may begin in infancy.[12]

Optic neuritis can herald MS in children just as in adults. One-half to three-quarters of children who present with optic neuritis develop MS.[13–15] Thus, head MRI should be considered in children with optic neuritis.

References

1. Gall JC, Hayles AB, Siekert RG, Keith HM. Multiple sclerosis in children: a clinical study of 40 cases with onset in childhood. *Pediatrics* 1958; **21**: 703–9.
2. Sindern E, Haas J, Stark E, Wurster U. Early onset MS under the age of 16: clinical and paraclinical features. *Acta Neurol Scand* 1992; **86**: 280–4.
3. Guilhoto LM, Osorio CA, Machado LR, *et al*. Pediatric multiple sclerosis report of 14 cases. *Brain Dev* 1995; **17**: 9–12.
4. Duquette P, Murray TJ, Pleines J, *et al*. Multiple sclerosis in childhood: clinical profile in 125 patients. *J Pediatr* 1987; **111**: 359–63.
5. Ebner F, Millner MM, Justich E. Multiple sclerosis in children: value of serial MR studies to monitor patients. *AJNR Am J Neuroradiol* 1990; **11**: 1023–7.
6. Hanefeld F, Bauer HJ, Christen HJ, Kruse B, Bruhn H, Frahm J. Multiple sclerosis in childhood: report of 15 cases. *Brain Dev* 1991; **13**: 410–16.
7. Osborn AG, Harnsberger HR, Smoker WR, Boyer RS. Multiple sclerosis in adolescents: CT and MR findings. *AJNR Am J Neuroradiol* 1990; **11**: 489–94.
8. Scaioli V, Rumi V, Cimino C, Angelini L. Childhood multiple sclerosis (MS): multimodal evoked potentials (EP) and magnetic resonance imaging (MRI) comparative study. *Neuropediatrics* 1991; **22**: 15–23.
9. Kolodny EH. Dysmyelination and demyelinating conditions in infancy. *Curr Opin Neurol Neurosurg* 1993; **6**: 379–86.
10. Valk J, van der Knaap MS. *Magnetic resonance of myelin, myelination, and myelin disorders*. New York: Springer-Verlag, 1989.
11. Glasier CM, Robbins MB, Davis PC, Ceballos E, Bates SR. Clinical, neurodiagnostic, and MR findings in children with spinal and brain stem multiple sclerosis. *AJNR Am J Neuroradiol* 1995; **16**: 87–95.
12. Shaw CM, Alvord EC Jr. Multiple sclerosis beginning in infancy. *J Child Neurol* 1987; **2**: 252–6.
13. Riikonen R, Ketonen L, Sipponen J. Magnetic resonance imaging, evoked responses and cerebrospinal fluid findings in a follow-up study of children with optic neuritis. *Acta Neurol Scand* 1988; **77**: 44–9.
14. Rizzo JF III, Lisle S. Risk of developing multiple sclerosis after uncomplicated optic neuritis: a long-term prospective study. *Neurology* 1988; **38**: 185–90.
15. Francis DA, Compston DA, Batchelor JR, McDonald WI. A reassessment of the risk of multiple sclerosis developing in patients with optic neuritis after extended follow-up. *J Neurol Neurosurg Psychiatr* 1987; **50**: 758–65.

13
Trauma

85. *Use CT to evaluate head injuries; skull radiographs are rarely indicated*

CT is the primary neuroimaging test for patients with acute head trauma. It enables detection of skull fractures, subdural hematomas, epidural hematomas, brain contusions, subarachnoid hemorrhages, pneumocephalus, penetrating objects (e.g. bullets), and diffuse axonal injury. Early recognition of intracranial injury is important to facilitate neurosurgical intervention in appropriate cases. Traumatic brain injuries may cause extra-axial or intra-axial lesions that develop at or after the time of the trauma.

Any victim of head trauma with a penetrating head wound, depressed skull fracture, decreased level of consciousness, or focal neurologic sign should have non-contrasted head CT. CT should not be performed in patients who are asymptomatic or have only a mild headache, mild dizziness, scalp hematoma, scalp laceration, superficial contusion, or abrasion.[1] Individual decisions need to be made for patients between these extremes.

Plain skull radiographs are inappropriate in the screening evaluation of a victim of head trauma.[2] If used, they may provide a false sense of security because of their low sensitivity. They are sometimes useful for defining skull fractures before neurosurgical intervention.

MRI is useful for identifying diffuse axonal injury and hematomas, but is usually impractical in the acute setting.

Skull fractures

Skull fractures can be linear, depressed, or compound. Linear skull fractures can be missed on CT, especially if the image orientation is in the same plane as the fracture. The presence of a linear fracture indicates a significant head trauma but does not correlate well with underlying brain injury. Thus, missing a linear fracture on CT is generally without consequence.

Depressed skull fractures are usually associated with brain injury (Fig. 85.1). Bone fragments below the inner table can cause dural tears and brain contusion, and be a source of infection. Depressed skull fractures are readily identified on CT bone window settings. Surgery is usually necessary.

Basilar skull fractures are associated with injuries to the internal carotid artery, cavernous sinus, or cranial nerves. Dural tears may produce a chronic CSF leak through the nose (rhinorrhea) or ear (otorrhea), increasing the risk of meningitis and intracranial abscess. Temporal bone fractures may result in facial paralysis, sensorineural hearing loss, vertigo, or nystagmus. Raccoon eyes, Battle's sign (postauricular ecchymosis), and hemotympanum are clinical signs of basilar skull fracture. CT with thin sections through the skull base usually demonstrates the fracture. Surface rendering (three-dimensional reconstruction) aids in detecting basilar skull fractures. Three-dimensional reconstruction is also valuable for facial and spine fractures. Treatment of a basilar skull fracture is usually conservative except for CSF leaks, which require surgical repair.

Separated (diastatic) sutures are well visualized on CT, especially with surface rendering. Traumatic suture separations with interposition of meninges leading to progressive expansion and leptomeningeal cysts may occur in children under 2 years old. Surgery is necessary for proper healing.

Epidural hematomas

See maxim 86.

Subdural hematomas

See maxim 87.

Diffuse axonal injury

Diffuse axonal injury (DAI) results from acceleration and deceleration trauma, which causes shearing of the white matter. The gray–white matter junction, corpus callosum, corona radiata, and internal capsule are the most commonly affected locations. MRI is best for detecting DAI, but evidence for this injury, especially when severe, is usually present on CT. DAI appears as patchy low-density areas without significant mass effect on CT and high T2 signal areas on MRI in the acute setting. Punctate hemorrhages indicate more severe damage. Diffuse brain swelling is ominous, especially if present on the initial study. Mild DAI may not be detectable on CT or MRI.

Cortical contusions

Contusions are usually located in the cortex and do not involve the deep white matter. They are most frequent in the inferior frontal or temporal lobes where the brain is adjacent to the rough basilar skull (Fig. 85.2). Contusions develop at the site of impact (coup) or directly opposite the impact (contrecoup) due to acceleration–deceleration forces.

Contusions are not independently correlated with impaired consciousness or poor long-term neurologic outcome, but they are often a companion of more significant brain injury, such as compound skull fractures and DAI. Contusions are frequently present in fatally injured head-trauma victims.

Acute hemorrhagic contusions appear as high-density foci on CT and are often surrounded by a rim of low-density edema. "Nonhemorrhagic contusions" appear as low-density regions and may be difficult to detect with CT because they can blend with the cortical sulci. Several days after the trauma, these lesions typically become more visible due to increasing edema, mass effect, and sometimes delayed hemorrhage. MRI often demonstrates cortical contusions better than CT, especially as

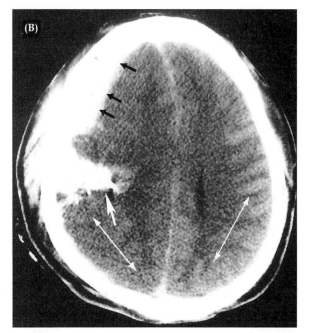

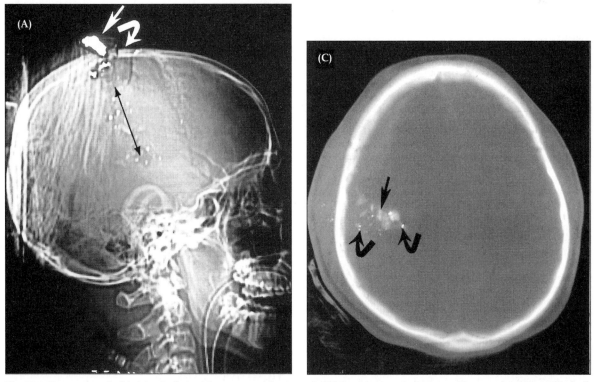

Fig. 85.1 Penetrating skull injury (bullet) with compound fracture. **A:** CT scout image of a patient with a gunshot injury. The bulk of the bullet is located in the subcutaneous tissue and skull (straight white arrow). Fragments of the bullet penetrate deep into the brain along the course of the double-headed black arrow. A compound fragmented skull fracture is present (curved white arrow). **B:** Axial image in the same study with soft-tissue (brain window) settings (window width 80, level 40). A subdural hematoma (small black arrows) and intraparenchymal hemorrhage (white arrow) are present in the right frontoparietal region. The gray–white differentiation is disrupted in the right hemisphere (compare regions with double-headed white arrows) and there is midline shift to the left indicative of mass effect. **C:** The same image as in B but with bone window settings (window width 3000, level 200). At this setting, it is possible to differentiate effectively between the bullet fragments, which are very dense (curved arrows), and bone fragments which are moderately dense (straight arrow). The blood and brain cannot be differentiated at this setting.

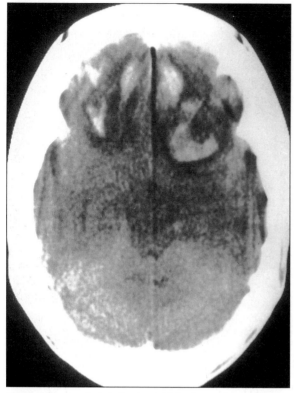

Fig. 85.2 Traumatic inferior frontal lobe hemorrhages (why to scan before tapping!). Axial noncontrasted CT in a victim of head trauma. Bilateral inferior frontal intracerebral hemorrhages are present with surrounding edema. This black man was brought to the emergency department by friends because of several days of a personality change. In the emergency department he had a fever to 39°C and hematuria. He was apathetic and offered no complaints. He had a "non-focal" neurologic examination according to the emergency department physicians. Because of the fever, he underwent a lumbar puncture. Neuroimaging was not performed before the lumbar puncture because there was "no evidence" of a mass lesion. Following the lumbar puncture, his level of consciousness diminished, and neurologic consultation and this CT were obtained. A brain herniation cascade developed (maxim 42) and he died 6 hours after the lumbar puncture despite aggressive treatment. Ultimately it was learned that he had been beaten with a baseball bat during an assault 2 days earlier. Superficial ecchymoses were not apparent because of his dark skin color. The hematuria and probably fever were the result of kidney trauma (a perirenal hematoma was present at autopsy). The apathy was the only evidence of the bifrontal lobe injury and should have been considered a "focal neurologic sign". Fortunately, this type of situation is uncommon; however, it illustrates the point that in unknown circumstances a head neuroimaging study should be obtained before lumbar puncture.

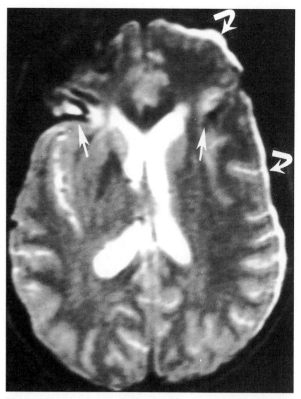

Fig. 85.3 Traumatic cerebral contusions. Axial T2-weighted image of a 26-year-old 2 weeks after a motor vehicle accident. Several contusions are present in the frontal lobes, consisting of high signal (extracellular methemoglobin) surrounded by low signal (hemosiderin) (straight arrows). A subacute left subdural hematoma is also present (curved arrows).

Deep-gray-matter injuries

Traumatic damage to the basal ganglia or thalamus is rare. When present, the morbidity and mortality rate are high. Deep-gray-matter injuries are typically hemorrhagic due to laceration of small perforating blood vessels. They appear as high-density foci on CT, but are better identified with MRI.

Brain-stem injury

Brain-stem injuries are poorly demarcated on CT owing to beam-hardening artifact in the posterior fossa. Traumatic brain-stem injury is usually indirect and accompanied by supratentorial DAI.

Pneumocephalus

The presence of intracranial air (black on CT and MRI) after head trauma indicates a dural rupture and requires that a source be identified (Fig. 61.1). It is often located in the anterior portion of the anterior fossa (paranasal sinus region) or adjacent to the sphenoid bone in the middle cranial fossa. Fractures in the posterior wall of the frontal sinus or ethmoid air cells frequently result in

significant artifact at the bone–brain interface does not occur on MRI. Bland traumatic injuries appear as high-signal foci on T2-weighted images. Acute and subacute hemorrhagic injuries are better evaluated on T1-weighted images (Table 60.1). MRI is a valuable tool in the subacute period because hematomas become isodense to brain on CT several days to weeks after the initial hemorrhage (Fig. 85.3).

pneumocephalus. Sphenoid sinus fractures are associated with parasellar air. Temporal bone fractures are usually not accompanied by pneumocephalus. The presence of pneumocephalus increases the likelihood of meningitis and brain abscess.

References

1. Masters SJ, McClean PM, Arcarese JS, *et al*. Skull x-ray examinations after head trauma: recommendations by a multidisciplinary panel and validation study. *N Engl J Med* 1987: **316**: 84–91.
2. Masters SJ. Evaluation of head trauma: efficacy of skull films. *AJNR Am J Neuroradiol* 1980; **1**: 329–37.

86. *Epidural hematomas may be associated with a lucid interval. If visible on CT, they indicate an emergency*

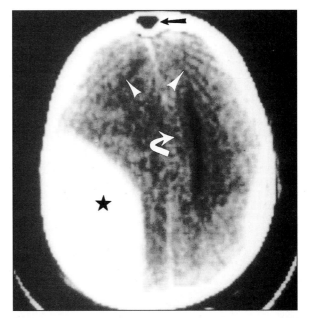

Fig. 86.1 Epidural hematoma. Noncontrasted axial CT discloses a large acute right-parietal epidural hematoma (star). Midline shift with subfalcial herniation is present (curved white arrow). The alternating white and black streaks are beam-hardening artifact (white arrowheads). The frontal sinus is visible at this level (black arrow).

Epidural hematomas are nearly always acute and are medical emergencies. Most are due to arterial bleeding, unlike subdural hematomas which usually have a venous source. Almost all epidural hematomas are associated with a skull fracture. They are uncommon in patients under 2 years or over 60 years of age. Epidural hematomas are usually located in the supratentorial compartment and are rare in the posterior fossa. The majority are the result of a tear in the middle meningeal artery. Less commonly a laceration in one of the other meningeal arteries or dural sinuses causes an epidural hematoma. Emergent surgical drainage is frequently life-saving.

Most small epidural hematomas (less than 1 cm) tend not to enlarge and can be followed without surgical treatment.[1] However, early CT can be misleading because delayed expansion of an epidural hematoma may occur. Epidural hematomas are one of the major reasons why serial neurologic examinations are necessary in patients with head injuries.[2–5]

The classic clinical course is initial loss of consciousness due to a concussion, followed by an awake (lucid) period lasting up to several hours, followed by a second loss of consciousness associated with expansion of the epidural hemorrhage. The lucid interval may be absent if the initial concussion results in a prolonged coma. Once an epidural hematoma enlarges, the neurologic examination usually reveals a hemiparesis which may be contralateral (due to direct compression of the adjacent cerebral hemisphere) or ipsilateral (due to mass effect with compression of the contralateral cerebral peduncle at the tentorial incisura). Either ipsilateral or bilateral mydriasis and ophthalmoparesis can result from uncal herniation compressing the third cranial nerve (maxim 42). The neurologic dysfunction is a direct consequence of mass effect from the hematoma. There is usually minimal initial underlying brain parenchymal damage. Most epidural hematomas, in contrast to subdural hematomas, enlarge rapidly, presumably due to the high pressure of the arterial source.

Epidural hematomas have a convex-lens shape and are located adjacent to the skull (Fig. 86.1). When the middle meningeal artery is the source, they are located over the frontotemporal convexity. Because they consist of acute blood, they are high density on CT.[6] Sometimes "hypodense bubbles" are present, suggesting fresh bleeding or venous sinus involvement.[7,8] The characteristic semicircular configuration is a result of dural attachments at the cranial sutures. The limited extension helps to differentiate an epidural from a subdural hematoma. Subdural hematomas course along the entire surface of a hemisphere (maxim 87).

It is worthwhile emphasizing the CT pattern of epidural blood because small epidural hematomas, before expansion with resulting clinical symptoms, may be obscured by the adjacent skull, especially if appropriate window and level settings are not used (maxim 61). The CT findings in epidural hematoma, subarachnoid hemorrhage (maxim 58) and obliteration of the fourth ventricle (maxim 43) are three of the most important CT patterns to recognize. Anyone who examines head CTs (including the emergency physician, internist, family practitioner, and general surgeon) needs to be able to identify even the subtlest of these patterns. Emergency treatment in these three conditions is often life-saving, and delay of treatment can be fatal.

References

1. Servadei F, Faccani G, Roccella P, *et al.* Asymptomatic extradural hematomas: results of a multicenter study of 158 cases in minor head injury. *Acta Neurochir* 1989; **96**: 39–45.

2. Smith HK, Miller JD. The danger of an ultra-early computed tomographic scan in a patient with an evolving acute epidural hematoma. *Neurosurgery* 1991; **29**: 258–60.

3. Di Rocco A, Ellis SJ, Landes C. Delayed epidural hematoma. *Neuroradiology* 1991; **33**: 253–4.

4. Milo R, Razon N, Schiffer J. Delayed epidural hematoma: a review. *Acta Neurochir* 1987; **84**: 13–23.

5. Bricolo AP, Pasut LM. Extradural hematoma: toward zero mortality. *Neurosurgery* 1984; **14**: 8–12.

6. Reider-Groswasser I, Frishman E, Razon N. Epidural hematoma: computerized tomography (CT) parameters in 19 patients. *Brain Inj* 1991; **5**: 17–21.

7. Sener RN. Acute epidural hematoma with hypodense bubbles on CT: a sign of dural sinus tear or solely fresh bleeding? *Pediatr Radiol* 1993; **23**: 628.

8. Chee CP, Habib ZA. Hypodense bubbles in acute extradural hematomas following venous sinus tear: a CT scan appearance. *Neuroradiology* 1991; **33**: 152–4.

87. *Contrast can make an "invisible" isodense subdural hematoma readily apparent on CT*

Most subdural hematomas result from direct trauma to the head which tears the bridging veins traversing the space between the brain and dura mater. They may be bilateral. Sudden acceleration–deceleration (whiplash) injuries may also produce a subdural hematoma, especially in a young child (maxim 81). A forceful sneeze or cough, or retching, may cause a subdural hematoma in patients with atrophy because the bridging veins are stretched between the dura and the atrophic brain. Minor trauma can result in a subdural hematoma in patients who have a coagulopathy or who are taking anticoagulants. A common complication of over-aggressive ventricular shunting is a subdural hematoma (Figs 78.1C and 87.1). In this situation, excessive shrinkage of the ventricles followed by shifting of the brain away from the calvarium may rip the bridging veins.

Subdural hematomas are crescent-shaped, extending over the entire surface of a cerebral hemisphere with margins at the interhemispheric fissure (falx cerebri) and tentorium cerebelli (Figs. 61.1 and 61.2). In contrast to epidural hematomas (maxim 86), subdural hematomas freely cross the cranial suture lines. To varying degrees, effacement of the cortical sulci and lateral fissure, obliteration of the ipsilateral lateral ventricle, and midline shift with subfalcial herniation accompany subdural hematomas. Small acute subdural hematomas are difficult to detect using the standard brain parenchymal CT

window and level settings. A wide window (150–200) and a level of 60–80 helps to separate the high-density bone from high-density hemorrhage (maxim 61). Fixed or transient focal neurologic deficits may develop with all but the smallest subdural hematomas. Focal seizures that sometimes secondarily generalize may occur.

Intracerebral hematomas in general are initially hyperdense to brain and become less dense as they age.[1] They are isodense 1–4 weeks after they develop and hypodense thereafter (maxim 56). In anemic patients (hemoglobin below 10 g/mL), acute hematomas may be isodense to brain (maxim 61). Hematomas may briefly increase in density during the first 3 days due to clot retraction. Subsequently they progressively decrease in density by approximately 1.5 Hounsfield units per day. (Pure acute blood registers at 80–85 Hounsfield units. However, most acute blood is mixed with brain and edema and the actual measured Hounsfield number is usually 60–80. Brain parenchyma registers at 25–40 Hounsfield units; maxim 1.) Absorption of a hematoma starts at the periphery and progresses towards the center.

When subdural hematomas are isodense to brain, they are difficult to identify on noncontrasted CT.[2,3] The only clue to their presence may be subtle mass effect on the cortical sulci, or mild midline shift.[4] Bilateral isodense subdurals are especially easy to overlook because no asymmetries may be present. However, with intravenous contrast, the brain parenchyma usually enhances sufficiently to make the distinction between the non-enhancing blood collection and brain tissue possible.[5,6] In addition, the cortical veins are usually visible with contrast in this situation because they are distant from the skull (Fig. 87.1). As the subdural hematoma ages, it becomes hypodense to the brain parenchyma on CT.

A subdural hematoma may have an atypical appearance and heterogeneous density. It may be located in the interhemispheric region or in any other location along the dura. The inhomogeneity is caused by intermixed CSF from an arachnoid defect, uneven metabolism of blood, or small recurrent hemorrhages interspersing fresh blood with decaying blood (Fig. 87.2).[7]

MRI can be used to definitively detect and age a subdural hematoma.[8] However, when no abnormalities are recognized on CT, for instance when an isodense subdural hematoma is present, MRI may not be obtained. Missed isodense subdural hematomas are especially frequent when they are bilateral.[9] Because CT is obtained in most situations in which a subdural hematoma might occur (maxim 85), overlooking isodense subdural hematomas is an ongoing problem. MRI is particularly useful in the chronic stage because it distinguishes a liquefied subdural hematoma, which is composed of proteinaceous fluid, from a traumatic CSF collection (hygroma). On CT, hygromas and chronic subdural hematomas in the subdural space are both hypodense.

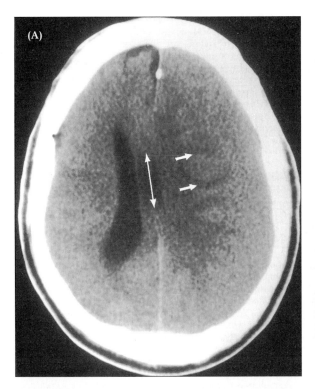

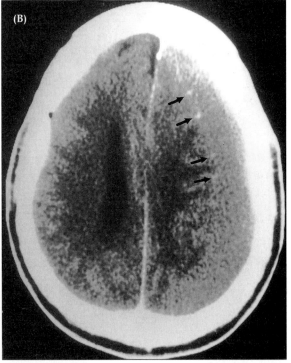

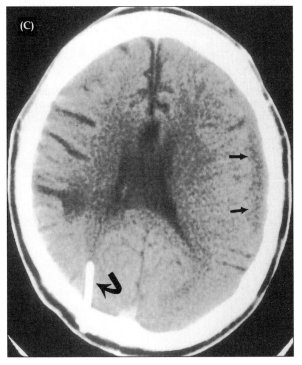

Fig. 87.1 Isodense subdural hematoma.

A: Noncontrasted axial CT at the level of the body of the lateral ventricle in a patient with a ventriculoperitoneal shunt (not shown). The body of the left lateral ventricle is effaced and the midline of the brain is shifted to the right (double-headed arrow in the center of the cavum septum pellucidum/vergae). No obvious mass lesion is present to account for the mass effect.

B: Contrast-enhanced axial CT in the same patient. The distinction between the cortex and the white matter is now obvious, and a left subdural hematoma isodense to the cortex is clearly discernible. Enhancing cortical veins line the surface of the brain (small black arrows). The cortical veins are not normally visualized because they are adjacent to the skull. Review of A reveals white-matter interdigitations (small white arrows) identifying the left hemisphere cortex and the isodense subdural hematoma.

C: Noncontrasted axial CT several months later reveals a residual hypodense subdural hematoma (straight arrows). The ventriculoperitoneal shunt is visible at this level (curved arrow). It had been surgically revised 2 months before A, resulting in excessive drainage and the subsequent development of the subdural hematoma.

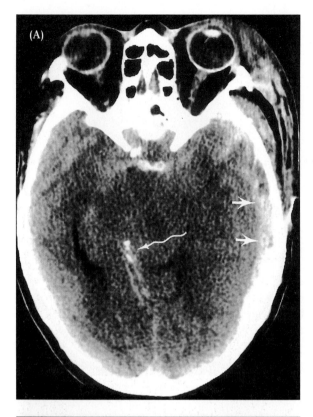

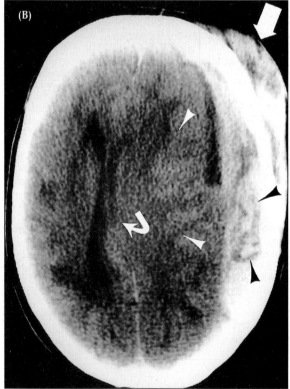

Fig. 87.2 Subdural hematoma with subfalcial and transtentorial (uncal) herniation. **A:** Noncontrasted axial CT at the level of the basal cisterns. The cisterns are obliterated and the brain stem is distorted (signs of uncal herniation). The pineal gland is shifted far to the right (wavy arrow). A left subdural hematoma with mixed densities is responsible (arrows). **B:** Noncontrasted axial CT of the same patient demonstrating two subdural hematomas with different density values and ages. The so-called "swirl sign" inside the lateral subdural hematoma (black arrowheads) represents active bleeding with unretracted liquid clot. The gray–white matter interface is displaced by a second large isodense subdural hematoma (arrowheads). Subfalcial herniation of the lateral ventricles is present (curved arrow). Extracranial soft-tissue swelling is indicated by the large white arrow.

References

1. Dolinskas CA, Bilaniuk LT, Zimmerman RA, Kuhl DE. Computed tomography of intracerebral hematomas. I. Transmission CT observations on hematoma resolution. *AJR Am J Roentgenol* 1977; **129**: 681–8.
2. Scotti G, Terbrugge K, Melancon D, *et al.* Evaluation of age of subdural hematomas by CT. *J Neurosurg* 1977; **47**: 311–15.
3. Weisberg LA. Analysis of the clinical and computed tomographic findings in isodense subdural hematoma. *Comput Radiol* 1986; **10**: 245–52.
4. Amendola MA, Ostrum BJ. Diagnosis of isodense subdural hematomas by computed tomography. *AJR Am J Roentgenol* 1977; **129**: 693–7.
5. Hayman LA, Evans RA, Hinck VC. Rapid-high-dose contrast computed tomography of isodense subdural hematoma and cerebral swelling. *Radiology* 1979; **131**: 381–3.
6. Tsai FY, Huprich JE, Segall HD, Teal JS. The contrast-enhanced CT scan in the diagnosis of isodense subdural hematoma. *J Neurosurg* 1979; **50**: 64–9.
7. Reed D, Robertson WD, Graeb DA, Lapointe JS, Nugent RA, Woodhurst WB. Acute subdural hematomas: atypical CT findings. *AJNR Am J Neuroradiol* 1986; **7**: 417–21.
8. Young IR, Bydder GM, Hall AS, *et al.* Extracerebral collections: recognition by NMR imaging. *AJNR Am J Neuroradiol* 1983; **4**: 833–4.
9. Greenhouse AH, Barr JW. The bilateral isodense subdural hematoma on computerized tomographic scan. *Arch Neurology* 1979; **36**: 305–7.

14
Spine and Spinal Cord: Congenital Malformations

88. *Mild Chiari I malformations are often discovered incidentally and are usually of no clinical significance*

The Chiari I malformation is a congenital hindbrain dysgenesis characterized by cerebellar tonsillar elongation and protrusion (ectopia) through the foramen magnum.[1] Since the advent of MRI, mild tonsillar ectopia is commonly identified on midsagittal MRI and is usually asymptomatic.

Cerebellar tonsils ascend (relative to the skull base) with increasing age. In the first decade the cerebellar tonsils may normally extend as low as 6 mm below the foramen magnum. During the second and third decades they should be less than 4 mm below the foramen magnum. In adolescents and adults, tonsillar ectopia of 5 mm or more is diagnostic of Chiari I malformation.

MR imaging is the best method of locating the cerebellar tonsils and to assess for a syrinx and determine its extent. Sagittal and coronal T1- and T2-weighted images readily demonstrate the relationship of the structures at the craniocervical junction. Fast T1-weighted sagittal images in flexion and extension may be used to determine whether functional cord compression is present. CT and conventional myelography are useful only in patients who cannot undergo MRI.

Symptoms of Chiari I malformation correlate with the extent of tonsillar ectopia. Patients with cerebellar tonsils 5–10 mm below the foramen magnum are often asymptomatic or minimally symptomatic.[2] When the tonsillar herniation exceeds 12 mm, symptoms are almost always present (Fig. 88.1). Manifestations of Chiari I malformation are due to stretching of the brain stem and spinal cord structures or to an associated syrinx. Symptoms include headache, neckache, vertigo, ataxia, nystagmus (especially vertical, downbeat), disequilibrium, myelopathy (with motor and sensory deficits), and asymmetric lower cranial nerve palsies (e.g. sensorineural hearing loss, dysphagia, dysarthria).[3,4] The pain classically worsens with straining or laughing.[5] Intermittent apnea may be present, especially in children. Symptomatic Chiari I malformations are somewhat more common in females. Surgical decompression of the base of the skull

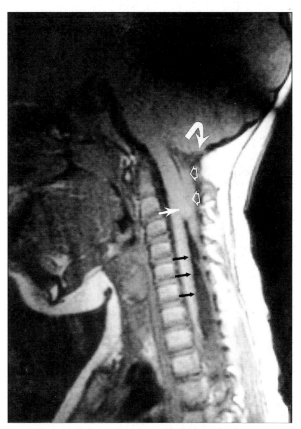

Fig. 88.1 Chiari I malformation. Off-midline sagittal T1-weighted image in a 3-year-old. The cerebellar tonsils are low-lying (better visualized on a more medial image; open white arrow). A cervicomedullar kink is present at the C2–C3 level (straight white arrow). The cisterna magna is obliterated (curved white arrow). Syringohydromyelia is present in the cervical spinal cord (small black arrows).

and spinal column relieves troubling symptoms and prevents progression in approximately three-quarters of cases.[6,7]

About 10% of patients with Chiari I malformation have mild-to-moderate hydrocephalus. Other brain anomalies are typically not components of the Chiari I malformation. However, Chiari I malformation is often associated with spinal cord and vertebral abnormalities (Fig. 78.5). One-half of patients with severe herniation have a syrinx (syringomyelia if the spinal cord alone is involved; syringobulbia if the brain stem is involved). Skeletal anomalies include basilar impression (platybasia), Klippel–Feil syndrome, and atlanto-occipital assimilation. Scoliosis may develop in patients with a syrinx.[8]

Chiari I malformations are due mainly to an undersized posterior fossa, which may be either congenital or acquired.[9,10] When the posterior fossa is too small, the growing brain expands its compartment by herniating the cerebellar tonsils through the foramen magnum. In some cases, the herniated tonsils, possibly through a ball-valve mechanism, disrupt CSF fluid dynamics and lead to syrinx formation. In Chiari I malformation, the syrinx is the result of expansion of the central canal and may or may not communicate with the fourth ventricle.[11] Chiari I malformations occur in rickets due to skull thickening.[12] Tonsillar herniation, meeting the criteria for a Chiari I malformation, occurs in the majority of children who have lumboperitoneal shunting, although it is symptomatic in less than 5% of cases.[13] Similarly, tonsillar herniation may be acquired after repeated lumbar punctures.[14]

References

1. Chiari H. Über Veränderungen des Kleinhirns infolge von Hydrocephalie des Grosshirns. *Dtsch Med Wochenschr* 1891; **17**: 1172–5.
2. Elster AD, Chin MY. Chiari I malformations: clinical and radiological reappraisal. *Radiology* 1992; **183**: 347–53.
3. Weber PC, Cass SP. Neurotologic manifestations of Chiari 1 malformation. *Otolaryngol Head Neck Surg* 1993; **109**: 853–60.
4. Sclafani AP, DeDio RM, Hendrix RA. The Chiari-I malformation. *Ear Nose Throat J* 1991; **70**: 208–12.
5. Pascual J, Oterino A, Berciano J. Headache in type I Chiari malformation. *Neurology* 1992; **42**: 1519–21.
6. Cristante L, Westphal M, Herrmann HD. Cranio-cervical decompression for Chiari I malformation: a retrospective evaluation of functional outcome with particular attention to the motor deficits. *Acta Neurochir* 1994; **130**: 94–100.
7. Nagib MG. An approach to symptomatic children (ages 4-14 years) with Chiari type I malformation. *Pediatr Neurosurg* 1994; **21**: 31–5.
8. Nohria V, Oakes WJ. Chiari I malformation: a review of 43 patients. *Pediatr Neurosurg* 1990–91; **16**: 222–7.
9. Stovner LJ, Bergan U, Nilsen G, Sjaastad O. Posterior cranial fossa dimensions in the Chiari I malformation: relation to pathogenesis and clinical presentation. *Neuroradiology* 1993; **35**: 113–18.
10. Pober BR, Filiano JJ. Association of Chiari I malformation and Williams syndrome. *Pediatr Neurol* 1995; **12**: 84–8.
11. Milhorat TH, Capocelli AL Jr, Anzil AP, Kotzen RM, Milhorat RH. Pathological basis of spinal cord cavitation in syringomyelia: analysis of 105 autopsy cases. *J Neurosurg* 1995; **82**: 802–12.
12. Caldemeyer KS, Boaz JC, Wappner RS, Moran CC, Smith RR, Quets JP. Chiari I malformation: association with hypophosphatemic rickets and MR imaging appearance. *Radiology* 1995; **195**: 733–8.
13. Chumas PD, Armstrong DC, Drake JM, *et al.* Tonsillar herniation: the rule rather than the exception after lumboperitoneal shunting in the pediatric population. *J Neurosurg* 1993; **78**: 568–73.
14. Sathi S, Stieg PE. "Acquired" Chiari I malformation after multiple lumbar punctures: case report. *Neurosurgery* 1993; **32**: 306–9.

89. *Chiari II malformations are always associated with malformations of the spinal cord*

The Chiari II malformation is complex, involving the hindbrain, spine, skull base, spinal canal, and spinal cord. Associated cerebral hemisphere anomalies are common. It typically presents in infancy or childhood and is also known as the Arnold–Chiari malformation. CT and MRI of the head and spine are needed to evaluate the many features of a Chiari II malformation. This syndrome can be detected *in utero* with MRI.[1]

The most significant radiologic finding is a shallow posterior fossa that is far too small to house the cerebellum and brain stem (Fig. 89.1). The cerebellum often towers upward through an enlarged incisura and compresses into a tentorium that has a low attachment. It may wrap around the brain stem. In all cases, the cerebellar tonsils extend downward through a widened foramen magnum. The medulla and inferior cerebellar vermis are also displaced into the cervical spinal canal in a cascading fashion, resulting in compression.[2] The herniated portion of the cerebellum may degenerate, in some cases leaving little cerebellum except the flocculus. Cervical-medullary kinks develop in three-quarters of affected patients because the spinal cord is relatively fixed by the dentate ligaments. The fourth ventricle is small in axial cross-sectional area, elongated, and displaced downward. When a "normal-sized" fourth ventricle is present, hydrocephalus is probable. Intravenous contrast is generally not needed, but if used enhancement of ectopic choroid plexus in the base of the caudally displaced fourth ventricle should not be misinterpreted.[3]

Multiple cerebral hemisphere anomalies usually coexist with the posterior fossa and spinal abnormalities in Chiari II malformation. In most affected patients, the corpus callosum is aplastic or hypoplastic and the rostrum is usually absent. Obstructive hydrocephalus resulting from posterior fossa compression or aqueduct stenosis is frequent. The caudate nuclei and massa intermedia tend to be prominent. The falx cerebri may be fenestrated, allowing gyri to cross the midline producing gyral interdigitations. The medial occipital lobes can appear polymicrogyric on the midsagittal MRI, although the cortex is histologically normal. This condition is termed stenogyria. Neuronal migration abnormalities with heterotopic gray matter are frequent. Tectal beaking is common and is best visualized on the midline sagittal MRI, as are many of the anomalies.[4]

The posterior arches of C1 and the lower cervical vertebrae may not be completely fused. The dens is often scalloped. Over one-half of affected patients have syringomyelia. Cervical diastematomyelia (split cord) and cervical arachnoid cysts may be present.[5] Spina bifida is common. A meningomyelocele, present at birth, is frequently the first sign of a Chiari II malformation. After surgical closure of the meningomyelocele, hydrocephalus

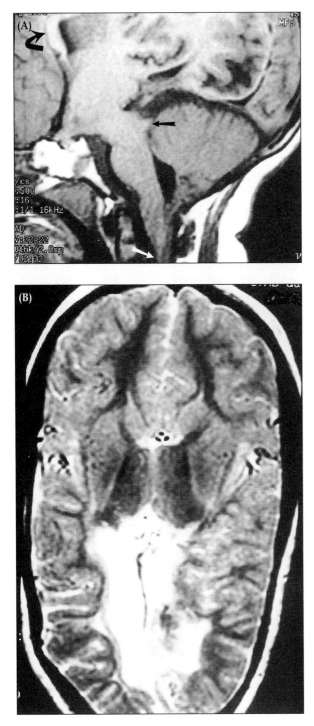

Fig. 89.1 Chiari II malformation. **A:** Sagittal T1-weighted image through the posterior fossa. A shallow posterior fossa and a low-lying fourth ventricle are present. The cerebellar tonsils descend through the foramen magnum (poorly visualized on this image). Tectal beaking is indicated by the straight black arrow. The corpus callosum is hypoplastic with absence of splenium and rostrum (the curved black arrow identifies the genu of the corpus callosum). A cervicomedullary kink is present at the C2 level (straight white arrow). **B:** Axial T2-weighted image through the basal ganglia in the same patient. Owing to the absent posterior corpus callosum, the occipital horns meet in the midline. The posterior falx cerebri is missing. The frontal lobe myelination pattern is normal for the patient's age, although visualization of the white-matter tracts is limited. The volume of white matter in the posterior temporal and occipital lobes is decreased.

play a role in the bulbar palsy.[6] Surgical decompression of the posterior fossa and spinal canal, before significant compression, can prevent much of the disability.[7–10]

Serial assessment of the ventricular size may be done with CT. However, repetitive shunt assessments using ionizing radiation (CT) can result in a substantial lifetime exposure. Cataracts induced by radiation are the most frequent consequence, but an increased risk of cancer is always a concern. MRI, which does not apparently have any associated long-term risks, is the safest serial neuroimaging method, but high cost limits its use and ventricular shunts are not well visualized.

References

1. Dinh DH, Wright RM, Hanigan WC. The use of magnetic resonance imaging for the diagnosis of fetal intracranial anomalies. *Childs Nerv Syst* 1990; **6**: 212–15.
2. Curnes JT, Oakes WJ, Boyko OB. MR imaging of hindbrain deformity in Chiari II patients with and without symptoms of brainstem compression. *AJNR Am J Neuroradiol* 1989; **10**: 293–302.
3. Stark JE, Glasier CM. MR demonstration of ectopic fourth ventricular choroid plexus in Chiari II malformation. *AJNR Am J Neuroradiol* 1993; **14**: 618–21.
4. Wolpert SM, Anderson M, Scott RM, Kwan ES, Runge VM. Chiari II malformation: MR imaging evaluation. *AJR Am J Roentgenol* 1987; **149**: 1033–42.
5. Naidich TP, McLono DG, Fulling KH. The Chiari II malformation: part IV. The hindbrain deformity. *Neuroradiology* 1983; **25**: 179–97.
6. Wolpert SM, Scott RM, Platenberg C, Runge VM. The clinical significance of hindbrain herniation and deformity as shown on MR images of patients with Chiari II malformation. *AJNR Am J Neuroradiol* 1988; **9**: 1075–8.
7. Pollack IF, Pang D, Albright AL, Krieger D. Outcome following hindbrain decompression of symptomatic Chiari malformations in children previously treated with myelomeningocele closure and shunts. *J Neurosurg* 1992; **77**: 881–8.
8. Vandertop WP, Asai A, Hoffman HJ, *et al.* Surgical decompression for symptomatic Chiari II malformation in neonates with myelomeningocele. *J Neurosurg* 1992; **77**: 541–4.
9. Dyste GN, Menezes AH, VanGilder JC. Symptomatic Chiari malformations: an analysis of presentation, management, and long-term outcome. *J Neurosurg* 1989; **71**: 159–68.
10. Eisenstat DD, Bernstein M, Fleming JF, Vanderlinden RG, Schutz H. Chiari malformation in adults: a review of 40 cases. *Can J Neurol Sci* 1986; **13**: 221–8.

requiring shunting frequently develops. After shunting, the medial walls of the trigones and the adjacent CSF space may appear abnormal.

Clinical presentation is usually the result of spinal cord or brain stem dysfunction. Compression of the medulla and spinal cord by the cerebellar tonsils at the foramen magnum and C1 level cause dysphagia, apnea, and stridor. Disorganization of the brain-stem nuclei may also

90. *Tethered spinal cords can be occult and occasionally become symptomatic during adulthood*

A tethered spinal cord is present in an adult when the distal tip of the conus medullaris is located at or below L2. The tethering lesion is usually a thickened filum terminale, intradural lipoma, intradural adhesions, or neural placode after surgical closure of a meningomyelocele. Tethered cords are associated with a terminal syrinx and occult spinal dysraphism in infants with an anorectal anomaly.[1,2] Most tethered cords are diagnosed in children but some present during adulthood. Symptoms typically develop during periods of rapid growth, especially during adolescence. In adults, additional tugging on a tight conus, narrowing of the spinal canal, the development of lumbar lordosis, or direct trauma to the back or buttocks may precipitate the symptomatic onset.[3]

In adults, the most common presenting symptom is diffuse, non-dermatomal leg pain, often referred to the anorectal region.[3] Lower extremity sensorimotor deficits (with weakness and hyperreflexia) and bladder and bowel dysfunction commonly develop. In children, progressive foot and spinal deformities occur. The leg symptoms are usually greatest after prolonged sitting, with movements requiring flexion, and after exercise, and are related to stretching and tugging of the distal spinal cord. Head flexion and the normal lengthening of the spinal column, from swelling of the intervertebral discs during sleep, may precipitate symptoms.

MRI, especially the T1-weighted sagittal acquisition, gives the best imaging (Fig. 90.1). When a tethered cord is diagnosed, the complete spine has to be examined to rule out associated pathology such as an intradural lipoma or diastematomyelia (split spinal cord). The fibrous filum terminale produces higher signal than CSF on T1-weighted images, enabling its visualization. The filum is thickened if its diameter is greater than 2 mm. Axial images are often better for examination of the filum and nerve roots. The distal conus is well visualized on both T1- and T2-weighted images. In most cases, MRI alone is adequate and myelography and postmyelography CT are not needed. To differentiate between lipoma and hematoma, fat-suppression MRI is sometimes required. Intravenous contrast rarely adds additional information.

Treatment of a tethered cord involves surgical release. Improvement in pain and weakness usually follows surgery, but bowel and bladder dysfunction often do not resolve. Improvement of certain types of scoliosis, kyphosis, and lordosis often follow release of the tethering.[4] Early diagnosis and surgery, before the development of irreversible lesions, is the key to successful treatment. Unfortunately the diagnosis of a tethered cord is often not considered in adults until after incorrect diagnoses are made, inappropriate treatments are attempted, and permanent damage has developed.[5]

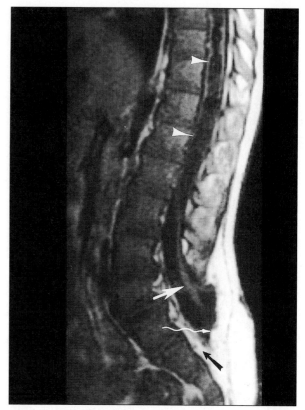

Fig. 90.1 Tethered spinal cord. Sagittal T1-weighted image through the lumbosacral spine in a child. A tethered spinal cord is present (straight white arrow). The tethering point into a lipoma (wavy white arrow) is at the L5–S1 level. A lipoma extends from the subcutaneous fat through a spina bifida defect into the ventral subarachnoid space (straight black arrow). The caudal spinal canal is expanded. An extensive associated syrinx is present (white arrowheads).

In infants, recurrent tethering is common after surgery because the neural contents come into contact with the posterior dura in the narrow spinal canal.[6]

References

1. Iskandar BJ, Oakes WJ, McLaughlin C, Osumi AK, Tien RD. Terminal syringohydromyelia and occult spinal dysraphism. *J Neurosurg* 1994; **81**: 513–19.
2. Davidoff AM, Thompson CV, Grimm JM, Shorter NA, Filston HC, Oakes WJ. Occult spinal dysraphism in patients with anal agenesis. *J Pediatr Surg* 1991; **26**: 1001–5.
3. Pang D, Wilberger JE Jr. Tethered cord syndrome in adults. *J Neurosurg* 1982; **57**: 32–47.
4. Reigel DH, Tchernoukha K, Bazmi B, Kortyna R, Rotenstein D. Change in spinal curvature following release of tethered spinal cord associated with spina bifida. *Pediatr Neurosurg* 1994; **20**: 30–42.
5. Kaplan JO, Quencer RM. The occult tethered conus syndrome in the adult. *Radiology* 1980; **137**: 387–91.
6. Zide B, Constantini S, Epstein FJ. Prevention of recurrent tethered spinal cord. *Pediatr Neurosurg* 1995; **22**: 111–14.

15
Spine and Spinal Cord: Tumors

91. *Most spinal cord neoplasms are gliomas (astrocytomas and ependymomas)*

Intraspinal lesions are classified according to the compartment involved as intramedullary, intradural extramedullary, and extradural. Identifying the compartment and the level usually narrows the differential diagnosis (Table 91.1). Tumors may arise in any compartment, originating from the nerve roots, meninges, and the spinal cord. Combining the information about location with the clinical characteristics often leads to the diagnosis. For instance, thoracic meningiomas tend to occur in middle-aged women.

MRI gives the best definition of intraspinal lesions, especially those that are intramedullary.[1] It enables noninvasive examination of the entire spinal cord (Fig. 92.1) and in most situations has replaced intrathecal contrast myelography (conventional and CT), which mainly evaluates the spinal cord indirectly. However, CT and plain radiographs are complementary to MRI in many situations because they highlight the bony detail.

Table 91.1 Common spinal canal lesions by location

Intramedullary	Intradural extramedullary	Extradural
Neoplasm (ependymoma, astrocytoma)	Neoplasm (nerve sheath tumor – schwannoma, meningioma)	Disc herniation
Syrinx (syringohydromyelia)	Metastasis (drop metastasis)	Osteophyte
Demyelinating disease or myelitis, e.g. viral, autoimmune	Arachnoiditis	Metastasis: direct to the epidural space or tumor from adjacent bone
Metastasis	Infection (subdural empyema)	Trauma (hematoma, bone fragment)
Ischemia, infarction	Arteriovenous malformation	Arachnoid cyst
Arteriovenous malformation		Infection (epidural empyema)

Almost all intramedullary tumors are gliomas. In adults, just under two-thirds are ependymomas and one-third astrocytomas. Astrocytomas are the most common spinal cord tumor in children. Occasionally oligodendrogliomas develop within the spinal cord. Ependymomas typically develop in the distal cord, particularly in the conus medullaris. Astrocytomas are more frequent at cervical and thoracic levels. On rare occasions metastases and schwannomas are intramedullary. Intramedullary spinal cord tumors may have an associated syrinx.

Spinal-cord ependymomas are slow-growing tumors that arise from the ependymal cells in the central canal or ependymal cell nests in the filum terminale. The peak incidence is in the fourth and fifth decade with a slight male predominance. Patients present with local or radicular pain, weakness, numbness, and incontinence. Intraspinal ependymomas are usually well circumscribed. In the filum region, the ependymoma initially displaces the adjacent nerve roots. Eventually the tumor adheres to and engulfs the nerve roots of the cauda equina. However, in the lower thoracic and lumbosacral region it is often difficult to determine whether a lesion originates in the conus medullaris, filum terminale, or cauda equina. Distant metastases from a small ependymoma are uncommon, although they may signal tumor recurrence following treatment.

Ependymomas are hypointense on T1-weighted images and hyperintense on T2-weighted images. Surrounding edema may make the lesion indistinct. They typically grow circumferentially into an oblong mass. Focal widening of the spinal cord is the major finding on myelography. Ependymomas may undergo cystic degeneration, resulting in heterogeneous signal on T2-weighted images. Most ependymomas enhance homogeneously with well-circumscribed margins. MRI detects almost all ependymomas, even when they are small. However, the imaging characteristics of ependymomas are not specific and biopsy is required for diagnosis.

Spinal cord astrocytomas are usually located in the cervical or thoracic regions. Only one-fifth are located in the conus medullaris and just 5% occur in the filum terminale.[2] They are twice as common in males. Spinal-cord astrocytomas tend to be less malignant than cerebral astrocytomas. They usually have a fibrillary or pilocytic histology. They are slow-growing and typically involve multiple segments at the time of diagnosis. Pain followed by weakness, incontinence, and numbness is the typical presentation.

Astrocytomas (like ependymomas) are hypointense on T1-weighted images and hyperintense on T2-weighted images. Astrocytomas have a tendency to involve the entire width of the cord. When the whole cord is involved, the bony spinal canal may enlarge to accommodate the neoplasm. In this circumstance, plain radiographs reveal separated pedicles that are thinned along their medial margins. The vertebral column may appear scalloped because the disc spaces are resistant to pressure from the tumor, preserving the posterior vertebral bodies. The bony changes are most common in the lumbosacral region.

Intravenous enhancement usually helps to separate the tumor from the surrounding edema. One-half of spinal-cord astrocytomas are cystic, resulting in a more heterogeneous appearance than in most ependymomas, especially on contrast-enhanced images. Nevertheless, there is a significant overlap in the neuroimaging appearance and the MRI enhancement pattern of ependymomas and astrocytomas. Intravenous enhancement is most useful in planning biopsy as it differentiates gross tumor from cystic regions.

References

1. Zimmerman RA, Bilaunik LT. Imaging of tumors of the spinal canal and cord. *Radiol Clin North Am* 1988; **26**: 965–1007.
2. Enzmann DR, DeLaPaz RL, Ruben JB. *Magnetic resonance in the spine.* St Louis: CV Mosby, 1990: 301.

92. *The most common malignant spinal column neoplasms are metastases*

Tumors affecting the spine are common. The following neoplasms may involve the spine:

- Breast
- Prostate
- Lung
- Renal cell
- Sarcoma
- Ewing's sarcoma
- Neuroblastoma
- Melanoma
- Lymphoma
- Leukemia
- Multiple myeloma

They may cause spinal cord and nerve root compression and often present as a neurologic emergency. The most common malignancies of the spine are metastases. The tumor, via hematogeneous spread, replaces the normal fatty bone marrow with nonfatty tumor cells. As it

progresses, it destroys bone trabeculae, usually without a periosteal response. However, Hodgkin's lymphoma, breast, and prostate cancer may cause a bony response, resulting in local sclerosis. The intervertebral discs are usually not involved in tumors of the spine. When the disc space is obliterated, infection is most likely.

Plain spine radiographs, to evaluate the bone integrity, should generally be the first radiologic study obtained on all patients with acute spinal cord syndromes. Further investigation usually requires intrathecal contrast myelography, CT, nuclear medicine bone scan, or MRI. In patients with spinal metastases, imaging (MRI or bone scan) of the entire vertebral column should be done before initiation of treatment because multilevel disease is common.

Plain radiographs and CT demonstrate bone marrow abnormalities by displaying the trabecular and cortical bone architecture. When the bone marrow changes are diffuse, CT may not differentiate between normal and abnormal marrow. Most tumors (exceptions include myeloma) exhibit increased activity on nuclear medicine bone scans, which serve as a good screening examination of the entire skeleton. MRI is the most sensitive imaging study for detecting bony metastases and primary bone tumors in the spine, and is superior to intrathecal contrast myelography. With MRI the entire spine can be imaged rapidly (sagittal images), images can be acquired in multiple planes for greatest lesion definition, and the vertebrae, epidural space, and spinal cord are simultaneously well visualized (Fig. 92.1). The diffuse bone marrow changes sometimes produced by leukemic infiltrates, lymphoma, and myeloma may be difficult to detect on all imaging modalities. However, MRI detects focal vertebral body lesions even in myeloma and lymphoma.

Vertebral body metastases are typically low signal on T1-weighted spin-echo images and high signal on T2-weighted images compared with the normal bone marrow. Bony sclerosis, however, produces low signal on both T1- and T2-weighted images. Although intravenous contrast helps with leptomeningeal tumor involvement and soft-tissue tumor, it may obscure vertebral body metastases because tumor enhancement may result in a signal identical to that of normal fat. This problem is solved by using a fat-suppression technique with intravenous contrast. Fat suppression eliminates the high background signal from the bone marrow fat, providing better visualization of enhancing lesions. Even with contrast, tumors may be difficult to detect in elderly patients because of heterogeneous bone marrow signal due to past vertebral body collapse. However, MRI is usually able to differentiate a collapsed vertebral body due to osteoporosis or trauma from malignant disease.[1]

The force needed to fracture a normal vertebral body is substantial, and traumatic vertebral body collapse in a normal person is usually associated with fragmentation of the bone or intervertebral disc. In contrast, osteoporotic and neoplastic vertebral body compressions may occur with minimal force (normal activity is often

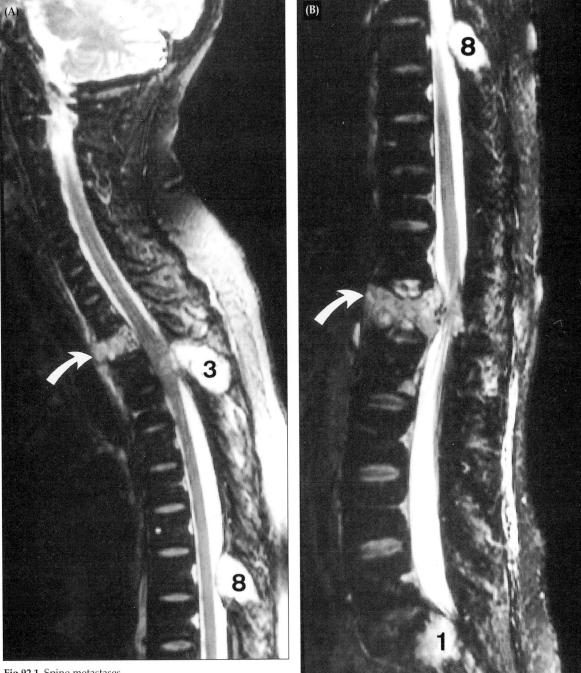

Fig 92.1 Spine metastases.

A and **B:** Sagittal T2-weighted images with fat suppression in a 43-year-old with melanoma. Diffuse metastatic involvement of the axial skeleton is present with tumor in the T2 and L1 vertebral bodies (curved arrows). Note the relative sparing of the intervertebral discs. (In contrast, infection often results in obliteration of the intervertebral discs; maxim 94.) Tumor is also present in the posterior elements of T3 (3) and T8 (8) and the sacrum (1). There is clear visualization of the metastases on T2-weighted images even without contrast. **C:** Axial T1-weighted gadolinium-enhanced image with fat suppression through the lesion at T2–T3 (note the image location in the inset at the lower right). Significant epidural extension of the bony metastasis is present with compressive anterolateral displacement of the spinal cord (arrow).

enough) due to underlying bone structure weakening. Because osteoporosis is caused by diffuse mineral loss with an otherwise normal bone marrow, a collapsed osteoporotic vertebral body should produce near normal

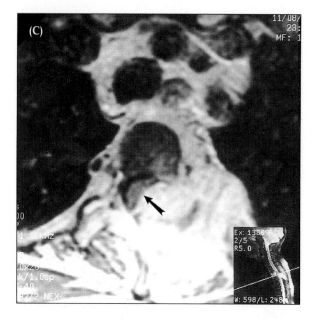

process.[2] Metastases weaken the vertebral bodies by bony destruction and replace the bone marrow with abnormal cells. Thus, abnormal marrow signal is expected on MRI in metastatic disease. A paraspinal mass suggests, but is not pathognomonic of, malignancy.

In patients with multiple myeloma, normal bone marrow MRI signal is frequently present even in collapsed vertebral bodies. Only one-half of patients with myeloma have abnormal bone marrow signals relative to other vertebral bodies.[3] Because patients with myeloma often also have osteoporosis, the cause of a collapsed vertebral body is often unclear.

The anatomical distribution of spine abnormalities may help in the differential diagnosis. Involvement of the pedicles and laminae without vertebral body involvement is only rarely due to metastases, whereas the presence of abnormal signal in both the vertebral body and posterior elements does not discriminate benign from malignant disease because osteoporotic changes may also occur in both locations.

References

1. Yuh WT, Zachar CK, Barloon TJ, Sato Y, Sickels WJ, Hawes DR. Vertebral compression fractures: distinction between benign and malignant causes with MR imaging. *Radiology* 1989; **172**; 215–18.
2. Horowitz SW, Fine M, Azar-Kia B. Differentiation of benign *versus* metastatic vertebral body compression fractures. *Neuroradiology* 1991; **33**(suppl): 114–16.
3. Libshitz HI, Malthouse SR, Cunningham D, MacVicar AD, Husband JE. Multiple myeloma: appearance at MR imaging. *Radiology* 1992; **182**: 833–7.

93. *Hemangiomas of the vertebral body are common but rarely symptomatic. However, thoracic hemangiomas in young women may cause pain and compressive myelopathy*

Hemangiomas are the most common benign tumors of blood vessels. They are slow-growing primary neoplasms of capillary, cavernous, or venous origin and are composed of large, mature, thin-walled, endothelial-lined, blood-filled vessels.[1] In the vertebral body the abnormal vessels replace the cancellous bones and may involve all or a portion of a vertebral body. The remaining trabeculae are thickened. Completely extraosseous hemangiomas are rare. Hemangiomas vary from predominantly fatty lesions to mainly vascular lesions with little or no fat.

Most hemangiomas are discovered incidentally and remain asymptomatic throughout life. Only a small percentage of patients present with pain or spinal cord or nerve root compression. Hemangiomas become symptomatic when pathologic vertebral compression fractures or epidural extension occurs. Symptomatic lesions are most frequent in the thoracic region in teenage girls and are unfortunately often discovered only after the onset of paraplegia. Hemangiomas may become symptomatic during pregnancy.[2,3]

Vertebral hemangiomas have a characteristic appearance on plain radiographs, CT, and MRI. On plain radiographs the trabeculae are thick and the vertebrae have vertical striations. A classic "polka dot" density in the bone marrow is present on CT. CT is useful to detect ossified extravertebral angiomatous masses.[4,5]

Intraosseous hemangiomas are typically mottled high-signal lesions on both T1- and T2-weighted MRI (a unique signal combination for a bone lesion).[6] High T1 signal is due to lipid (Fig. 93.1). The cause of the high T2 signal is not known but may be related to increased water content. Fat suppression is useful for defining fatty hemangiomas and distinguishing them from metastases. The vascular variety enhance with intravenous contrast.

Fatty hemangiomas are almost always asymptomatic. The vascular variety, which demonstrate soft-tissue density on CT and low signal on MRI, have a greater potential to become symptomatic and result in spinal cord compression.[7] Other features that are associated with symptomatic hemangiomas include a location between T3 and T9, involvement of the entire vertebral body, extension to the neural arch, an expanded cortex with indistinct margins, and an irregular honeycomb pattern.[8]

In patients with symptomatic hemangiomas, selective spinal angiography usually reveals hypervascularity with dilated feeding arteries without shunting or early venous drainage. The hypervascularity may extend into the spinal canal. Only minimal vascularity is present in the fatty form of hemangiomas.[7]

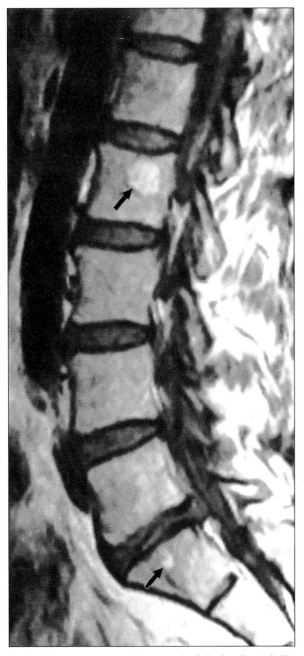

Fig. 93.1 Hemangioma of the vertebral body. Sagittal T1-weighted image. Well-circumscribed round lesions of high signal intensity relative to normal bone marrow are present in the L2 and S1 vertebral bodies (arrows). These lesions were also high signal on the T2-weighted images (not shown). This pattern is typical for vertebral body hemangiomas.

Symptomatic vertebral hemangiomas are treated with selective angiography and embolization (using methyl methacrylate), surgery, radiation therapy, and combinations of these methods with good success.[9–15]

References

1. Yochum TR, Lile RL, Schultz GD, Mick TJ, Brown CW. Acquired spinal stenosis secondary to an expanding thoracic vertebral hemangioma. *Spine* 1993; **18**: 299–305.
2. Redekop GJ, Del Maestro RF. Vertebral hemangioma causing spinal cord compression during pregnancy. *Surg Neurol* 1992; **38**: 210–15.
3. Tekkok IH, Acikgoz B, Saglam S, Onol B. Vertebral hemangioma symptomatic during pregnancy: report of a case and review of the literature. *Neurosurgery* 1993; **32**: 302–6.
4. Schnyder P, Fankhauser H, Mansouri B. Computed tomography in spinal hemangioma with cord compression: report of two cases. *Skeletal Radiol* 1986; **15**: 372–5.
5. Yu R, Brunner DR, Rao KC. Role of computed tomography in symptomatic vertebral hemangiomas. *J Comput Tomogr* 1984; **8**: 311–15.
6. Ross JS, Masaryk TJ, Modic MT, Carter JR, Mapstone T, Dengel FH. Vertebral hemangiomas: MR imaging. *Radiology* 1987; **165**: 165–9.
7. Laredo JD, Assouline E, Gelbert F, Wybier M, Merland JJ, Tubiana JM. Vertebral hemangiomas: fat content as a sign of aggressiveness. *Radiology* 1990; **177**: 467–72.
8. Laredo JD, Reizine D, Bard M, Merland JJ. Vertebral hemangiomas: radiologic evaluation. *Radiology* 1986; **161**: 183–9.
9. Nicola N, Lins E. Vertebral hemangioma: retrograde embolization: stabilization with methyl methacrylate. *Surg Neurol* 1987; **27**: 481–6.
10. Smith TP, Koci T, Mehringer CM, *et al.* Transarterial embolization of vertebral hemangioma. *J Vasc Interventional Radiol* 1993; **4**: 681–5.
11. Fox MW, Onofrio BM. The natural history and management of symptomatic and asymptomatic vertebral hemangiomas. *J Neurosurg* 1993; **78**: 36–45.
12. Raco A, Ciappetta P, Artico M, Salvati M, Guidetti G, Guglielmi G. Vertebral hemangiomas with cord compression: the role of embolization in five cases. *Surg Neurol* 1990; **34**: 164–8.
13. Nicola N, Lins E. Vertebral hemangioma late results of retrograde embolization: stabilization with methyl methacrylate in two cases. *Surg Neurol* 1993; **40**: 491–4.
14. Baker ND, Klein MJ, Greenspan A, Neuwirth M. Symptomatic vertebral hemangiomas: a report of four cases. *Skeletal Radiol* 1986; **15**: 458–63.
15. Faria SL, Schlupp WR, Chiminazzo H Jr. Radiotherapy in the treatment of vertebral hemangiomas. *Int J Radiat Oncol Biol Phys* 1985; **11**: 387–90.

16
Spine and Spinal Cord: Infections, Back Pain, and Myelopathy

94. *Consider Pott's disease (tuberculosis of the spine) in middle-aged patients with lesions consistent with spinal metastases but no primary malignancy*

Although the incidence of pulmonary tuberculosis has decreased in recent decades, tuberculosis involving the spine has become more common. Immunosuppression, alcoholism, drug abuse, and HIV infection predispose patients to tuberculosis. Tuberculosis tends to occur in children in the developing countries and in middle-aged adults in North America and Europe.

Tuberculosis of the spine develops insidiously compared with tuberculosis infection in other organs. There may be a history of several years of symptoms. The lower thoracic and lumbosacral spine are the most commonly affected sites. Cervical spine involvement is rare, but does occur.[1,2] When diagnosed, nearly 90% of patients have at least two vertebral bodies involved and half have three or more involved. "Skip" lesions (infection in non-adjacent vertebrae) and a prominent paraspinal abscess, larger than in most pyogenic infections, are common.

Tuberculous spondylitis starts in the anterior, inferior portion of the vertebral body and spreads beneath the longitudinal ligaments. Disc space narrowing occurs secondarily with resultant vertebral body fusion and eventual gibbous formation. Flattened vertebrae may develop (vertebra plana). However, in some cases, the intervertebral discs are relatively preserved, probably because of a paucity of proteolytic enzymes in mycobacteria. In contrast, pyogenic spine osteomyelitis usually results in obliteration of the disc space (Fig. 94.1).

Tuberculosis is one of the great imitators of other diseases (maxim 75). When posterior element involvement is present, differentiating between infection and tumor in the spine may be difficult.[3,4] This is especially true when the disc space is relatively preserved (usually an indication of neoplasm and not infection). Noncontiguous involvement of multiple vertebral bodies increases the likelihood that spinal tuberculosis will be misdiagnosed as metastases on neuroimaging. Intravenous contrast may suggest tuberculosis if the epidural and subligamentous spaces enhance or if an enhancing paraspinal abscess is present.

The bony abnormalities and disc space disease, if present, are usually apparent on plain radiographs and CT. Conventional myelography only indirectly demonstrates cord compression or spinal canal obstruction. CT myelography and spine MRI are the best studies, but neither is definitively diagnostic. CT may reveal characteristic fragmentary bone destruction with bone fragments present in two-thirds of epidural soft-tissue masses. Osteolytic, subperiosteal, or well-defined lytic lesions with sclerotic margins may also occur.[5] MRI typically demonstrates slightly narrowed disc spaces, low T1 signal and high T2 signal within the marrow of at least two adjacent vertebrae, subligamentous or epidural soft-tissue masses with high T2 signal, and cortical bone erosion.[6] Biopsy is usually required for diagnosis.

Low back pain is the most common presenting complaint in adults. Kyphosis may occur before or after treatment because of spinal fusion.[7] Paraparesis, fever, sensory disturbance, and bowel and bladder dysfunction may develop. Treatment consists of appropriate antimicrobials for 1 year. Surgical decompression is needed in patients with neurologic symptoms.[8] The neuroradiologic features may progress despite clinical improvement during the initial months of therapy.[9]

References

1. Wurtz R, Quader Z, Simon D, Langer B. Cervical tuberculous vertebral osteomyelitis: case report and discussion of the literature. *Clin Infect Dis* 1993; **16**: 806–8.
2. Slater RR Jr, Beale RW, Bullitt E. Pott's disease of the cervical spine. *South Med J* 1991; **84**: 521–3.
3. Smith AS, Weinstein MA, Mizushima A, *et al*. MR imaging characteristics of tuberculous spondylitis *vs* vertebral osteomyelitis. *AJR Am J Roentgenol* 1989; **153**: 399–405.
4. Monaghan D, Gupta A, Barrington NA. Case report: tuberculosis of the spine: an unusual presentation. *Clin Radiol* 1991; **43**: 360–2.
5. Jain R, Sawhney S, Berry M. Computed tomography of vertebral tuberculosis: patterns of bone destruction. *Clin Radiol* 1993; **47**: 196–9.
6. Thrush A, Enzmann D. MR imaging of infectious spondylitis. *AJNR Am J Neuroradiol* 1990; **11**: 1171–80.
7. Pun WK, Chow SP, Luk KD, Cheng CL, Hsu LC, Leong JC. Tuberculosis of the lumbosacral junction: long-term follow-up of 26 cases. *J Bone Joint Surg [Br]* 1990; **72**: 675–8.
8. Nussbaum ES, Rockswold GL, Bergman TA, Erickson DL, Seljeskog EL. Spinal tuberculosis: a diagnostic and management challenge. *J Neurosurg* 1995; **83**: 243–7.
9. Boxer DI, Pratt C, Hine AL, McNicol M. Radiological features during and following treatment of spinal tuberculosis. *Br J Radiol* 1992; **65**: 476–9.

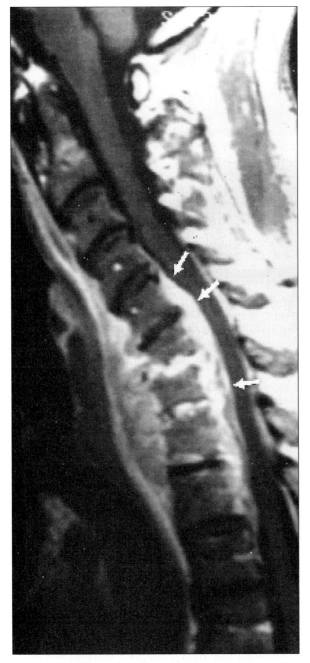

Fig. 94.1 Infection of the spine. Sagittal contrast-enhanced T1-weighted image of a 56-year-old man. An enhancing soft-tissue mass with an epidural component is present (arrows). Several vertebral bodies and especially the disc spaces are also involved and enhance with contrast. *Pseudomonas aeruginosa* was cultured from an aspirate. The predilection for disc space involvement is typical of spinal infections (tumors tend to spare the discs).

95. *Epidural lipomatosis is usually a consequence of steroid therapy and may cause myelopathy or radiculopathy*

An abnormal accumulation of normal fat in the epidural space is termed epidural lipomatosis. Rarely, the epidural lipomatosis becomes extensive enough to compress the spinal cord or cauda equina and cause myelopathy or radiculopathy requiring surgical decompression.[1-8] Symptomatic epidural lipomatosis occurs primarily in patients receiving long-term steroid therapy, although it may happen within a few months of initiating steroids. It may also develop in Cushing's syndrome, massive obesity, or without apparent cause.[9,10] No underlying preexisting abnormal fatty lesion is typically present.[11] This condition is reported in patients with rheumatoid arthritis, asthma, and after organ transplant who were receiving only modest doses of steroids.[12]

Epidural lipomatosis is readily identifiable on MRI or CT with intrathecal contrast. MRI is preferable as the entire spine can be surveyed in a single study and use of a fat-suppression technique will confirm the nature of the tissue. Epidural lipomatosis is most common in the thoracic region where the abnormal epidural fat usually accumulates more prominently posterior to the spinal cord. In the lumbar region, it is more likely to surround the thecal sac. It may also be focal and asymmetric, although the diffuse circumferential form is most common.

On axial images, the thecal sac often has a characteristic stellate appearance with three rays emanating from a central core, producing a trifid shape resembling the letter Y or a cloverleaf.[13] On midline sagittal T1-weighted image, the width of the accumulated fat usually measures at least 6 mm before it causes cord compression, although this diameter may be smaller if spinal stenosis is present.[14] Conventional myelography is inadequate in this condition because all it shows is an epidural lesion and it provides no information to help differentiate epidural lipomatosis from abscess, hematoma, or tumor.

Treatment of epidural lipomatosis often requires decompressive surgery with removal of the fat. If the diagnosis is made before significant symptoms have developed, an attempt to treat by discontinuation of steroids or by dramatic weight loss can be considered.[15]

References

1. Russell NA, Belanger G, Benoit BG, Latter DN, Finestone DL, Armstrong GW. Spinal epidural lipomatosis: a complication of glucocorticoid therapy. *Can J Neurol Sci* 1984; **11**: 383–6.
2. Randall BC, Muraki AS, Osborn RE, Brown F. Epidural lipomatosis with lumbar radiculopathy: CT appearance. *J Comput Assist Tomogr* 1986; **10**: 1039–41.
3. Soloniuk DS, Pecararo SR, Munschauer FE. Myelopathy secondary to spinal epidural lipomatosis. *Spine* 1989; **14**: 119–22.
4. Chapman PH, Martuza RL, Poletti CE, Karchmer AW. Symptomatic spinal epidural lipomatosis associated with Cushing's syndrome. *Neurosurgery* 1981; **8**: 724–7.

5. Kaplan JG, Barasch E, Hirschfeld A, Ross L, Einberg K, Gordon M. Spinal epidural lipomatosis: a serious complication of iatrogenic Cushing's syndrome. *Neurology* 1989; **39**: 1031–4.

6. Fessler RG, Johnson DL, Brown FD, Erickson RK, Reid SA, Kranzler L. Epidural lipomatosis in steroid-treated patients. *Spine* 1992; **17**: 183–8.

7. Roy-Camille R, Mazel C, Husson JL, Saillant G. Symptomatic spinal epidural lipomatosis induced by a long-term steroid treatment: review of the literature and report of two additional cases. *Spine* 1991; **16**: 1365–71.

8. Haddad SF, Hitchon PW, Godersky JC. Idiopathic and glucocorticoid-induced spinal epidural lipomatosis. *J Neurosurg* 1991; **74**: 38–42.

9. Stern JD, Quint DJ, Sweasey TA, Hoff JT. Spinal epidural lipomatosis: two new idiopathic cases and a review of the literature. *J Spinal Disorders* 1994; **7**: 343–9.

10. van Rooij WJ, Borstlap AC, Canta LR, Tijssen CC. Lumbar epidural lipomatosis causing neurogenic claudication in two obese patients. *Clin Neurol Neurosurg* 1994; **96**: 181–4.

11. Buthiau D, Piette JC, Ducerveau MN, Robert G, Godeau P, Heitz F. Steroid-induced spinal epidural lipomatosis: CT survey. *J Comput Assist Tomogr* 1988; **12**: 501–3.

12. Lee M, Lekias J, Gubbay SS, Hurst PE. Spinal cord compression by extradural fat after renal transplantation. *Med J Aust* 1975; **1**: 201–3.

13. Kuhn MJ, Youssef HT, Swan TL, Swenson LC. Lumbar epidural lipomatosis: the "Y" sign of thecal sac compression. *Comput Med Imaging Graph* 1994; **18**: 367–72.

14. Quint DJ, Boulos RS, Sanders WP, Mehta BA, Patel SC, Tiel RL. Epidural lipomatosis. *Radiology* 1988; **169**: 485–90.

15. Beges C, Rousselin B, Chevrot A, *et al*. Epidural lipomatosis: interest of magnetic resonance imaging in a weight-reduction treated case. *Spine* 1994; **19**: 251–4.

96. *A patient with acute myelopathy and normal neuroimaging of the spinal cord probably has transverse myelitis or a spinal cord infarction*

A patient who presents with acute or subacute myelopathy requires emergent neuroimaging with MRI, CT, or conventional intrathecal contrast myelography. The neuroimaging procedure is performed immediately to evaluate for a treatable lesion such as an epidural abscess, tumor (maxims 91 and 92), or hematoma. A negative study in conjunction with clinical features of spinal cord dysfunction suggests the diagnosis of transverse myelitis or transverse myelopathy. In this situation, high-dose pulse steroids may be warranted.

Acute transverse myelitis is not a single disease but a clinical syndrome with a wide variety of causes including:

- Multiple sclerosis
- Parainfection: viral (cytomegalovirus, herpes simplex virus, Epstein–Barr virus, human immunodeficiency virus); bacterial (neuroborreliosis – Lyme disease, syphilis, mycoplasma)
- Vaccination
- Connective tissue disease (systemic lupus erythematosus, Sjögren's syndrome)
- Granulomatous disease (neurosarcoidosis)
- Vasculitis (maxim 53)
- Paraneoplastic syndrome.

Causes of transverse myelopathy include:

- Spinal cord infarction
- Spinal arteriovenous malformation
- Trauma (especially in patients with osteophytes and spinal stenosis)
- Compressive lesion (tumor, hematoma, abscess)
- Intramedullary lesion (tumor, syrinx)
- B12 deficiency (combined systems degeneration)
- Amyotrophic lateral sclerosis.

In many patients the cause is never determined. Acute transverse myelitis is classically a monophasic illness, but may be recurrent in some conditions (e.g. multiple sclerosis, Sjögren's syndrome, idiopathic).[1] In practice, the presence of CSF pleocytosis (in the absence of a compressive lesion or an obvious vascular cause) is used to classify the condition as a transverse myelitis (inflammatory) as opposed to a transverse myelopathy, but this differentiation really requires pathologic examination of the spinal cord.

Acute transverse myelitis is characterized by rapidly progressive motor, sensory, and autonomic dysfunction of the spinal cord developing over hours or days. Sphincter dysfunction is usually prominent. Horner's syndrome may be present if the cervical cord is affected. Transverse myelitis/myelopathy may develop asymmetrically, predominating in any spinal cord distribution:

1. *Complete cord*
   - Paraplegia or quadriplegia – corticospinal tracts
   - Sensory level – lateral spinothalamic tracts and dorsal columns.
2. *Brown–Sequard (hemicord syndrome)*
   - Ipsilateral spastic paresis – corticospinal tract
   - Ipsilateral lower motor neuron weakness at the level of the lesion – anterior horn
   - Ipsilateral epicritic (vibration, proprioception) sensory loss – dorsal column
   - Contralateral protopathic (pain and temperature) sensory loss – lateral spinothalamic tract
   - Ipsilateral Horner's syndrome (with cervical lesions)
3. *Brown–Sequard plus*
   - Hemicord syndrome plus variable sensory and motor dysfunction from partial involvement of the other half of the spinal cord
4. *Anterior spinal artery*
   - Symmetric paresis or plegia – bilateral corticospinal tract
   - Symmetric protopathic sensory loss – bilateral lateral spinothalamic tract
   - Sparing of epicritic sensory function – dorsal columns supplied by posterior spinal artery
5. *Posterior spinal artery*
   - Epicritic sensory loss – bilateral dorsal columns

6. *Central cord*
  - Spastic lower extremity paresis – bilateral corticospinal tract
  - Flaccid upper extremity paresis – bilateral anterior horns
  - Variable sensory deficits.

Either a complete level or a Brown–Sequard plus syndrome typify most parainfectious transverse myelitis presentations. A vascular cause usually presents with an anterior spinal artery syndrome or, very rarely, a posterior spinal artery syndrome. Most commonly, the thoracic cord is affected, but any single level or multiple levels may be involved. Traumatic myelopathies associated with arthritic disease (spondylosis) usually occur at a lower cervical level and present with a central cord syndrome. Because the spinal cord ends at approximately the L1 vertebral body, examination of a patient with a myelopathy should concentrate on the cervical and thoracic spine (not the lumbosacral region unless a tethered cord is present; maxim 90).

MRI with gadolinium enhancement is the method of choice for examination of the spinal cord. Not only is rapid evaluation of the complete spinal cord possible with MRI (using phase array coils with a large field of view; Fig. 92.1), but it is the method most likely to detect and characterize a pathologic process affecting the spinal cord. Approximately one-half of patients with transverse myelitis have an abnormal MRI, most commonly with enlargement (swelling) of the spinal cord.[2,3] Intramedullary high T2 signal and enhancement on T1 images may be present (Fig. 96.1).[4] The findings are best visualized on sagittal T2-weighted images with 3–4 mm sections and a 512 matrix (not the conventional sagittal T1). The clinical outcome of transverse myelitis with an abnormal MRI is usually worse than in cases with no lesion on MRI.[5] Spinal cord atrophy may eventually develop in patients with persistent deficits.[6]

Conventional myelography and CT have a very limited role in the diagnosis of transverse myelitis, except to rule out compressive disease, although they occasionally detect swelling of the spinal cord.

Although a definitive diagnosis is not possible with neuroimaging, certain MRI features suggest the diagnosis of multiple sclerosis (MS). A single lesion, especially when it involves the C2–C3 level, is much more common in MS than in other diseases. An MS plaque tends to have an oval shape and extends longitudinally along the cord. The region involved is usually asymmetric and does not have associated spinal cord swelling.[7] Over three-quarters of patients with such a lesion will develop MS.[2] Patients with negative MRI and asymmetric clinical features are also prone to develop MS. In contrast, patients with multiple segment involvement rarely develop MS. Many patients with spinal MS have brain lesions consistent with MS on MRI (maxims 19 and 20) and CSF IgG oligoclonal bands. Patients with parainfectious, vascular, or traumatic causes for a transverse myelopathy usually do not have brain lesions or oligoclonal bands.

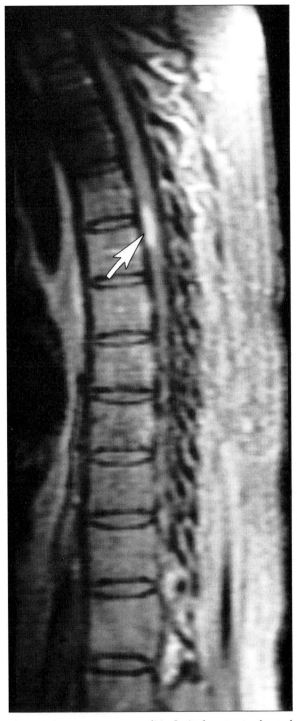

Fig. 96.1 Acute transverse myelitis. Sagittal contrast-enhanced T1-weighted image with fat suppression through the thoracic spinal cord in a patient with transverse myelitis. An enhancing area, involving almost the entire diameter, is present in the midcord without significant swelling (arrow). The enhancement disappeared after symptoms improved.

References

1. Tippett DS, Fishman PS, Panitch HS. Relapsing transverse myelitis. *Neurology* 1991; **41**: 703–6.
2. Campi A, Filippi M, Comi G, *et al*. Acute transverse myelopathy: spinal and cranial MR study with clinical follow-up. *AJNR Am J Neuroradiol* 1995; **16**: 115–23.
3. Austin SG, Zee CS, Waters C. The role of magnetic resonance imaging in acute transverse myelitis. *Can J Neurol Sci* 1992; **19**: 508–11.
4. Gero B, Sze G, Sharif H. MR imaging of intradural inflammatory diseases of the spine. *AJNR Am J Neuroradiol* 1991; **12**: 1009–19.
5. Scott T, Weikers N, Hospodar M, Wapenski J. Acute transverse myelitis: a retrospective study using magnetic resonance imaging. *Can J Neurol Sci* 1994; **21**: 133–6.
6. Shen WC, Lee SK, Ho YJ, Lee KR, Mak SC, Chi CS. MRI of sequela of transverse myelitis. *Pediatr Radiol* 1992; **22**: 382–3.
7. Jeffery DR, Mandler RN, Davis LE. Transverse myelitis: retrospective analysis of 33 cases, with differentiation of cases associated with multiple sclerosis and parainfectious events. *Arch Neurol* 1993; **50**: 532–5.

97. *Asymptomatic disc herniations and disc bulges are common radiologic findings*

Low back pain is a leading reason for visits to physicians and work-related disability in developed countries. The most common causes are paraspinal muscle strain and nerve root irritation (sciatica). Paraspinal muscle strain is characterized by pain localized to the back that is exacerbated with contraction (bending toward the affected side against resistance). Sciatica is characterized by pain radiating from the lower back around the hip and into the leg. Sciatica is usually exacerbated with sitting, a Valsalva maneuver, or stretch (e.g. straight leg raise test). Sciatic pain worsens with deep palpation of the buttocks in the sciatic notch. Pressure from a wallet in this region may be the sole precipitating cause of sciatica. Most patients with low back pain improve with a few days of bed rest followed by progressive mobilization or light activity.

Surprisingly, approximately two-thirds of adults who have never experienced back pain have abnormal MRIs of the lumbosacral spine.[1] One-half of these asymptomatic patients have disc bulges (circumferential symmetric disc extension beyond the interspace) and one-quarter have disc protrusions (focal or asymmetric disc extension beyond the interspace; Fig. 97.1). Over one-third have abnormalities at multiple levels. Schmorl's nodes (herniation of the disc into the adjacent vertebral body), annular

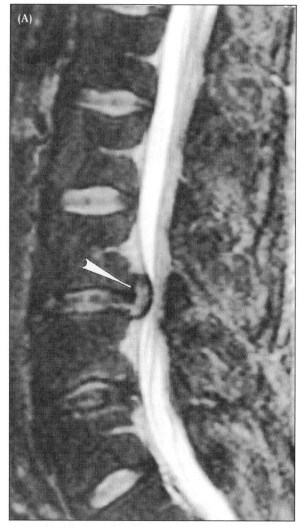

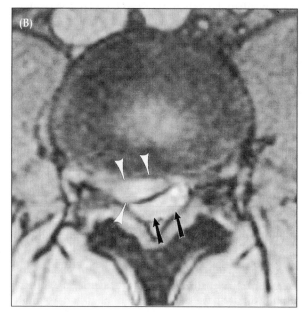

Fig. 97.1 Intervertebral disc herniation (protrusion) of the nucleus pulposus. **A:** Sagittal T2-weighted fat-suppressed image. A herniated intervertebral disc with resulting spinal stenosis is present at the L3/L4 interspace. **B:** Axial T2-weighted gradient-echo image. The herniated intervertebral disc (white arrowheads) is easily differentiated from the thecal sac (black arrows), revealing the extent of the disc herniation (protrusion) and the degree of stenosis.

defects, or posterior articular process arthropathy are present in 10–20%. Conventional intrathecal contrast myelography, CT, and discograms are similarly abnormal in a significant percentage of asymptomatic adults. The extremely high frequency of these neuroimaging findings in asymptomatic individuals makes their presence of limited importance in the initial evaluation of a patient with back pain.

The only intervertebral disc neuroimaging abnormality that may correlate with symptomatic radiculopathy is a disc extrusion defined as an extreme extension of the disc beyond the interspace. The base of a disc extrusion, against the disc of origin, is narrower than the depth of the extruding material or there is no connection between the extrusion and the disc of origin at all.[1]

Radiologic evaluation should be reserved for patients being considered for surgery. These patients include those with pain who have failed conservative therapy (bedrest, physical therapy, and anti-inflammatory medications) for several weeks (usually at least six). Because only a portion of patients with sciatic pain improve with surgery, the decision regarding surgery to treat the symptom of pain should be made with care. Muscle weakness or diminished reflexes in the distribution of a nerve root implies potentially permanent damage and mandates early radiologic evaluation for surgery. Early surgery in patients who are becoming weak usually reverses the deficit. The presence of deep pain with percussion directly over the spine raises the possibility of a vertebral body malignancy or infection and warrants at least plain spine radiography. Upper motor neuron deficits (spasticity, hyperreflexia, extensor plantar responses) indicate spinal cord or brain involvement and require investigation above the lumbosacral region because the spinal cord ends at approximately L1 (except with a tethered cord; maxim 90). Use of lumbosacral spine radiographs to evaluate a patient with myelopathic signs confined to the lower extremities is a common error made by inexperienced physicians.

The observation that disc herniations improve in a significant number of patients whose symptoms resolve suggests that symptomatic patients do have visible disc abnormalities.[2–5] The problem is that it is generally not possible to distinguish a symptomatic from an incidental disc on any neuroimaging study currently available. Radiologic tests also do not predict which patients with sciatica will benefit from conservative treatment and which will require surgery. In patients with low back pain, neuroimaging studies should be reserved for those who have failed conservative treatment or have weakness or atypical features.[6]

References

1. Jensen MC, Brant-Zawadzki MN, Obuchowski N, Modic MT, Malkasian D, Ross JS. Magnetic resonance imaging of the lumbar spine in people without back pain. *N Engl J Med* 1994; **331**: 69–73.
2. Bush K, Cowan N, Katz DE, Gishen P. The natural history of sciatica associated with disc pathology: a prospective study with clinical and independent radiologic follow-up. *Spine* 1992; **17**: 1205–12.
3. Thelander U, Fagerlund M, Friberg S, Larsson S. Describing the size of lumbar disc herniations using computed tomography: a comparison of different size index calculations and their relation to sciatica. *Spine* 1994; **19**: 1979–84.
4. Dullerud R, Nakstad PH. CT changes after conservative treatment for lumbar disk herniation. *Acta Radiol* 1994; **35**: 415–19.
5. Delauche-Cavallier MC, Budet C, Laredo JD, et al. Lumbar disc herniation: computed tomography scan changes after conservative treatment of nerve root compression. *Spine* 1992; **17**: 927–33.
6. Deyo A. Magnetic resonance imaging of the lumbar spine: terrific test or tar baby? *N Engl J Med* 1994; **331**: 115–16.

98. *Differentiating a postoperative scar from a herniated disc is usually possible*

Over 200 000 lumbar disc operations are done annually in the USA and many more are performed worldwide. Approximately 15% of patients operated on have recurrent back pain due to either residual disc fragments or recurrent herniation;[1,2] these patients often improve with repeat surgery. Another 15% have recurrent back pain but only have scar tissue and do not respond to reoperation to remove the scar. In these patients with "failed back surgery syndrome", distinguishing between those who have recurrent or residual herniation and those with exclusively postsurgical scar tissue is important in deciding which patients are candidates for repeat surgery.

After lumbar disc surgery, granulation tissue (early scar) develops within the spinal canal.[3] The size of the early scar does not correlate with outcome.[4] As the collagen and other fibrosing material ripen and the water content decreases during the initial postoperative months, the granulation tissue shrinks to form a mature scar. The location of the scar within the epidural space affects its evolution. Anterior and lateral recess scars are more heterogeneous than are posterior scars and may never fully mature. Posterior epidural scars usually reach maturity within 4 months.

The appearance of the scar on MRI and the enhancement pattern are a direct consequence of the physiology of healing. There are significant differences between disc material, young scar (granulation tissue), and mature scar on MRI.[5] To a lesser extent, these features may be visualized on intravenous contrast-enhanced CT, but MRI permits better definition of tissue margins and composition than does CT. In contrast, a postoperative scar appears identical to disc material on myelography (both conventional and CT) and noncontrasted CT.

On unenhanced MRI, the presence of mass effect, lesion location adjacent to the disc space, and signal isointense to the disc suggest that the material is herniated nucleus pulposus,[6] although up to one-third of scars also have these features. Enhancement of the material with intravenous contrast strongly suggests that the lesion is a scar and not a herniated disc. Early scars enhance because of increased vascularity and widened extravascular spaces, and mature scars may enhance because of loose endothelial gap junctions.[7] Eventually the enhancement of most mature scars tends to fade. On occasion, enhancing granulation tissue forms around herniated disc material associated with a tear in the annulus fibrosus and may erroneously be labeled a scar following surgery.[8]

The MRI signal intensity of scars tends to be different for scars in different locations. The anterior and lateral epidural scars have higher T2 signal intensity due to larger extracellular spaces.[7] Scars in these locations usually vaguely enhance.

Total normalization of the posterior extraspinal structures generally follows surgery. Gas, visible with CT or MRI, persists in the disc space for over 3 years after surgery in nearly one-half of patients.[9] CT findings do not predict which patients have clinical symptoms.[9,10]

Intrathecal contrast myelography indirectly visualizes the spinal canal and cord by demonstrating the subdural space. Extradural lesions producing impression on the thecal sac, root sleeve defects, and root swelling, the classic signs of disc herniation, may also be caused by scarring. Intrathecal contrast myelography has little role in the postoperative evaluation of back pain. Contrast-enhanced CT shows changes similar to those on MRI, but does not differentiate tissue as well.[11] CT discography (CT following injection of contrast into the disc space) is not as effective at identifying disc herniations as CT myelography.[12]

References

1. Rish BL. A critique of the surgical management of lumbar disc disease in a private neurosurgical practice. *Spine* 1984; **9**: 500–4.
2. Rish BL. A critique of posterior lumbar interbody fusion: 12 years' experience with 250 patients. *Surg Neurol* 1989; **31**: 281–9.
3. Dina TS, Boden SD, Davis DO. Lumbar spine after surgery for herniated disk: imaging findings in the early postoperative period. *AJR Am J Roentgenol* 1995; **164**: 665–71.
4. Jensen TT, Overgaard S, Thomsen NO, Kramp S, Petersen OF, Hansen JH. Postoperative computed tomography three months after lumbar disc surgery: a prospective single-blind study. *Spine* 1991; **16**: 620–2.
5. Ross JS, Masaryk TJ, Schrader M, Gentili A, Bohlman H, Modic MT. MR imaging of the postoperative lumbar spine: assessment with gadopentetate dimeglumine. *AJNR Am J Neuroradiol* 1990; **11**: 771–6.
6. Bundschuh CV, Modic MT, Ross JS, Masaryk TJ, Bohlman H. Epidural fibrosis and recurrent disk herniation in the lumbar spine: MR imaging assessment. *AJR Am J Roentgenol* 1988; **150**: 923–32.
7. Bundschuh CV, Stein L, Slusser JH, Schinco FP, Ladaga LE, Dillon JD. Distinguishing between scar and recurrent herniated disk in postoperative patients: value of contrast-enhanced CT and MR imaging. *AJNR Am J Neuroradiol* 1990; **11**: 949–58.
8. Ross JS, Modic MT, Masaryk TJ. Tears of the annulus fibrosus: assessment with Gd-DTPA-enhanced MR imaging. *AJR Am J Roentgenol* 1990; **154**: 159–62.
9. Heilbronner R, Fankhauser H, Schnyder P, de Tribolet N. Computed tomography of the postoperative intervertebral disc and lumbar spinal canal: serial long-term investigation in 19 patients after successful operation for lumbar disc herniation. *Neurosurgery* 1991; **29**: 1–7.
10. Tullberg T, Rydberg J, Isacsson J. Radiographic changes after lumbar discectomy: sequential enhanced computed tomography in relation to clinical observations. *Spine* 1993; **18**: 843–50.
11. Sotiropoulos S, Chafetz NI, Lang P, *et al.* Differentiation between postoperative scar and recurrent disk herniation: prospective comparison of MR, CT, and contrast-enhanced CT. *AJNR Am J Roentgenol* 1989; **10**: 639–43.
12. Hodge JC, Ghelman B, Schneider R, Rappoport LH, O'Leary PF, Cammisa FP Jr. Recurrent disk *versus* scar in the postoperative patient: the role of computed tomography (CT)/diskography and CT/myelography. *J Spinal Disorders* 1994; **7**: 470–7.

99. *Irreversible neurologic injury is commonly present when spinal stenosis is diagnosed*

Congenitally small spinal canals (the result of short pedicles) in the lumbosacral and cervical regions are present in a significant portion of the population. Symptomatic spinal stenosis usually develops when a patient with a congenitally small canal develops a superimposed process such as spondylosis (degenerative disc disease), degenerative facet disease, disc bulges or herniations, ligament degeneration (especially hypertrophy of the ligamentum flavum), or spondylolisthesis (maxim 79).[1] Almost all cases of spinal stenosis develop insidiously, and by the time of presentation produce significant neurologic deficits. The natural history of spinal stenosis is not well studied, but many patients improve or stabilize with conservative measures.

Cervical stenosis usually presents as a chronic myelopathy. Gait difficulty due to spasticity is the most common presentation. Urinary dysfunction, sensory changes, weakness, and hyperreflexia are often present. In the upper extremities, radicular pain and flaccid paresis with evidence of denervation (atrophy, fasciculations) may occur due to concurrent cervical radiculopathies. Neck range of motion is usually limited, although neck pain occurs in less than 10% of cases. A combination of direct compression, repeated trauma from normal neck movements, and vascular compromise probably causes the myelopathy, but the exact mechanism of the cervical spinal cord injury is not known. Surgery to open the canal and relieve the compression in cervical stenosis rarely results in dramatic improvement, but some improvement and prevention of further deterioration is typical as long as other major diseases are not prominent.[2,3]

Lumbar stenosis classically presents with pseudoclaudication (pain in the buttocks or legs on standing or walking that is relieved by resting with the trunk flexed). Pseudoclaudication usually does not occur on walking upstairs or uphill (because the trunk is flexed) and may not improve with rest unless the trunk is flexed, differentiating it from arteriosclerotic claudication. Pseudoclaudication may be unilateral, but is most commonly bilateral. Other symptoms of a cauda equina syndrome including radicular pain, weakness, sensory impairment, and bladder dysfunction may be present. Pseudoclaudication does not occur with lumbar disc disease. Most patients with lumbar stenosis respond to conservative therapy, but surgery is usually successful at relieving pain if symptoms are severe, except in patients with coexisting spondylolisthesis.[4,5]

Thoracic spinal stenosis is uncommon.[6] When spinal stenosis develops in the thoracic spine, it is usually confined to a single level. Clinical symptoms include pseudoclaudication and myelopathic features.

The high frequency of asymptomatic degenerative spine disease in the general population requires that other diseases be considered before attributing a patient's symptoms to spinal stenosis. Patients with apparent symptomatic cervical stenosis should be evaluated for B12 deficiency (combined systems degeneration; pernicious anemia) because a combined peripheral neuropathy and myelopathy may occur with both and the hematologic abnormalities may develop only after significant neurologic deficits have developed. Other causes of myelopathy are listed in maxim 96. Neuroimaging of the cervical and sometimes the lumbar spine should be considered before making the diagnosis of amyotrophic lateral sclerosis (ALS) (especially when bulbar symptoms are not prominent) because combined cervical and lumbar stenosis may also cause a mixture of upper and lower motor neuron dysfunction in all extremities.

Because spinal stenosis has many causes, a variety of radiographic appearances are possible (Fig. 99.1). CT and MRI help to differentiate the various types of spinal stenosis and are invaluable for preoperative planning. Most patients with borderline congenital spinal stenosis (with short pedicles) do not become clinically symptomatic until an acquired spinal stenosis develops. Increased disc and facet joint instability results in

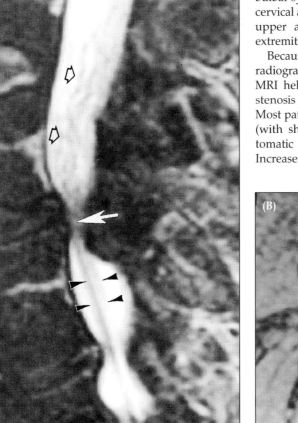

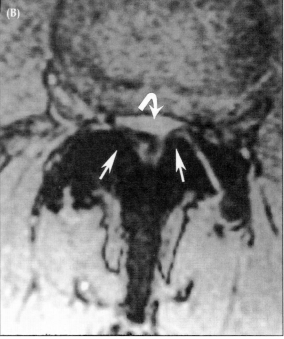

Fig. 99.1 Lumbar spinal stenosis. **A:** Sagittal T2-weighted image through the lumbosacral spine. Significant spinal stenosis is present at the L2–L3 level. The CSF signal intensity is lower above the stenosis because of increased motion compared with the relatively stationary CSF below the stenosis. At the site of stenosis, the dural sac appears as a low signal intensity because of a paucity of CSF (white arrow). Directly above the stenotic region, the nerve roots are bunched and curled in the middle of the thecal sac because they are not free to slide through the narrow space (open black arrows). Distal to the stenosis, the roots are straight and stretched (black arrowheads). **B:** Axial T2-weighted gradient-echo image through the stenotic level. The canal compromise is primarily caused by degenerative disease of the facets (low signal; straight arrows) which are encroaching on the thecal sac (curved arrow).

abnormal motion and ligamentous laxity, leading to a cascade of progressive degenerative changes affecting the lateral recesses, neural foramina, and central spinal canal. In a broad sense, spinal stenosis includes lateral recess stenosis and neural foramen stenosis in addition to central canal stenosis.

Plain spine radiographs, conventional intrathecal myelography, CT, CT myelography, and MRI all demonstrate the reduced diameter of the spinal canal and thecal sac in spinal stenosis. The degree of stenosis may be overestimated on high field strength MRI because of pulsation of the CSF and truncation artifact.[7] However, measurements of the size of the bony spinal canal do not always predict which patients will develop symptoms. MRI is the best method for demonstrating multiple osteophytes, bulging or herniated discs, neural foramina stenosis, and lateral recess stenosis, although contrasted CT usually also reveals these abnormalities.[8]

Sagittal T2-weighted images reveal extradural defects from osteophytes and protruding discs compressing the high-signal CSF in the subarachnoid space. At the level of the stenosis, the dural sac has low signal because little CSF is present at the stenotic location. In severe stenosis, the ventral subarachnoid space is completely effaced. The CSF signal may be lower above the stenosis because of fluid motion and higher below the stenosis where the fluid is relatively stationary. The T1-weighted sagittal image is less helpful because the disc material, ligaments, cortical bone and CSF all produce low signal.

In cervical stenosis, secondary spinal cord changes, manifested as abnormal intramedullary high T2 signal, may develop at the constricted levels. Myelomalacia consisting of neuronal loss, demyelination, and chronic infarction are present on pathologic examination. In lumbar stenosis, the nerve roots may be clumped and curled and may become edematous because they cannot slide normally (Fig. 99.1A). This appearance can be misdiagnosed as a vascular malformation because the compressed roots may enhance with intravenous contrast.

Degenerative changes with bony overgrowth into the bony spinal canal, degenerative facet joint disease, and hypertrophy of the ligamentum flavum all contribute to acquired spinal stenosis (Fig. 99.1B). Neural foramen narrowing may cause root symptoms. The neural foramen normally has the shape of an inverted pear in the sagittal plane. The nerve root exits immediately underneath the pedicle where the neural foramen is widest. Degenerative changes in the superior articular process of the facet joint may constrict the foramen at this location. However, because the nerve root and ganglion occupy only a small portion of the neural foramen, even severe foramen stenosis may be asymptomatic. Hypertrophy of the inferior articular facet is more likely to result in central canal stenosis.

References

1. Weisz GM, Lee P. Spinal canal stenosis. Concept of spinal reserve capacity: radiologic measurements and clinical applications. *Clin Orthop* 1983; **179**: 134–40.
2. Seifert V, van Krieken FM, Zimmermann M, Stolke D, Bao SD. Microsurgery of the cervical spine in elderly patients. Part 1: Surgery of degenerative disease. *Acta Neurochir* 1994; **131**: 119–24.
3. Snow RB, Weiner H. Cervical laminectomy and foraminotomy as surgical treatment of cervical spondylosis: a follow-up study with analysis of failures. *J Spinal Disorders* 1993; **6**: 245–50.
4. Tuite GF, Doran SE, Stern JD, *et al.* Outcome after laminectomy for lumbar spinal stenosis. Part II: Radiographic changes and clinical correlations. *J Neurosurg* 1994; **81**: 707–15.
5. McCullen GM, Bernini PM, Bernstein SH, Tosteson TD. Clinical and roentgenographic results of decompression for lumbar spinal stenosis. *J Spinal Disorders* 1994; **7**: 380–7.
6. Yamamoto I, Matsumae M, Ikeda A, Shibuya N, Sato O, Nakamura K. Thoracic spinal stenosis: experience with seven cases. *J Neurosurg* 1988; **68**: 37–40.
7. Reul J, Gievers B, Weis J, Thron A. Assessment of the narrow cervical spinal canal: a prospective comparison of MRI, myelography and CT–myelography. *Neuroradiology* 1995; **37**: 187–91.
8. Schnebel B, Kingston S, Watkins R, Dillin W. Comparison of MRI to contrast CT in the diagnosis of spinal stenosis. *Spine* 1989; **14**: 332–7.

17
Epilogue

100. *Behold the future: automated structural and functional diagnosis, desktop three-dimensional surgical planning and radiologic anatomy teaching, and, of course, no more film*

Routine daily use of telemedicine technology will be a fact of life within the next one or two decades. All images (radiologic and other), medical records, and other patient information will be stored in a digital format on computer disks (probably optical because of their resistance to deterioration and properties of rapid random access). Within 25 years, the term "x-ray film" is likely to be as archaic as the concept of four humors causing medical disease is today. The notion of patients carrying a credit card-sized computer disk with all their medical information is hardly science fiction. The hardware already exists for these applications; only the software needs to be developed. Future improvements in hardware will certainly further increase speed and decrease cost, but the major current obstacle is software, including standardization of storage formats.

Digitalization of medical images opens a new world of image postprocessing. Automated segmentation and feature extraction methods with comparisons of imaged structures to a normative database for determination of pathology are already being developed.[1] Coregistration of the different imaging methods for multispectral analysis will result in high sensitivity and specificity of medical image interpretation.

The current manual method of extracting data, relying on a human being (radiologist/physician) to analyze two-dimensional images of a three-dimensional structure, will initially be augmented over the coming decades and eventually replaced (probably within a century) by automated computed medical image analysis. Radiology will be an interventional specialty in the 22nd century.

As imaging technology advances, microscopic views will be possible. Soon, radiologists will indeed be pathologists. Spectroscopic MRI techniques, and neurotransmitter receptor and antibody radioactive-labeled agents for PET and SPECT, will dramatically improve our understanding of pathophysiology and initiate a new field of dynamic radiologic physiology.

Within the next decade or so, three-dimensional visualization (using stereoscopic or holographic displays) of medical images will be present on every physician's desktop. Interactive software will enable physicians to plan and practice surgery while sitting in their offices. Medical students will learn anatomy by repeatedly dissecting the computed images from a wide variety of viewpoints. The lay public will employ these and similar scientific "virtual reality" devices as computer games, offering the scientific community a new opportunity to narrow the knowledge gap.

Advances in functional neuroimaging, primarily with functional MRI but also with PET and SPECT, will create a large specialty field of radiologic psychophysics. Teams of researchers and clinicians will be required to make full use of this technology.[2]

Interactive medical imaging will be in routine use in the operating room within the next few years. Structural anatomy will be coregistered with functional features (probably mainly determined with functional MRI) during neurosurgical procedures to optimize preservation of eloquent brain regions. Improvements in directed-beam therapy are likely to obviate the need for conventional surgery in many situations.

One ultimate goal of engineers during the next century will be development of a tool similar to the medical tricorder that Drs Crusher, McCoy, and Pulaski use on *Star Trek*. Such a versatile, yet compact, hand-held device is within the realm of possibility. This device will probably combine magnetic resonance, ultrasound, and nuclear medicine technology. Because of advances in genetics and imaging resulting from engineering advances and computerization, medicine will be as different a century from now as it is today compared with the dark ages a millennium ago.

References

1. Ashton EA, Berg MJ, Parker KJ, Weisberg J, Chen CW, Ketonen L. Segmentation and feature extraction techniques, with applications to MRI head studies. *Mag Reson Med* 1995; **33**: 670–7.
2. Thatcher RW, Hallet M, Zeffiro T, John ER, Huerta M. (eds), *Functional neuroimaging technical foundations*. New York: Academic Press, 1994.

Index